Nature,
Human Nature, &
Human Difference

Nature, Human Nature, & Human Difference

Race in Early Modern Philosophy

Justin E. H. Smith

PRINCETON UNIVERSITY PRESS
PRINCETON AND OXFORD

Published by Princeton University Press, 41 William Street,
Princeton, New Jersey 08540
In the United Kingdom: Princeton University Press, 6 Oxford Street,
Woodstock, Oxfordshire OX20 1TR

press.princeton.edu

Cover Art: Cover of the first edition of Williem
Piso's *Historia Naturalis Brasiliae*, 1648.

First paperback printing, 2017
Paperback ISBN: 978-0-691-17634-5

The Library of Congress has cataloged the cloth edition as follows:
Smith, Justin E. H.
Nature, human nature, and human difference : race in early
modern philosophy / Justin E. H. Smith.
pages cm
Includes bibliographical references and index.
ISBN 978-0-691-15364-3 (hardback : acid-free paper)
1. Race—Philosophy. 2. Ethnicity—Philosophy. 3. Philosophy of nature.
4. Science—Philosophy. 5. Evolution (Biology) I. Title.
GN269.S65 2015
305.8001—dc23
2014048310

British Library Cataloging-in-Publication Data is available

This book has been composed in Sabon LT Std

Printed on acid-free paper. ∞

Printed in the United States of America

Again, we may decry the color prejudice of the South, yet it remains a heavy fact. Such curious kinks of the human mind exist and must be reckoned with soberly.

—*W.E.B. DuBois, The Souls of Black Folk (1903)*

I am black. At the same time, I have always held the religion of science to be the only true one, the only one worthy of the constant attention and the infinite devotion of any man who allows himself to be guided by nothing but free reason. How could I reconcile the conclusions that seem to be drawn from this same science against the aptitudes of the Blacks with this passionate and deep veneration that is for me a profound need of the intellect?

—*Anténor Firmin, On the Equality of the Races (1885)*

Whatever is killed and can be killed, necessarily lives.

—*Anton Wilhelm Amo, On the Impassivity of the Human Mind (1734)*

Contents

Acknowledgments

This book was written with the support of a research grant for the period 2009–12 from the Social Sciences and Humanities Research Council of Canada. It came together in its present form during a research sabbatical as a member at the Institute for Advanced Study, Princeton, in the winter and spring of 2011. Parts of chapter 3 previously appeared in Marcelo Dascal and Victor Boantza (eds.), *Controversies within the Scientific Revolution* (John Benjamins, 2011); parts of chapter 4 previously appeared in Charles T. Wolfe and Ofer Gal (eds.), *The Body as Object and Instrument of Knowledge: Embodied Empiricism in Early Modern Science* (Springer, 2010); parts of chapter 7 previously appeared in *History of Science* 50, 4 (December 2012). Material that later found its way into the book has been presented in more forums than I can document, including the Committee for Interdisciplinary Science Studies at the CUNY Graduate Center; the University of South Florida; the Humboldt University of Berlin; Université Paris Diderot–Paris 7; the École Normale Supérieure, Paris; the University of Helsinki; the Institute for Advanced Study, Princeton; Villanova University; and the Montreal Interuniversity Workshop in the History of Philosophy. The present version of the book has benefitted from inspiring discussions with, and from the careful readings and helpful comments of, Delphine Antoine-Mahut, Matthew Barker, Magali Bessone, D. Graham Burnett, Nathaniel Adam Tobias Coleman, Harvey Cormier, James Delbourgo, Claude-Olivier Doron, François Duchesneau, Beth Epstein, Victor Emma-Adamah, Dan Garber, Bryce Huebner, Julie Klein, Antoine Leveque, David Livingstone Smith, Koffi Maglo, Stephen Menn, Adriaan Neele, Anne-Lise Rey, Jason Stanley, Tzuchien Tho, Catherine Wilson, Charles T. Wolfe, and two anonymous and extremely insightful referees for Princeton University Press. Particular gratitude and admiration are reserved for Adina Ruiu, from whom I learn so much. The employees of La Présence Africaine bookstore in the rue des Écoles have generously helped me over the past year to discover many remarkable books and authors. In so doing they have greatly expanded the scope of my reading, and helped me, at just the right time, to range still further, here, from the intellectual village in which I grew up.

This book is dedicated, "hook, line, and sinker," to the memory of Kyle "Tracker" Brown (1971–2007).

—Paris, September 2014

A Note on Citations and Terminology

Unless otherwise indicated, translations from original-language texts are our own. In cases where there is an authoritative or influential English edition of a given text, this text is cited rather than the original non-English edition. Thus for example Pierre Gassendi, François Bernier, and J. F. Blumenbach are cited from period translations (or, in the last case, a later nineteenth-century translation), and thus we are reading them more or less in the same words an English-language reader would have read them in the era of their greatest influence. An attempt has been made to avoid anachronism and to stay true to actors' categories, both in our own translations of citations and in the analysis that follows. There are important debates in and beyond scholarly circles about the suitability of the terms "indigenous," "New World," and "native" for describing the Americas and their original inhabitants. With sensitivity to the issues raised in these debates, these and related terms have nonetheless been retained, in the hope of using them as a conceptual conduit to the way the encounter with this half of the world was experienced and thought by early modern Europeans. One area in which the respect for actors' categories poses considerable difficulties is in the discussion of racial categories. Terms such as "Negro" and "Moor" have been retained in citations, and when they occur in subsequent analysis they are wrapped in quotation marks that simultaneously function as scare quotes as well as indicating a classic use/mention distinction: these are *never* our own terms, but always refer back to a historical author under discussion. As for "black," "white," and related terms, these are used without quotation marks, though with a certain degree of discomfort. Since a key argument of the book concerns the historical contingency of these terms, scare quotes may be understood as implied throughout.

Nature,
Human Nature, &
Human Difference

Introduction

I.1. Nature

In 1782, in the journal of an obscure Dutch scientific society, we find a relation of the voyage of a European seafarer to the Gold Coast of Africa some decades earlier. In the town of Axim in present-day Ghana, we learn, at some point in the late 1750s, David Henri Gallandat met a man he describes as a "hermit" and a "soothsayer." "His father and a sister were still alive," Gallandat relates, "and lived a four-days' journey inland. He had a brother who was a slave in the colony of Suriname."[1] So far, there is nothing exceptional in this relation: countless families were broken up by the slave trade in just this way. But we also learn that the hermit's soothsaying practice was deeply informed by "philosophy." Gallandat is not using this term in a loose sense, either. The man he meets, we are told, "spoke various languages—Hebrew, Greek, Latin, French, High and Low German; he was very knowledgeable in astrology and astronomy, and a great philosopher."[2] In fact, this man, we learn, "had been sent to study at Halle and in Wittenberg, where in 1727 he was promoted to Doctor of Philosophy and Master of Liberal Arts."[3]

On a certain understanding, there have been countless philosophers in Africa, whose status as such required no recognition by European institutions, no conferral of rank.[4] On a narrower understanding, however, Anton Wilhelm Amo may rightly be held up as the first African philosopher in modern history. Gallandat tells us that after the death of Amo's "master," Duke August Wilhelm of Braunschweig-Wolfenbüttel, the philosopher-slave grew "melancholy," and "decided to return to his home country."

[1] *Verhandelingen uitgegeven door het Zeeuwsch Genootschap der Wetenschappen te Vlissingen*, Negende Deel, Middelburg: Pieter Gillissen, 1782, 19–20.

[2] *Verhandelingen*, 19–20. "[H]y sprak verscheiden taalen, Hebreeuws, Grieks, Latyn, Fransch, Hoog- en Nederduitsch; was zeer kundig in de Astrologie en Astronomie, en een groot Wysgeer."

[3] *Verhandelingen*, 20.

[4] For some reflections on the character and history of indigenous African philosophy, see, among many other works, Placide Tempels, *La philosophie bantoue*, Elisabethville: La Présence Africaine, 1945; Paulin J. Hountondji, *Sur la "philosophie africaine,"* Paris: Maspéro, 1976; Alexis Kagame, *La philosophie Bantu comparée*, Paris: Présence Africaine, 1976; Robin Horton, *Patterns of Thought in Africa and the West: Essays on Magic, Religion, and Science*, Cambridge: Cambridge University Press, 1993; Kwasi Wiredu (ed.), *A Companion to African Philosophy*, London: Blackwell, 2004; Emmanuel Chukwudi Eze, *African Philosophy: An Anthology*, London: Wiley-Blackwell, 1998. See also Justin E. H. Smith, *A Global History of Philosophy*, Princeton: Princeton University Press, forthcoming.

What Gallandat fails to mention is that between the time of August Wilhelm's death and Amo's departure from Germany, a scurrilous *Spottgedicht*, a satirical and libelous poem, was published in 1747 by a certain Johann Ernst Philippi. It is not clear whether the events described in the poem ever took place, but this is a question of secondary importance. Amo is accused in the poem of falling in love with a certain Mademoiselle Astrine, a German brunette. At some point the goddess Venus comes to resolve the problematic case, judging unsympathetically that "a Moor is something foreign to German maidens."[5] She then condemns him to a life of sorrow:

> You, Amo, are mistaken; with your vile nature
> Your heart will never be content.[6]

The goddess's judgment, or rather that of the author who created her, marks a sharp contrast with everything we know about Amo's earlier life. In his university education, in the appraisal of his works of philosophy and jurisprudence, in the esteem in which he was held as a teacher at Wittenberg and Jena, there is no evidence whatsoever of a perception on the part of the Germans that his African origins, let alone his skin color and other physical features, should serve as an impediment to his leading a full and rewarding life as a thinker and as a human being. To put this point somewhat differently, there was no evidence that Amo's "nature" (*Wesen*) could be judged to be something distinct from his "heart" (where this is understood as a poetic substitution for "soul" or "self," and not, obviously, as a bodily organ). How exactly a person's nature came to be associated with his or her external physical features, and how therefore a person's nature could be said to be vile simply as a result of the conformation, color, size, or shape of these features, has much to do with the very history of philosophy to which Amo devoted much of his life. The question of how "race" came to mean "nature" will be the principal focus of this book.

I.2. Historical Ontology

This book aims to do more than one thing. It is however, first and foremost, a work of historical ontology, in Ian Hacking's sense.[7] That is, it

[5]Johann Ernst Philippi, *Belustigende Poetische Schaubühne, und auf derselben I. Ein Poßirlicher Student, Hanß Dümchen aus Norden, nebst Zwölf seiner lustigen Cameraden. II. Die Academische Scheinjungfer, als ein Muster aller Cocketten. III. Herrn M. Amo, eines gelehrten Mohren, galanter Liebes-Antrag an eine schöne Brünette, Madem. Astrine. IV. Der Mademoiselle Astrine, Parodische Antwort auf solchen Antrag eines verliebten Mohren*, Cöthen: in der Cörnerischen Buchhandlung, 1747, n.p.

[6]Philippi, *Belustigende Poetische Schaubühne*, n.p. "Du irrest, Amo, dich; bey ihrem schnöden Wesen / Wird dein Hertz nie genesen."

[7]See Ian Hacking, *Historical Ontology*, Cambridge, MA: Harvard University Press, 2002.

aims to show how kinds of things, and kinds of people, that appear to be carved out within nature itself in fact come into being in the course of human history as a result of changes in the ways human beings conceptualize the world around them. As Hacking writes, "My historical ontology is concerned with objects or their effects which do not exist in any recognizable form until they are objects of scientific study."[8] Here Hacking emphasizes science, while at other times he allows historical ontology to extend to the emergence of "discourses" in general. But in the example that interests us either the narrower or the broader conception of the project will serve just as well, for the modern discourse of race runs in near perfect parallel to the rise of race as a scientific problem. In some of his most influential work, Hacking has taken the categorization of mental illness as an illustrative case study in the generation of historical, or, as he sometimes calls them, "transient," kinds.[9] Yet he emphasizes that transience is not the same thing as illusoriness, that at the very least it serves us well to take seriously and to investigate the reasons why a culture generates new kinds from one era to another, and perhaps even, following upon our investigation, to retain them.[10] Hacking's approach, which owes a deep debt to Michel Foucault's work on the history of sexuality and its focus—inter alia—on the construction in relatively recent history of the homosexual as a kind of person,[11] is certainly fruitful far beyond the study of mental illness.

In recent years there has been a great deal of very important research in genetics and related disciplines on the limits and problems of racial classifications, and in philosophy there has been an intense engagement with this research both as a problem in the philosophy of science, as well as with respect to its implications for the social ontology of race. We will be considering some of this research in the following chapter, but at the

[8] Hacking, *Historical Ontology*, 11.

[9] See Ian Hacking, *Mad Travelers: Reflections on the Reality of Transient Mental Illness*, Charlottesville: University of Virginia Press, 1998. For an interesting critique of Hacking's argument, see Rachel Cooper, "Why Hacking Is Wrong about Human Kinds," *British Journal for the Philosophy of Science* 55, 1 (2004): 73–85. For a more recent engagement by Hacking specifically with the question of race as it relates to his broader concern about kinds of people, see Ian Hacking, "Genetics, Biosocial Groups and the Future of Identity," *Daedalus* 135, 4 (Fall 2006): 81–95.

[10] Related to this, Naomi Zack has written of race: "By now, science has moved on, but common sense and humanistic scholarship lag by over a century. . . . *It is the taxonomy of human races that science fails to support, not any one or even many of the hereditary traits that society deems racial.*" Naomi Zack, *The Philosophy of Science and Race*, London: Routledge, 2002, xi.

[11] See Michel Foucault, *Histoire de la sexualité*, vols. 1–3, Paris: Gallimard, 1976–84. For a treatment of the significance of Foucault's late-period philosophy for thinking about race, particularly in a colonial context, see Ann Laura Stoler, *Race and the Education of Desire: Foucault's* History of Sexuality *and the Colonial Order of Things*, Durham, NC: Duke University Press, 1995.

outset it is worth noting that most of it neglects to take seriously the way in which categories of any sort that are deployed by human beings are deeply embedded in history, and require historical research (and ideally also cross-cultural comparison) in order to be fully understood. This is of course not the first book to suggest that the category of race—both the particular racial categories into which we divide the human species today, as well as the very idea that the human species can be so divided—might appropriately be seen as a consequence of concrete changes in modern European discourses, not least scientific discourse, about human diversity. Yet there has been little sustained study of the intellectual history of the period in which the categories of race as we understand them took shape.[12] In part, this absence can be explained by the reasonable perception that the most important factors in the shaping of these categories were not "intellectual" at all, but economic, and indeed that it might even be offensive to think of the history that left us with the horrid legacy of slavery and systemic racism as having anything intellectual about it at all. Accordingly, excellent and abundant scholarship has been produced on the economic and social history of slavery.[13] In this scholarship, in general, the things that people implicated in this history came to tell themselves and others about what kinds of people there are have been seen, not unreasonably, as at best a posteriori rationalization or coming to terms with a world economic system that had taken shape not as a result of any innovations on the plane of ideas, but as a result of the sum total of practices out of which that economic system emerged. A vivid example of this sort of thinking is offered by the colonial governor in

[12]Some, but by no means all, notable recent exceptions include Katherine M. Faull (ed.), *Anthropology and the German Enlightenment: Perspectives on Humanity*, *Bucknell Review* 38, 92 (1995); Robert Bernasconi and Sybol Cook (eds.), *Race and Racism in Continental Philosophy*, Bloomington: Indiana University Press, 2003; Robert Bernasconi and Tommy Lee Lott (eds.), *The Idea of Race*, Indianapolis: Hackett, 2000; Sara Eigen and Mark Larrimore (eds.), *The German Invention of Race*, Albany: State University of New York Press, 2007; Andrew S. Curran, *The Anatomy of Blackness: Science and Slavery in an Age of Enlightenment*, Baltimore: Johns Hopkins University Press, 2011; Joan Pau Rubiés, "Christianity and Civilisation in Sixteenth-Century Ethnological Discourse," in Henriette Bugge and Joan Pau Rubiés (eds.), *Shifting Cultures: Interaction and Discourse in the Expansion of Europe*, Münster: Lit Verlag, 1995, 35–60; Rebecca Earle, *The Body of the Conquistador: Food, Race and the Colonial Experience in Spanish America, 1492–1700*, Cambridge: Cambridge University Press, 2012.

[13]Obviously, we can barely begin to summarize this vast literature here, but a few representative works include Ira Berlin, *Many Thousands Gone: The First Two Centuries of Slavery in North America*, Cambridge, MA: Harvard University Press, 1998; David Brion Davis, *Inhuman Bondage: The Rise and Fall of Slavery in the New World*, Oxford: Oxford University Press, 2006; Patrick Manning (ed.), *Slave Trades, 1500–1800: Globalization of Forced Labour*, Brookfield, VT: Variorum, 1996; Seymour Drescher, *Abolition: A History of Slavery and Antislavery*, Cambridge: Cambridge University Press, 2009.

Jamaica, Edward Long, whose 1774 *History of Jamaica* amounts to a long-winded defense of the plantation system: "When we reflect on the nature of [black people]," Long writes, "and their dissimilarity to the rest of mankind, must we not conclude that they are a different species of the same genus? . . . Every member of the creation is wisely fitted and adapted to their certain uses, and confined within their certain bounds."[14] The uses for which Long saw Africans as suited were plantation labor and servitude, and it is reasonable to suppose that his polygenism, or the theory of separate creations for the different races, offered merely ad hoc theoretical support of an economic order he would have supported no matter what the evidence about human origins seemed to reveal.

The explanatory priority of the economic over the intellectual will not be disputed here. The infamous *Code noir* was established throughout the French colonies, enshrining in law the inequality of both African slaves as well as of free blacks, in 1685, just one year after François Bernier published his "New Division of the Earth," reducing humanity to four or five biogeographically defined "races or species."[15] It is hard not to see Bernier's proposal not as pure disinterested theorizing, but rather as a distillation of the emerging political preoccupation of his time and place: to provide an articulation in words that could impose some sense and legitimacy on an increasingly harsh and unequal system of labor extraction. Yet we will proceed here in the conviction that there is always a complex interplay between what is said, what is believed, and what is done, and that at least part of the study of the history of modern racism must consist in accounting for the way in which early modern thinkers conceptualized and talked about human diversity. There is a problem of philosophical anthropology that would have been there had there been no New World encounter with Native Americans, and had there been no transatlantic slave trade. These were the two events, the one after the other, that directly stimulated the reflections on philosophical anthropology in Europe and its colonies that are of interest in the present study, but the problem would have been there had modern history unfolded very differently, and this is something that it is in many respects easier to see when we begin by focusing our attention on the earliest history, in the sixteenth and seventeenth centuries, when slavery had not yet taken on a rigidly racial dimension for Europeans, but was as likely to be associated

[14]Edward Long, *The History of Jamaica*, 3 vols., London, 1774, 2:356, 374–75, 53, cited in Paul Stock, "'Almost a Separate Race': Racial Thought and the Idea of Europe in British Encyclopaedias and Histories, 1771–1830," *Modern Intellectual History* 8, 1 (2011): 3–29, 13.

[15]François Bernier, "Nouvelle division de la terre," *Journal des Sçavans* (April 24, 1684): 133–40, 133.

in their minds with the Ottoman slave trade or with the economic system of ancient Rome.[16]

By the time Thomas Jefferson wrote his 1787 *Notes on the State of Virginia*, the transatlantic slave trade had come to be seen as something entirely new and unprecedented in global history, rather than simply a later development of a trade that had previously been focused elsewhere and concentrated on different ethnic groups. Thus Jefferson compares the issues involved in Roman manumission of slaves with those in present-day America: "Among the Romans emancipation required but one effort. The slave, when made free, might mix with, without staining the blood of his master. But with us, a second is necessary, unknown to history. When freed, he is to be removed beyond the reach of mixture."[17] For Jefferson, ancient history could provide no key for dealing with the problems of modern American slavery, since this latter social institution was unique in history, and had no precedent, in view of its racial dimensions. But the Portuguese slave traders who had plied the coast of West Africa a few centuries earlier had by no means conceived of their activity as a radical break with past practices. Somehow, between the 1500s and Jefferson's 1787 work, modern slavery in the Atlantic world had come to be seen as fundamentally racial, as grounded in racial difference rather than simply rationalized post hoc in terms of it.

The function of polygenesis as a theory of human origins would also change dramatically from the sixteenth to the eighteenth century. The doctrine of polygenesis had its first expression in the sixteenth century in the radical freethinker Lucilio Vanini, who would as a result of this idea have his tongue torn out by the Inquisition. In his mouth, the theory that the Native Americans had a separate and natural origin, that they were earth-born rather than divinely created, let alone descended from Adam, can be seen as a sort of attempt at a thoroughgoing naturalism about humanity's origins and connection to the rest of the natural world. This thesis is familiar at least since Richard Popkin's great work on Isaac La Peyrère and the doctrine of pre-Adamism:[18] in the early modern pe-

[16]For a classic study of the institutions of slavery in Europe that preceded the Atlantic slave trade, see Charles Verlinden, *L'esclavage dans l'Europe médiévale, vol. 1: Péninsule ibérique—France*, Bruges: De Tempel, 1955; *L'esclavage dans l'Europe médiévale, vol. 2: Italie—colonies italiennes du Levant—Levant latin—Empire byzantin*, Ghent, 1977.

[17]Thomas Jefferson, *Notes on the State of Virginia* (1787), in Emmanuel Chukwudi Eze (ed.), *Race and the Enlightenment: A Reader*, London: Blackwell, 1997, 103.

[18]Richard H. Popkin, *Isaac La Peyrère (1596–1676): His Life, Work and Influence*, Leiden: Brill, 1987. For a similar thesis, see Giuliano Gliozzi, *Adamo e il nuovo mondo. La nascità dell'antropologia come ideologia coloniale: dalle genealogie bibliche alle teorie razziali (1500–1700)*, Florence: Franco Angeli, 1977. For another excellent more recent study of the long-term impact of La Peyrère's work and of the doctrine of pre-Adamism, see David Livingstone, *The Pre-Adamite Theory*, Philadelphia: American Philosophical Society, 1992.

riod, attributing separate origins to separate human groups often had more to do with questioning the authority of scripture (and, we might add, with exploring proto-evolutionary ideas that ultimately stem from Arabic-Aristotelian accounts of spontaneous generation) than with excluding certain human groups from a full share in humanity. And yet, undeniably, the two go together in surprising ways: when John Locke argues a century after Vanini that sub-Saharan Africans are the product of long, promiscuous hybridism with apes, he is *both* revisiting a Vaninian theme, promoting the non-separateness of human beings from the rest of living nature, *and* engaging in obvious apologetics for the system of slavery which he had extensive personal interests in maintaining. This can't mean that Locke's nominalism is *just* an apology for slavery; after all, he also adduces the example of the cat-rat, alongside the ape-man, in order to make the case that nature itself does not care about the boundaries between "species."[19]

That we are limiting ourselves to looking at thinkers here—and in most cases at thinkers who were able to live a life of thought in large part because the category of race was not disadvantageous to them—further distinguishes our focus not only from those who hold that economic history tells us the deepest story, but also from those who believe, again not unreasonably, that it is the muffled voices of those whom this new way of talking about kinds of people silenced and enchained that most need to be recovered by scholars today. But Isaac La Peyrère, G. W. Leibniz, Carl Linnaeus, and J. F. Blumenbach were not just mouthpieces of power. They were also heirs to a scientific and philosophical tradition that made it possible to say some things, and not others, about human nature and human difference, and what exactly they were able to say made a tremendous difference for the perception of the legitimacy of the racist institutions that were in the course of emerging in their era. How and why European authors came to have the beliefs that they had concerning race, in the period around the end of the seventeenth and the beginning of the eighteenth century, will be the focus of this study.

What we want to understand is the full intellectual background that made this realism possible in general, and that also made the particular racial distinctions between, for example, blacks and whites, possible. It is our hypothesis that a crucial feature of the emergence of the modern race concept was the collapse of a certain universalism about human

See also Livingstone, *Adam's Ancestors: Race, Religion, and the Politics of Human Origins*, Baltimore: Johns Hopkins University Press, 2008.

[19]On the relationship between Locke's entanglements in the institution of slavery and his philosophy, see Robert Bernasconi and Anika Maaza Mann, "The Contradictions of Racism: Locke, Slavery, and the *Two Treatises*," in Andrew Valls (ed.), *Race and Racism in Modern Philosophy*, Ithaca, NY: Cornell University Press, 2005, 89–107.

nature, which had been sustained by a belief in the transcendent essence of the human soul, and this belief's gradual but steady replacement over the course of the early modern period by a conception of human beings as *natural* beings, and thus as no less susceptible to classification in terms of a naturalistic taxonomy than any other natural being, plant or animal or mineral. The peculiarly modern ontologization of human difference, then, was made possible by the rejection of human nature, and the parallel insertion of humans into nature.

According to a certain, broadly Foucauldian view, kinds of people are in no small measure artifacts of social practices—they are, to put it somewhat crudely, written into existence through these practices. At the most general level, we may conjecture that the individual human being is itself such an artifact, that what it is to be an individual human in the modern world is to be registered as such, in a church registry of baptisms or in a file in the department of vital statistics. On this view, the reason why you can drive right past a dead dog on the side of the road, whereas a dead human would require you to stop, is that in the case of the human, unlike the dog, there is paperwork to be completed; the life cannot be closed off without a notation in the vital statistical records.[20] In this way, it is not that a being gets the status of a legal being in recognition of its prior status as a moral being, but rather it is precisely the other way around: beings come to have a certain moral charge to them—they may not be arbitrarily killed, and if they are dead their corpses must be treated according to a set of rules, for example—because they are classified as legal individuals.

Of course, the interplay between these two is complex, and here we are emphasizing the priority of the legal over the moral for the sake of argument. To consider another example, in the abortion debate, opponents of the practice have a prior commitment to the moral status of fetuses, and so seek to win a legal status for these entities that would reflect the moral one. But it is quite likely, given the evidence from similar cases, that if fetuses did have a long-established legal status, many who do not in our own culture believe that abortion is a moral issue would think differently. We already do think very differently about our moral commitments to different classes of animals (pests, livestock, vermin, wildlife, zoo animals, to name a few), where clearly the only basis for distinction is a legal or social one, stemming from the position they occupy in human society,

[20]For an account of the way certain administrative practices (in the event, legal ones) affect the perception of the moral status of animals, see Justin E. H. Smith, "The Criminal Trial and Punishment of Animals: A Case Study in Shame and Necessity," in Andreas Blank (ed.), *Animals: New Essays*, Munich: Philosophia Verlag, forthcoming.

and has nothing at all to do with differences in their internal capacities, their neurophysiology, or the like.

The connection to the problem of race should be obvious: kinds of people are to no small extent administered into being, brought into existence through record keeping, census taking, and, indeed, bills of sale. A census form asks whether a citizen is "white," and the possibility of answering this question affirmatively helps to bring into being a subkind of the human species that is by no means simply there and given, ready to be picked out, prior to the emergence of social practices such as the census. Censuses, in part, bring white people into existence, but once they are in existence they easily come to appear as if they had been there all along. This is in part what Hacking means by "looping": human kinds, in contrast with properly natural kinds such as helium or water, come to be what they are in large part as a result of the human act of identifying them as this or that. Two millennia ago no one thought of themselves as neurotic, or straight, or white, and nothing has changed in human biology in the meantime that could explain how these categories came into being on their own. This is not to say that no one *is* melancholic, neurotic, straight, white, and so on, but only that how that person got to be that way cannot be accounted for in the same way as, say, how birds evolved the ability to fly, or how iron oxidizes.

In some cases, such as the diagnosis of mental illness, kinds of people are looped into existence out of a desire, successful or not, to help them. Racial categories seem to have been looped into existence, by contrast, for the facilitation of the systematic exploitation of certain groups of people by others. Again, the categories facilitate the exploitation in large part because of the way moral status flows from legal status. Why can the one man be enslaved, and the other not? Because the one belongs to the natural-seeming kind of people that is suitable for enslavement. This reasoning is tautological from the outside, yet self-evident from within. Edward Long, as we have seen, provides a vivid illustration of it in his defense of plantation labor in Jamaica. But again, categories cannot be made to stick on the slightest whim of their would-be coiner. They must build upon habits of thinking that are already somewhat in place. And this is where the history of natural science becomes crucial for understanding the history of modern racial thinking, for the latter built directly upon innovations in the former. Modern racial thinking could not have taken the form it did if it had not been able to piggyback, so to speak, on conceptual innovations in the way science was beginning to approach the diversity of the natural world, and in particular of the living world.

This much ought to be obvious: racial thinking could not have been biologized if there were no emerging science of biology. It may be worthwhile to dwell on this obvious point, however, and to see what more unexpected insights might be drawn out of it. What might not be so obvious, or what seems to be ever in need of renewed pointing out, is a point that ought to be of importance for our understanding of the differing, yet ideally parallel, scope and aims of the natural and social sciences: the emergence of racial categories, of categories of kinds of humans, may in large part be understood as an *overextension* of the project of biological classification that was proving so successful in the same period. We might go further, and suggest that all of the subsequent kinds of people that would emerge over the course of the nineteenth and twentieth centuries, the kinds of central interest to Foucault and Hacking, amount to a further reaching still, an unprecedented, peculiarly modern ambition to make sense of the slightest variations within the human species as if these were themselves species differentia. Thus for example Foucault's well-known argument that until the nineteenth century there was no such thing as "the homosexual," but only people whose desires could impel them to do various things at various times. But the last two centuries have witnessed a proliferation of purportedly natural kinds of humans, a typology of "extroverts," "depressives," and so on, whose objects are generally spoken of as if on an ontological par with elephants and slime molds. Things were not always this way. In fact, as we will see, they were not yet this way throughout much of the early part of the period we call "modern."

I.3. The History of Science and the History of Philosophy

It is no exaggeration to say that early modern globalization was one of the most important impetuses behind the radical transformations that occurred within philosophy between the sixteenth and the eighteenth centuries. To some extent, the importance of intercultural encounters, both the real ones that were related by travel writers with varying degrees of veracity, as well as the fictional ones inspired by these real ones, has been duly acknowledged by political philosophers. Most of us know, for example, that Locke could not have developed his theory of private property in precisely the way he did if he had not had the example of Native Americans, who supposedly lived without private property, as a comparison class.[21] Social contract theories, it is generally recognized, drew heavily on shoddy ethnographic evidence, though ethnographic evidence nonetheless, from supposedly more primitive groups of people in recently encountered parts of the world. The significance of early modern globalization for the history of moral philosophy has been widely acknowl-

[21] See John Locke, *The Second Treatise of Government*, New York: Macmillan, 1987.

edged, too. Michel de Montaigne's argument for relativism based on the supposed cultural practices of New World natives is by now a commonplace.[22] The impact beyond political and moral philosophy, however, in epistemology, metaphysics, and natural philosophy, remains dramatically less important among scholars of the history of philosophy.[23]

In the present study attention will be paid to both sorts of impact, already considered above, on philosophical reflection concerning human nature and human difference in early modern Europe: taxonomy, or theory of classification, as it developed in European natural philosophy from the Renaissance on, on the one hand; and, on the other, the ethnographic reports that were coming back to Europe from around the world throughout the age of exploration. These might appear to be two very disparate sources, but they are actually interwoven with one another in significant ways. For many years now historians of science have emphasized the crucial importance of colonialism for the growth of scientific knowledge in the early modern period. Ideas about the number and variety of plant species, for example, and thus ideas about the scope of the undertaking of botanical taxonomy, emerged in the course of what Londa Schiebinger has very helpfully dubbed "colonial bioprospecting":[24] the task of searching the world for useful plants to be brought back to Europe for a profit. In this respect, historians of science have compellingly argued, we cannot at all adequately understand natural philosophy and natural history as these developed in Europe if we do not look at them as a regional inflection of global developments. One such crucial development was the transformation of the knowledge project of biological taxonomy into a properly global endeavor.

It should not be surprising, in turn, to find that alongside bioprospecting we are able to identify a parallel activity of "ethnoprospecting": the effort to carry out an exhaustive global survey of human diversity. It should also not be surprising that, like bioprospecting, this parallel project ended up changing the way European thinkers perceived the relevant

[22]See Frank Lestringant, *Le Brésil de Montaigne. Le nouveau monde des* Essais *(1580–1592)*, Paris, 2005.

[23]This lacuna is significantly less noticeable outside the Anglophone world, and outside the discipline of philosophy narrowly conceived. See, for example, Sergio Landucci, *I filosofi e i selvaggi, 1580–1780*, Bari, 1972; Tzvetan Todorov, *La conquête de l'Amérique: la question de l'autre*, Paris: Seuil, 1982; Stuart B. Schwartz (ed.), *Implicit Understandings: Observing, Reporting, and Reflecting on the Encounters between Europeans and Other Peoples in the Early Modern Era*, Cambridge: Cambridge University Press, 1994.

[24]See Londa Schiebinger, *Plants and Empire: Colonial Bioprospecting in the Atlantic World*, Cambridge, MA: Harvard University Press, 2007. For an excellent study of the connection among early modern commerce, colonialism, and science, see Harold J. Cook, *Matters of Exchange: Commerce, Medicine, and Science in the Dutch Golden Age*, New Haven: Yale University Press, 2007.

variety of diversity—human or plant—itself. That is, going out into the world, and undertaking an exhaustive survey of the variety of kinds, could not but impact the way in which people interested in this project thought about the nature of kinds themselves. In particular belief in the fixity and discreteness of biological species grew increasingly difficult to support in the face of massive new information about the plenitude of kinds in nature, which strongly suggested the possibility of finding intermediate kinds between any two given kinds. This possibility, in turn, strongly motivated a view of the variety of kinds on which any given kind is a variation on other kinds, rather than a sharply cordoned off, eternally fixed class.

Thus, we will be looking at early modern taxonomy, along with the project of bioprospecting that feeds into it, as an endeavor with significant similarities to and connections with early modern travel writing, conceived in part as the project of "ethnoprospecting." It will be argued that in order to understand the forces that shaped thinking about racial difference in early modern philosophy, we must look to the philosophers' own interest in the projects of scientific classification and physical anthropology, with an eye to the way these projects were influenced by early modern globalization and by the associated projects of global commerce, collection, and systematization of the order of nature.

So far, we have claimed to be principally interested in the history of philosophy, yet have also been moving fairly nonchalantly between this history and the history of science. It is easy to explain why we are permitting ourselves to move so freely in these two worlds, without much attention to the boundary that supposedly divides them: methodologically, the approach here is one that takes actors' categories seriously, and as a matter of simple historical fact early modern philosophers conceived of the domains of scientific inquiry mentioned above as very much part of the project of philosophy. They did not call these domains of inquiry "science," let alone "biology" or, more narrowly, "taxonomy." They called them, variously, "natural philosophy" and "natural history," the former implying a search for the general principles of nature, the latter usually involving a cataloguing of particulars. Both of these terms were rich with significations that have been lost in the intervening centuries, and part of the aim of this book is to reconstruct these significations to the point where we become able to see how and why thinkers from Pierre Gassendi and Gottfried Wilhelm Leibniz up through Johann Friedrich Blumenbach and Immanuel Kant were able to conceptualize questions of what we would call "physical anthropology" as part and parcel of the project of philosophy.

As we will see, it is not simply that they took an interest in questions of measurable human variation as Renaissance men, they did not see these questions as side interests or digressions from their main project as metaphysicians, any more than Aristotle saw his own interest in, say, marine invertebrates as a hobby rather than as a constitutive part of his overarching philosophical project. Physical anthropology, taxonomy, and natural history (conceived, with Leibniz, as the surveying and cataloguing of "singular things") simply *were* philosophy for the people we all recognize today as the great modern philosophers, and to neglect to study these areas of inquiry *as* philosophy is to fail to take an interest in the scope and aims of these thinkers' philosophical projects. Moreover, even if respect of actors' categories is not part of your own scholarly approach to the history of philosophy, it is undeniable that the physical anthropology, taxonomy, and other endeavors of the philosophers we are considering were in fact integrated with the concerns of theirs that we are all still perfectly capable of recognizing as philosophical. Most pertinently, all of these domains of inquiry were centrally focused on the problem of human nature—*Quid sit homo*?—and the philosophers who pursued them simply took for granted that the study of physical anthropology and related areas had an important role to play in the answering of this fundamental philosophical question.

Once we recognize that inquiry into the natural world was in part constitutive of early modern philosophy, it becomes easier to map this inquiry in relation to other concerns of early modern philosophers. Some of these concerns will be easily recognizable from today's perspective as philosophical, but others, like much of natural science itself, will have also, in the intervening centuries, lost their evident philosophical relevance. One of these faded concerns of early modern philosophy, whose relevance we will attempt to recover here, indeed lies very close to natural history on the conceptual map of early modern philosophy: the humble genre of travel writing. In fact, far from being the mere early modern equivalent of tourist guides, travel reports are the closest thing we have from that era to what would later develop into comparative ethnography. Exposure to other cultures, as well as the attempt, though often crude and judgmental, to make sense of their beliefs and practices, was an extremely important source of reflection on the nature of morality, reason, and other key notions of Western philosophy. With very few exceptions, early modern philosophers were explicitly and deeply interested in figuring out the implications of contemporary encounters with other cultures—both with their "folk science" or knowledge systems, as well as with their customs and morality—for the understanding of basic philosophical questions. Consider, for example, two of the most significant texts of Renaissance and early modern philosophy: Thomas More's *Utopia*,[25] and a century later

[25]Thomas More, *De optimo rei publicae statu deque nova insula Utopia*, Leuven, 1516.

Francis Bacon's *New Atlantis*,[26] both of which construct their visions of the philosophical project around the fictional artifice of an ethnographic encounter with inhabitants of islands, each said to be located somewhere in the New World. The few cases in which philosophers exhibit no such interest, such as that of René Descartes, far from being the norm, amount to exceptions that are themselves in need of explanation.[27]

It is at least true that for the most part, sixteenth- and seventeenth-century naturalists gave short shrift to the variety of cultures encountered throughout the world, preferring instead to devote their attention to the botanical and geographical features (and, to a lesser extent, to the zoological varieties) they came across. When "natives" figure at all in utopian fictions, they are generally so removed from the ethnographic information that was already available as to warrant the conclusion that they are intended simply as projections of European aspirations, and not as in any way modeled after the New World inhabitants themselves. Indeed, most typically the inhabitants of distant islands off the coast of South America were imagined as descending from ancient seafarers who originated in the Old World, and thus who had access to ancient learning, as well, perhaps, to ancient revelation. But this fictional imagining of origins also reflected a quite legitimate theory of human monogenesis current in early modern ethnography. Native Americans triggered a lively debate in sixteenth-century Spain and the Spanish colonies concerning whether or not Americans are natural slaves in Aristotle's sense, or whether they are fully equal to Europeans.[28] *Libres esprits* of the same era, such as Montaigne, were, again, interested in the figure of the New World "cannibal" as a challenge to the universality of moral principles.

In the early to mid-seventeenth century, for some philosophers the very existence of cultural difference seems to have presented a troubling obstacle to their own claim on universality. Whether or not one is correct in describing the early seventeenth-century as a "counter-Renaissance," in which humble curiosity about particulars—including particular cultural practices—is replaced by an ambitious quest for an absolute and universal certainty that would not mesh well with any interest in cultural

[26]Francis Bacon, *Nova Atlantis. Mundus alter et idem, sive Terra australis antehac semper incognita: longis itineribus peregrini Academici nuperrime lustrata Authore Mercurio Britanico . . .* , Utrecht: Apud Joannem a Waesberge, 1643 [1624].

[27]Franklin Perkins has interestingly contrasted Descartes's style and aims as a philosopher with those of Leibniz, in particular with an eye to explaining why in the former case we find scarcely any mention of the non-European world (or, one might add, any particular interest in European customs, tradition, etc., as worthy of attention). See Perkins, *Leibniz and China: A Commerce of Light*, Cambridge: Cambridge University Press, 2005.

[28]See Anthony Pagden, *The Fall of Natural Man: The American Indian and the Origins of Comparative Ethnology*, Cambridge: Cambridge University Press, 1986.

diversity,[29] the absence of references to non-European cultures in figures such as Descartes is at least noteworthy. The world outside Europe, Descartes may well have thought, could only provide complicating and messy evidence against the universality of his claims, and, more damagingly, against the a priori method of producing claims about what sort of entity a human being is. Descartes was not a skeptic (other than in a methodological sense), and perhaps correspondingly he was defiantly uninterested in the social world beyond the Republic of Letters. Those, in contrast, who believed in the importance of the accumulation of particulars, rather than in adherence to "notions" or concepts of reason, were intensely interested in gathering ethnographic information from explorers, and in acquiring natural knowledge indirectly from the subjects of ethnography themselves.

Scientific institutions such as the Royal Society of London—whose members were all, by seventeenth-century standards, philosophers—were by contrast intensely interested in collecting data on the flora and fauna, climates and cultures, of faraway places. And Locke, returning to the same sort of ethnographic evidence to which Montaigne had appealed a century before, makes the case for the absence of universal human essence on the grounds that cultural norms are radically different throughout the world. "There is scarce that Principle of Morality to be named," he writes in the *Essay Concerning Human Understanding* of 1690, "or Rule of Vertue to be thought on which is not, somewhere or other, slighted and condemned by the general Fashion of whole Societies of Men."[30] There seems indeed to be a direct correlation between the basic epistemological position taken up by seventeenth-century thinkers as to the usefulness of knowledge of particulars in contrast with universal principles, on the one hand, and on the other those thinkers' interest in drawing upon examples from the nascent (and largely fantastical) ethnographic literature for the purpose of illustrating some point of practical or theoretical philosophy.

Even in the case of Descartes, we are not dealing with someone who simply has no interest in the study of other cultures. Rather, the philosopher initially takes a deep interest in cultural difference, but consciously determines not to place it at the center of his philosophical inquiry, precisely because, in its variability, it seems unable to offer any answers to the universal questions that are of interest to him. Thus Descartes writes in chapter 1 of the *Discourse on Method*,

[29]Such an account was offered in Stephen Toulmin, *Cosmopolis: The Hidden Agenda of Modernity*, Chicago: University of Chicago Press, 1990.

[30]John Locke, *Essay Concerning Human Understanding*, book I, chap. 3, 16, in *The Works of John Locke Esq; In Three Volumes*, vol. 1, London: John Churchill, 1714.

> It is true that, while busied only in considering the manners of other men, I found here, too, scarce any ground for settled conviction, and remarked hardly less contradiction among them than in the opinions of the philosophers. So that the greatest advantage I derived from the study consisted in this, that, observing many things which, however extravagant and ridiculous to our apprehension, are yet by common consent received and approved by other great nations, I learned to entertain too decided a belief in regard to nothing of the truth of which I had been persuaded merely by example and custom.[31]

It is thus not that Descartes's deductive method, his approach to philosophy as an activity for the cloistered meditator, is in no way a result of a lack of interest in human cultural diversity, but rather of a concern, not at all unfounded, that the human species is simply too diverse to yield up answers to fundamental questions about human nature through the study of this diversity.

Some philosophers might object that to focus on the way race occurs in historical philosophical texts not centrally concerned with race is to mistake contingent or trivial features of a work for its actual substantial content. This is a common complaint leveled by philosophers against scholars who engage with philosophical texts from a position outside of philosophy narrowly conceived. Thus Martha Nussbaum has criticized Judith Butler's view that J. L. Austin heterosexualizes social bonds by taking "I thee wed" as the paradigmatic example of a speech act. Nussbaum comments, "It is usually a mistake to read earth-shaking significance into a philosopher's pedestrian choice of examples. Should we say that Aristotle's use of a low-fat diet to illustrate the practical syllogism suggests that chicken is at the heart of Aristotelian virtue?"[32]

Yet, pace Nussbaum, one might also reasonably suggest that the significance of examples should be judged on a case-by-case basis. The stock examples of chalk, tables, and desks that philosophy instructors employ in teaching in fact say quite a bit about our ontology, and it is not at all a disruption of philosophical reflection, but indeed a deepening of it, to ask what the taking of these mundane classroom objects as paradigmatic instances of thing says about our ontological commitments. Aristotle and Leibniz both, for example, would have seen chairs and chalk as each in need of a different sort of ontological analysis, whereas we tend to run them together. When it comes to examples touching upon race or gender, philosophers tend to be somewhat more tolerant than Nussbaum is here of Butler, and indeed many people working very much from within the

[31] René Descartes, *Discours de la méthode*, in *Oeuvres de Descartes*, ed. Charles Adam and Paul Tannery, Paris: Léopold Cerf, 1902, 6:10.

[32] Martha Nussbaum, "The Professor of Parody," *New Republic*, February 22, 1999.

discipline of philosophy, rather than picking away at it from outside, as Nussbaum supposes Butler to be doing, have acknowledged that examples drawn from racial or ethnic others in the history of philosophy may be more relevant to understanding this history than chicken is to Aristotle's virtue theory. Charles W. Mills, for example, argues that the offhand comments about race in classical philosophical works, far from being tangential to the main concerns of the authors, in fact express the overall systematic aim of these works: to perpetuate a social contract that promotes the interests of whites over other people.[33] We do not need to scrutinize his argument in order to appreciate that at the very least there are serious philosophers who are rather more interested in drawing out the significance of examples than Nussbaum thinks we ought to be.

In the present study, we are approaching the history of the concept of race as a problem for philosophy itself, and are taking treatments of race in philosophical texts in history as significant, even where race is not the principal topic of explicit concern for the author. It is hoped that in so doing we are not working against the texts, or proceeding deconstructively or antiphilosophically or anything like that, but only trying to understand the complex history of a difficult problem by paying attention to all the traces it leaves. As the nineteenth-century Haitian philosopher Anténor Firmin wrote of the science of anthropology (in which he included the historically grounded defense of the equality of races): "No study was ever more complex. Here, one must reason with confidence about all subjects, whether they are concerned with mind or with matter . . . , with phenomenon and with noumenon, following the terminology of Kant. Not everyone has it in them."[34]

I.4. Aims and Outline

Until now, we have painted a fairly detailed picture of many of the theoretical concerns in the background of the present study. We have yet to make our positive theses explicit, and we have yet to spell out how, concretely, we will go about arguing for them.

There are in fact three interwoven argumentative aims of this book. The first is that, whatever its philosophical shortcomings and whatever ill consequences its corollary theses had, metaphysical dualism in fact

[33] Charles W. Mills, *The Racial Contract*, Ithaca, NY: Cornell University Press, 1997.

[34] Anténor Firmin, *De l'égalité des races (anthropologie positive)*, Paris: Librairie Cotillon, 1885, 4. "Jamais étude ne fut plus complexe. Là, il faut raisonner avec assurance sur tous les sujets, qu'ils relevent de l'esprit ou de la matière; il faut envisager le monde et la pensée, le phénomène et le noumène, suivant la terminologie de Kant. Cela n'est pas de la force de chacun."

served as an important bulwark against the rise of essentialist thinking about human racial diversity, against the possibility of taking, as Philippi did in the case of Amo, the physical body for the "nature." That is—and this will be shown through a number of historical case studies—so long as the human soul was thought to be something fundamentally independent of the body, physical differences between human beings could not be taken as markers of essential difference.

Correlatively, a **second** thesis is that it was the naturalization of the human being, the discovery of the possibility of the study of human beings as natural entities, that made dualism a moribund research program by the end of the eighteenth century, and that also, simultaneously, made essentialist racial thinking possible. This thinking tracked in unmistakable ways the methods and systems that were actively being applied for the study of other, nonhuman domains of nature, particularly other domains of the living world, and in this respect we may say that the naturalization of the human being and the corollary rise of racial thinking involved an overextension of systematic scientific thought, in particular of taxonomic methods fruitfully applied in botany and to a lesser extent in zoology, chemistry, and other domains, to the human species. To put this second argument somewhat more succinctly: when human beings were fully inserted into nature, the unity of the human species was lost, and different groups were now held, themselves, to have different "natures." Humanity became, to use what is now a very polysemous word, as "diverse" as nature itself.

Third and finally, it will be argued that, while a commitment to the existence of a body-independent rational soul was in certain important respects a bulwark against the rise of modern racism, the interactions between this commitment and the rising naturalism of the modern period helped over the *longue durée* to generate the modern,[35] racially charged dichotomy between two basic varieties of people: the people of reason and the people of nature, so to speak. This is where the role of philosophy in the history of the concept of race is particularly important, and very much in need of excavation and analysis, for it was a conceit of the philosophical project itself, a project conceived as central to the identity of Europeans and generally conceived as nontransferable and unshareable with non-Europeans, that helped to strengthen the appearance of a fundamental difference between "the West" and "the rest." This ignoble contribution of philosophy was not in itself racial, but it helped to

[35]This term is of course most closely associated with Fernand Braudel and the Annales school of history, whose members sought to emphasize long-term processes in historical scholarship over the study of singular events or *histoire événementielle*. See in particular Braudel, "Histoire et sciences sociales: la longue durée," *Annales. Histoire, Sciences Sociales* 13, 4 (October–December 1958): 725–53.

ground in something apparently eternal and foundational the parallel project of natural science, which in the same period was actively seeking to establish essential differences between human groups by appeal to ever more fine-grained descriptions of minor, which is to say nonessential, differences. This contribution, moreover, is most evident in modern philosophy's engagement, or, more often, failure to engage, with the indigenous knowledge systems of the non-European peoples who had, since the end of the fifteenth century, been thrust into complicated, and generally disadvantageous, forms of contact and exchange with Europeans.

These theses are not entirely new. In a quite different and perhaps more appropriate scholarly idiom, Denise Da Silva has forcefully argued in her 2007 book, *Toward a Global History of Race*, that the purported universalism of the modern idea of "humanity" has been compromised all along by the fact that it is in its essence a racial idea. She argues, in particular, that from the seventeenth to the nineteenth centuries, science and philosophy were committed to "protecting *man*" by "the writing of the subject as a historical, self-determined thing—a temporary solution consolidated only in the mid-nineteenth century, when man became an object of scientific knowledge."[36] She goes on to show how the sciences of man dealt with the fundamental ontological problem of "humanity" by "deploy[ing] racial difference as a constitutive human attribute."[37] Da Silva's argument is subtle and impossible to summarize here, but from what has already been mentioned, the overlap with the present project should be clear. One significant point of disagreement, however, is that while she sees the fragmentation of the species into races as an extreme measure taken to "protect man," here it will be argued that this fragmentation was itself a consequence of a forceful rejection of an older and time-honored conception of humanity.

We are covering a very broad time period and a great number of authors, and are moving fairly freely between these. Such an approach is of course discouraged for many scholarly purposes. Our period extends, roughly, from the early sixteenth to the late eighteenth century, or, to put it handily, from Columbus to Kant. Other than for the middle of this span, namely, the second half of the seventeenth century and the beginning of the eighteenth, and in particular the work of Leibniz and his contemporaries, we can claim to offer no truly original scholarship, and are instead relying on the tremendous amount of interpretative work that is already available. But if our approach to these periods is avowedly derivative, it is

[36]Denise Da Silva, *Toward a Global History of Race*, Minneapolis: University of Minnesota Press, 2007, xiii.

[37]Da Silva, *Toward a Global History of Race*, xiii.

nonetheless justified in relation to the methodological aims and objectives of the present study.

This is so for at least three reasons. First, by taking the wide-focused perspective, it becomes possible to make out the true nature of the detailed contours of the period we know best and are most concerned to interpret correctly—namely, again, the period marked out by the mature work of Leibniz and various of his contemporaries, *the* period, as will be shown, in which the most important developments in the history of the concept of race were occurring. Second, the majority of scholarship on the history of the concept of race has taken the eighteenth century as the most pivotal period for understanding the concept's legacies today. By anchoring ourselves here in the seventeenth century, and erring into the high Enlightenment principally in order to make sense of *longue durée* developments that gave their first signs far earlier, we are helping to reorient the discussion of the history of the race concept chronologically, and in particular are helping to decouple it from the handful of authors, such as Denis Diderot, David Hume, Blumenbach, and Kant, who are generally taken to have contributed significantly, for better or worse, to the concept's legacy, rather than being taken as having jumped into the discussion of a topic that already had a long independent history before them. Finally, it is only by sampling freely from such a wide array of anecdotes, examples, and arguments that we gain the sort of clear picture, following a sort of Baconian method, that enables us to move to the level of generalities, that justifies our attempts to identify broad historical trends. This is certainly not how all scholarship proceeds, and narrow and focused studies are certainly of value. But the survey has its place as well.

In the first chapter, we will engage with current philosophical accounts of racial categories, paying particular attention to the relevance for recent philosophy of both social constructionism as well as the cognitivist approach in understanding categorial schemes such as that of modern racial classification. Our principal concern in this chapter is to show that, while much of the recent literature in fields as diverse as cognitive anthropology, analytic philosophy of race, and postcolonial theory has been tremendously useful in clarifying the precise nature and function of racial categories, and in accounting for their tenaciousness in a world in which they are recognized to be of little scientific value (with a few caveats, as we will see), nonetheless these approaches must be complemented by a deepened understanding of the historical development of the categories they call into question, and of the way current thinking about race is shaped and also constrained by a past of which we remain largely unaware.

In the second chapter we will consider a cluster of thorny methodological problems in any undertaking to examine the historical and philosophical background of the modern concept of race. In particular, we will highlight the following problem: if race is acknowledged to not exist in any transhistorical sense, and to have a vague and constantly mutating referent throughout history, then how can we know we are picking out the right textual sources that will tell us what people thought about race and when? A particularly vivid example, as will be shown, may be drawn from early modern medical writing: blackness, here, often has entirely nonracial connotations, and yet often these connotations become overlain or mingled with later strata of racial thought, and it is extremely difficult, as an interpretive matter, to determine in many cases whether there is a racial dimension to a given reference to blackness or not. In this chapter we will also be arguing for the necessity of a sort of casuistical approach, which does not eschew the study of particular cases, anecdotes, offhand examples or remarks that occur in the course of making other philosophical arguments. All of these, it will be argued, should be taken seriously in any attempt to give a general account of the functions race fulfills in the history of modern philosophy.

In the third chapter, with the bulk of our theoretical and methodological concerns by now spelled out, we will turn to the history of the race concept more narrowly speaking. We will first consider the early development of thinking about human diversity and human origins in the context of the Renaissance. In important respects, as we will see, later reflections in European philosophy echo debates that played out a century earlier within the Ibero-American world, largely as a result of the fact that the Iberians were the earliest Europeans to have significant encounters with non-European peoples in the modern era. We will focus in particular on those sixteenth-century engagements with the *novissima americana*, the latest news from the Americas, that dealt with the question of the origins and nature of biological kinds in the New World, and particularly with the origins and nature of New World peoples. One thing that will become clear in the discussion in this and the following chapter is that the possibility of radical or essential difference between different human groups, an essential difference that would be traced back to the different groups' distinct origins, was understood in the Renaissance and at the beginning of the early modern period to be a radically libertine view, and all but the most extreme thinkers held to some sort of conception of all human beings as fundamentally the same in virtue of the fact that they all descend from Adam and all possess saveable souls.

In the fourth chapter, we will follow the development of Renaissance debates about the nature of human diversity into the following century, and particularly as they played out in France and England. We will focus in particular on so-called polygenesis, as represented most famously by Isaac La Peyrère's pre-Adamism, according to which different human groups were created at different moments. On the most common version of this view, some or all of the peoples of non-Abrahamic faiths were created—and given an oblique and widely interpretable mention in scripture—prior to Adam. We will further consider some of the elaborate attempts to account for the migration of human beings throughout the world in terms of diffusionist models, focusing in particular on the work of the seventeenth-century English jurist Matthew Hale, and arguing that such attempts played an important role in the increasing naturalization of modern paleoanthropology, and therefore also of modern accounts of the nature of human diversity.

In the fifth chapter, we will focus on those early modern accounts of human phenotypic diversity (to use today's terminology) that do not resort to claims of essential difference, but instead appeal to some form or other of degeneration to account for human diversity. Degenerationism is the view that there was an original, ideal type of the human species (and generally also of animal species), but that different groups have deviated from this perfect state as a result of migration, changes in diet and in climate, and hybridity with other species. We will see that degenerationist accounts of human variety are particularly interesting in the way they conflate descriptive and normative claims. This is particularly clear in the work of the mid-seventeenth-century English Baconian philosopher John Bulwer, who effectively accounts for *all* human diversity in terms of harmful cultural practices. We will also consider in some detail the place of apes, and in particular of higher primates, in degenerationist reflections on the lower limits of the human species.

In the sixth chapter, we will turn to François Bernier's contribution to the history of racial thinking. This French physician and traveler is often credited with being the key innovator of the modern race concept. While some rigorous scholarship has recently appeared questioning Bernier's significance, his racial theory is seldom placed in his context as a Gassendian natural philosopher who was, in particular, intent to bring his own brand of modern, materialistic philosophy to bear in his experiences in the Moghul Empire in Persia and northern India. It will be argued that Bernier's principal innovation was to effectively decouple the concept of race from considerations of lineage, and instead to conceptualize it in biogeographical terms in which the precise origins or causes of the original differences of human physical appearance from region to region remain underdetermined.

In the seventh chapter, we will undertake an extensive treatment of the place of Leibniz in the history of the concept of race. In particular the significant points of difference between his own view, on the one hand, and Bernier's biogeographical view, on the other, will be drawn out. It will be shown, in fact, that Leibniz remains thoroughly committed to a conception of race that is rooted in earlier ideas about the temporal succession of members of a family or lineage. It will be shown, moreover, in what way his analysis of race may be seen as a concrete application of his very deepest philosophical commitment, according to which the order of the world amounts to a multiplicity that is underlain by unity. What we will see, in effect, is Leibniz's application of this philosophical commitment in his own account of human diversity, in particular through an analysis of his view that the human species is best divided up not according to physical differences from one region to the next, but rather according to differences of language, which itself is nothing other than the expression of a universal human reason.

In the eighth chapter, we will focus on the life and work of Anton Wilhelm Amo, to whom we have already been introduced above, who was active in Germany in the period between Leibniz and Kant. We will be particularly interested to see how Amo's identity as an African in Europe helped to shape both his philosophy and its reception, and what lessons may have been drawn in the era for thinking about the relationship between human racial diversity, on the one hand, and the universality of human reason, on the other. It will be argued, finally, that the position occupied by Amo in the philosophical landscape of early eighteenth-century Germany reveals the likely influence of Leibniz, who had provided a model for a nonracial philosophical anthropology for which he has generally not been given much credit.

In the ninth chapter, finally, we will undertake a survey of some of the more important developments in the history of the concept of race in eighteenth-century Germany. There was an interesting inconsistency, we will see, between the desire to make taxonomic distinctions and a hesitance to posit any real ontological divisions within the human species. This inconsistency was well represented in the physical-anthropological work of Blumenbach, one of Kant's great intellectual adversaries and, in many respects, the most important eighteenth-century theorist of human difference. Johann Gottfried Herder, a contemporary of Blumenbach's, was intensely interested in human diversity, but saw this diversity as entirely based in culture rather than biology, and saw cultural difference as an entirely neutral matter, rather than as a continuum of higher and lower. Herder constitutes an important link, it will be argued, between early modern universalism of the sort represented by Leibniz, on the one hand, and on the other the ideally value-neutral project of cultural anthropology as it would begin to emerge in the nineteenth century.

For the most part, the chapters may be read separately, though where fitting there are indications in the text for where to find related discussions in other parts of the book. The aim is to make the case for the three theses identified above not by a single, cumulative argument, but by following these theses like leitmotifs through a wide variety of texts and authors.

Chapter 1

Curious Kinks

1.1. Essence

To be a realist about race is to hold that racial categories pick out real kinds in nature. The philosophical notion of kind is difficult to summarize, and appears to be inherently unclear, but a few things may be said about it at the outset. To hold that something is a natural kind, in the philosophical tradition extending back to John Stuart Mill, is to hold that two individuals of that kind have properties in common that lie deeper than the properties in virtue of which we human beings group them together. The examples Mill invokes to clarify his distinction between groupings that are based on observable properties, on the one hand, and groupings based on natural-kind membership, on the other, are "white things" for the former, and animal, vegetable, and chemical kinds for the latter. Mill emphasizes that white things have no common properties other than their color, while "an hundred generations have not exhausted the common properties of animals or of plants, of sulphur or of phosphorus."[1] It is crucial for Mill that ultimately whether a given grouping picks out a natural kind or not is a question that can be resolved only empirically. The only way to determine whether white people constitute a real kind, in a way that white things do not, is through further study. This approach extends for Mill, not least, to the question of race: "To the logician, if a negro and a white man differ in the same manner (however less in degree) as a horse and camel do, that is, if their differences are inexhaustible, and not referable to any common cause, they are different species, whether they are descended from common ancestors or not. But if their differences can all be traced to climate and habits, or to some one or a few special differences in structure, they are not, in the logician's view, specifically distinct."[2] Mill is, in an important sense, right: there is no prima facie reason why there

[1] John Stuart Mill, *System of Logic: Ratiocinative and Inductive*, in John M. Robson (ed.), *The Collected Works of John Stuart Mill*, vol. 7, Toronto: University of Toronto Press, 1974, 156. For a useful discussion of this important text, see Ian Hacking, "Why Race Still Matters," *Daedalus* 134, 1 (Winter 2005): 102–16.

[2] Mill, *System of Logic*, 158.

should not be subspecies differences within *Homo sapiens*, nor is there any reason why these differences should not mark out "logical species," that is, again, kinds that are such that their differences from other kinds are not exhausted by the properties in virtue of which they are classed together. But of course the fact that race's status as a natural kind is a matter to be resolved empirically by no means prevents race as a social kind from having a life and history of its own. And part of the life of race as a social kind involves a certain resistance to empirical evidence.

Throughout this book, we will be using "realism" and "essentialism" more or less interchangeably. There are of course subtle distinctions between these two technical terms in philosophy, indeed each has a long and storied history, and each has been used in many different, though usually related, ways. In the history of philosophy, "essentialism" has generally described the view that any given being will have an essence, in virtue of which this being may be said to be the sort of being it is. Thus a particular being is held to be a cat because it shares in the essence of cathood. Usually implicit in this view is the idea that the abstract essence of a being is always going to amount to a more perfect or complete manifestation of the sort of being in question than will the particular being that shares in it. Thus there will always be at least some respects in which any particular cat fails to correspond to the ideal image of the cat, and according to essentialism it is this ideal image that tells us most definitively what it is to be a cat. Dan Sperber notes compellingly that precisely what enables us to pick out "defective" representatives of a given biological kind as representatives of that kind is that we interpret the supposed defect as a lack, and we mentally fill in what is missing. A three-legged cat, is missing *its* leg; to speak somewhat paradoxically, there *is* a leg, or so we imagine, that needs to be put back in our own cognition if in fact it happens to be missing in reality. Cats simply are four-legged, no matter how many actually existing legs any given cat will have.[3] To use the example discussed by Sperber, one can hold simultaneously that "birds have feathers" and that "this bird has no feathers," whereas one cannot hold that "this husband has no wife." Thus a key feature of essentialism is that it holds that there is some abstract standard by which to measure an individual being's degree of success or failure in its membership in a kind; the kind in turn is held to be something at least conceptually distinct from its individual members. For the most part, today, science strives to avoid essentialism. In biology, in particular, scientists prefer one version or another of the view that, typically, there need not be any fact of the

[3] See Dan Sperber, "Why Are Perfect Animals, Hybrids, and Monsters Food for Symbolic Thought?," *Method & Theory in the Study of Religion* 8, 2 (1996 [1975]): 143–69.

matter about the necessary and sufficient criteria for a particular being's membership within that population.

One interesting thing to note here is that, on this conception of essentialism, there has been virtually no explicit defense in the history of Western philosophy and science of such a view as applied to races. The explanation for this is simple: almost no one was ready to suppose that there is a positive essence in which all members of a nonwhite race share. That is, almost no one who has held that the human species naturally divides into distinct, real subgroups has been willing to admit that these subgroups are, so to speak, separate but equal, and that there is a way for a black person to be fully and perfectly a black person, alongside a white person who is fully and perfectly a white person, in the way that, say, a dog and a cat next to each other may be said to be perfectly adequate representatives of their respective kinds. Instead, the vast majority of racial theorists have supposed that it is only one subgroup of human beings that represents humanity par excellence—and in modern history, this subgroup is always constituted by white people—while other groups are defined in terms of the degree to which they approximate the standard set by this privileged group. It would be a strange sort of essentialism that, by comparison, held that there really is an essence of doghood, and cats represent particularly poor instantiations of it.

How, then, are we to make sense of racial essentialism, if it is so obviously not essentialist in the standard way, insofar as racial thinkers tend to suppose that there is only one human essence, and that different races succeed or fail in differing degrees to instantiate this essence? One possible explanation of what is happening is that in racial thought "human" comes to represent a higher-order taxon, akin to, say, "bird" rather than to "cat." Here, we might well find folk taxonomies holding that, say, an eagle is a more complete realization of the essence of birdhood than, say, a penguin. One might believe that there is also an essence of penguinhood, but this latter essence would not enter into consideration when considering how well or poorly the penguin manages to instantiate birdhood relative to the eagle. It is not however clear that this is what is happening in racial thinking. For the most part, as we have already begun to see, racial theorists have, to appropriate a theory developed by Thomas Laqueur for the history of gender,[4] held to what might be called a "one-race model" of human diversity: all human beings are held up to the ideal of the white man, and nonwhite men and women are seen as mere approximations of this ideal. The failure or misfire is typically accounted for in terms of "degeneration" (as we will discuss in detail below): the original stock of

[4] See Thomas Laqueur, *Making Sex: Body and Gender from the Greeks to Freud*, Cambridge, MA: Harvard University Press, 1990.

the human species started out with a European phenotype, and drifted off course as a result of climatic influence and diet, and also possibly as a result of lax morals and hybridism.

And yet, as much recent work indicates, it may be that we are looking in the wrong place if we try to find racial essentialism in any explicit or coherent theory of essences. Instead, we find it in the relatively unreflected responses of children and adults to questions about particular cases: a casuistic essentialism, perhaps, however paradoxical this may sound. And here, often, the evidence strongly suggests that people do think in an essentialist way, pretheoretically, about racial difference. As Sally Haslanger has helpfully summed up the state of scholarship within philosophy on race, an earlier generation of scholars sought to examine in an a priori way what is meant by race,[5] while more recently scholars have started to investigate in an empirical vein the ways in which race shows up in folk theories of human diversity.[6] Haslanger herself, along with Jennifer Saul,[7] has begun to argue that as philosophers we should not only analyze the concept or its empirical applications, but also investigate what we *should* mean when we deploy racial terms.[8]

Yet left out among these options is the historical reconstruction of the origins and development of the concept of race, which has the greatest affinity with the empirical approach Haslanger cites, to the extent that it is interested in providing an account of the ways in which people in fact do apply or have applied the concept of race. But the historical approach goes further, insofar as it aims, through a reconstruction of the way in which the concept came into being, to gain a sense of just how contingent and dispensable, or, on the contrary, how deep-rooted and tenacious, it might prove to be. Such an account, then, is necessarily complementary to

[5] See J. L. A. Garcia, "Current Conceptions of Racism: A Critical Examination of Some Recent Social Philosophy," *Journal of Social Philosophy* 28 (1997): 5–42; J. L. A. Garcia, "Philosophical Analysis and the Moral Concept of Racism," *Philosophy and Social Criticism* 25 (1999): 1–32; Lawrence Blum, *I'm Not a Racist, But . . . : The Moral Quandary of Race*, Ithaca, NY: Cornell University Press, 2002; Michael O. Hardimon, "The Ordinary Concept of Race," *Journal of Philosophy* 100 (2003): 437–55; Ron Mallon, "Passing, Traveling and Reality: Social Constructionism and the Metaphysics of Race," *Nous* 38 (2004): 644–73.

[6] See in particular Joshua Glasgow, *A Theory of Race*, New York: Routledge, 2009; Joshua Glasgow, Julie L. Shulman, and Enrique Covarrubias, "The Ordinary Conception of Race in the United States and Its Relation to Racial Attitudes: A New Approach," *Journal of Cognition and Culture* 9 (2009): 15–39; Luc Faucher and Edouard Machery, "Racism: Against Jorge Garcia's Moral and Psychological Monism," *Philosophy of the Social Sciences* 39 (2009): 41–62.

[7] Jennifer Saul, "Philosophical Analysis and Social Kinds: Gender and Race," *Proceedings of the Aristotelian Society* 80 (suppl., 2006): 119–44.

[8] See Sally Haslanger, "Language, Politics and 'the Folk': Looking for 'the Meaning' of 'Race,'" *Monist* 93, 2 (April 2010): 169–87.

the sort of approach Haslanger is advocating, to the extent that, if we are going to entertain proposals about how race should be talked about, it behooves us to come to understand also the way in which history shapes and perhaps constrains the possibilities of our current and future race talk.

1.2. Race and Cognition

It has been suggested above that newly coined categories must build upon ones that are already in place. If this is correct, it should not be surprising to learn that the racial categories that appeared over the course of the early modern period built upon a somewhat earlier, but still thoroughly modern, scientific project of biological (to use our own, anachronistic terminology) classification. These commitments together make the work of cognitively oriented anthropologists and historians of science of natural interest to the current study. Without having to affirm some of the more dogmatic aims of the former tendency, we may still recognize as plausible the cognitivist view that categorial systems throughout history and throughout the world amount to variations within certain parameters that are fixed by human biology. On this understanding, there is a limited range of ways in which different cultures can carve the world up, and each way is going to bear important, deep similarities to each other way, notwithstanding superficial differences. Some cognitivist researchers studying folk taxonomy have suggested that this should be understood on the model of Chomsky's theory of universal grammar: they hold, namely, that there is an evolved capacity that human beings have qua human beings, and the natural languages or, in our case, the natural categorial systems, amount in the end only to different inflections of this shared, human capacity. The task of the cognitivist anthropologist, on this understanding, is to penetrate through the local variations in order to uncover and describe the universal capacity, and the ultimate parameters it establishes for the range of possible local variations.

For some scholars, the universal scheme for classifying determines the shape and character not only of folk-taxonomical systems, but also of scientific taxonomy in the West from Aristotle through Linnaeus and beyond. Scott Atran, for example, has given a compelling cognitivist account of the development of Western taxonomy from Andrea Cesalpino to Carolus Linnaeus, arguing that the system they gave us amounts not so much to a radical break with the folk taxonomy, as to a further development or accrual of such taxonomical knowledge.[9] On this understanding,

[9]See in particular Scott Atran, *The Cognitive Foundations of Natural History: Towards an Anthropology of Science*, Cambridge: Cambridge University Press, 1990; Scott Atran, *The Native Mind and the Cultural Construction of Nature*, Cambridge, MA: MIT Press, 2008.

science does not initiate the project of taxonomy, so much as it brings folk-scientific knowledge into sharper focus by writing it down, systematizing it, making it explicit. In this respect, taxonomy fundamentally differs as a modern scientific enterprise from, say, mechanical physics, which is a form of knowledge that may with much greater justice be said to have been brought into existence only at the moment it became an active scientific research program.

Now if scientific taxonomy builds on folk taxonomy, and if racial classification builds on this in turn, there might be some basis for supposing that something about the modern habit of distinguishing between human groups on racial grounds is more deep-seated than we have acknowledged it to be. There is, certainly, a great deal of scholarly disagreement, as to how far back in at least Western history something we can positively identify as racism is to be found. Almost without exception, scholars have focused on antiblack racism to the exclusion of other possible forms of racism, and the scholarly consensus tends to be that this form of racism took shape over the course of the modern period in parallel to the growth of the institution of slavery. By and large this is also the view of the present study, though it is worth noting that some dissenting authors have made compelling cases for antiblack racism in classical antiquity, and particularly in classical Jewish and Islamic civilization.[10]

Prior to the supposedly modern invention of race, human beings may have practiced what could be called, on analogy to "folk taxonomy," "folk racial science": making distinctions between different groups, implicitly rooted in essences and indexed to perceived physiological traits, but without any explicit theory of the root causes of these distinctions. The early Greek historian and ethnographer Herodotus provides a good example of this. His distinctions blend the physiological, the linguistic, and the cultural without any concern to keep these separate. Thus for example in describing the various peoples of Scythia, he writes, "The Budinians . . . differ from the Gelonians in both language and lifestyle. The Budinians, who are nomadic, are the indigenous inhabitants of the

[10]See in particular David M. Goldenberg, *The Curse of Ham: Race and Slavery in Early Judaism, Christianity, and Islam*, Princeton: Princeton University Press, 2003. See also Raphael Jospe, "Teaching Judah Ha-Levi: Defining and Shattering Myths in Jewish Philosophy," in Raphael Jospe (ed.), *Paradigms in Jewish Philosophy*, London: Associated University Presses, 1997, 87–111. For the boldest statement of the view that there was nothing like modern antiblack racism in Greco-Roman antiquity, see Frank M. Snowden, *Blacks in Antiquity: Ethiopians in the Greco-Roman Experience*, Cambridge, MA: Harvard University Press, 1971; Snowden, *Before Color Prejudices: The Ancient View of Blacks*, Cambridge, MA: Harvard University Press, 1983. For a significant counterpoint to this view, see Benjamin Isaac, *The Invention of Racism in Classical Antiquity*, Princeton: Princeton University Press, 2004. See also Christopher Tuplin, "Greek Racism? Observations on the Character and Limits of Greek Ethnic Prejudice," in Gocha R. Tsetskhladze (ed.), *Ancient Greeks West and East*, Leiden: Brill, 1999, 47–75.

country, and they are the only race there to eat lice, whereas the Gelonians are farmers, grain-eaters, and gardeners; moreover, the two sets of people are altogether dissimilar in appearance and coloring."[11] In Tacitus, some centuries later, we see a remarkable attempt to root physiological features—in the event, the much-vaunted robustness of the Germans—in cultural practices:

> In every house the children grow up, thinly and meanly clad, to that bulk of body and limb which we behold with wonder. Every mother suckles her own children, and does not deliver them into the hands of servants and nurses. No indulgence distinguishes the young master from the slave. They lie together amidst the same cattle, upon the same ground, till age separates, and valor marks out, the free-born. The youths partake late of the pleasures of love, and hence pass the age of puberty unexhausted: nor are the virgins hurried into marriage; the same maturity, the same full growth is required: the sexes unite equally matched and robust; and the children inherit the vigor of their parents.[12]

Both Herodotus and Tacitus are writing in a vein that perhaps has its earliest expression in the Hippocratic treatise *Airs, Waters, Places*, which describes the relationship of populations to their natural environments, as well as the way these environments shape them physically and psychologically, yet generally with an eye to answering questions in the domain of what we would consider physical geography rather than physical anthropology (though as late as Kant we see very clearly that the question of "race" could in the modern period still at times be included under the banner of geography). In the classical period, this conformation of the body to the environment is typically described in neutral terms, or even in terms that celebrate the adaptability and variety of the human species. In the early modern period in contrast, as we will see, such change in the appearance of a population, as a result of climatic or other environmental factors, is almost without exception conceptualized as change for the worse, as degeneration.

Another striking difference between ancient and early modern accounts of human biogeographical variety has to do with the different conceptions in these periods of change over time. Typically, ancient authors are much less concerned to account for patterns of past migration that might have resulted in the current arrangement of different, physically diverse populations in different parts of the known world. Nowhere

[11]Herodotus, *The Histories*, trans. Robin Waterfield, Oxford: Oxford University Press, 1998, 109 [270].

[12]Tacitus, *A Treatise on the Situation, Manners and Inhabitants of Germany*, ed. and trans. John Aikin. Oxford: W. Baxter, 1823, 54.

does Hippocrates say that the people who are now Europeans arrived in Europe and became bellicose as a result of environmental conditions; he says only that Europeans are bellicose. It would not be unreasonable to suppose that the new concern with change over time as a result of change of habitat, whether this is conceived as adaptation or as generation, was a response to the increasing dispersion of Europeans throughout the globe in the early modern period, and to the increasing concern about the long-term effects on European populations of this dispersion. The rising racial essentialism of the period may, in turn, be seen as a way of securing the stability of the population through change in habitat by positing traits that are, somehow, resistant to any environmental influence.

It is important to distinguish the question of whether racism is a reflection of a universal feature of human cognition from the much more local question of the history of European, or Jewish or Muslim, perceptions of sub-Saharan Africans. If racism is rooted in a universal tendency of human cognition of the social world, then the contingent facts about what phenotypic variation within the human species there is, whether there are people with the phenotypes typical of sub-Saharan Africa or not, is perfectly irrelevant. If cognition of human difference as essential were innate, there would still be what we are prepared to identify as racism even in a world that consisted of only, say, Kazakhs and Mongols.

This last example, in fact, is borne out by empirical work. One controversial but intriguing set of studies from the cognitive anthropologist Francisco Gil-White has compellingly argued (on the basis mostly of field data among Kazakhs and Mongols in Central Asia) that ethnic groups are innately disposed to perceive neighboring ethnic groups as essentially different, where this difference is understood on the model of biological species difference.[13] The possibility that racism of this sort—which sees ethnic difference as essential difference—is deep-seated is not at all at odds with the view that there are numerous distinct features of modern racism. The comparison of modern racism to, say, the conceptualization of ethnic difference among the Mongolian nomads studied by Gil-White gives us something roughly analogous to the comparison between, say, modern entomological classifications and the folk entomology of a native Amazonian people. Modern racial science aspires to account for racial differences at a global level. Although it is positively laden with distinctions of value, it has at least a pretense of naturalism, purporting to simply describe real

[13] See Francisco Gil-White, "Are Ethnic Groups Biological 'Species' to the Human Brain? Essentialism in Our Cognition of Some Social Categories," *Current Anthropology* 42, 4 (2001): 515–54; Gil-White, "The Cognition of Ethnicity: Native Category Systems under the Field Experimental Microscope," *Field Methods* 14, 2 (2002): 170–98. For a strong critique of Gil-White's work, see Tim Ingold, "Commentary on Gil-White," in *Current Anthropology* 42 (2001): 541–42.

divisions as they present themselves in nature (traditional perceptions of essential ethnic differences, by contrast, would have been unconcerned with keeping these two domains separate). Finally, while both traditional ethnic essentialism and modern racism take ethnic difference as akin to biological species difference, modern racism attempts to offer an explicit theory of what this biological difference is, and does so by attaching itself, at least rhetorically, to the best taxonomical science of the day.

To sum up then: on the approach we will be taking here, realism or essentialism about ethnic difference may very well extend back vastly further in human history than the modern period, and it is admittedly important to study the very most deep-seated forms of perception of human difference that are given in human cognition. Yet this does not compromise the suggestion that there is an interesting account to be given concerning the emergence of distinct and novel ways of thinking about human difference that appear in the modern period, in part as a side effect or discursive echo of social and economic history, particularly the history of slavery, but in part also as a result of new developments in natural scientific reflection on the variety of nature.

1.3. Race without a Theory of Essences; or, Liberal Racism

It is hard to find anyone who advocates a thoroughgoing realist or essentialist theory of races. As Hacking writes of the notorious experimental psychologist J. Philippe Rushton, who in his lifetime published several studies purportedly proving the inferior intelligence of people of African descent, "although Rushton stands up and says the most amazing things in public, even he does not say, 'I am an essentialist about race.' "[14] Indeed, in the early modern period as well, at the very moment we begin to see modern racial realism emerging, we find virtually no one explicitly taking up that banner. In fact, virtually all explicit claims to the effect that there is a real, permanent, essential difference between different groups, rather than a temporary, reversible, contingent one, are made by radical freethinkers, such as Voltaire, in the aim of shocking contemporary sensibilities, rather than by conservative thinkers aiming to defend the status quo.

This is one of the most significant lessons of our historical investigation of the concept of race: racial realism has been sustained, from Blumenbach to Rushton, not as a result of any deep ontological commitment on the part of scientists and other theorists to a real, essential difference

[14] Ian Hacking, *The Social Construction of What?*, Cambridge, MA: Harvard University Press, 1999, 18.

between different human groups. Instead, it is sustained by a practice of quantifying small differences, which are then inflated in importance by a broader culture desirous of scientific legitimation for a racial realism to which it is committed on largely independent, nontheoretical grounds. Authors who produce the scientific data are generally at least circumspect when it comes to making strong claims concerning the reality of racial divisions, or they explicitly deny that racial boundaries mark any real or permanent divisions in nature (e.g., Blumenbach). It thus appears that it is the legitimizing effect itself of scientific research on a domain of nature, such as the diversity of the human species, that in turn leads to an essentialization of the categories deployed in the research. It matters little if a researcher's deeper commitments support the unity of the human species; if that researcher is at the same time using distinctions such as "Caucasian" and "Mongoloid," and identifying even minor points of difference between the members of these categories, the effect will be to bring these categories into existence, as social kinds that appear to be natural kinds.

If modern racism has been linked to real, measurable differences only in an ad hoc and exaggerated way, then why, we must still ask, is it so very tenacious? What is it about the modern world, or about human habits of thinking, or about the combination of these two, that inclines so many people to conceptualize the human species as consisting in a few basic, fixed, and biologically and socially significant subkinds? Analyzing the modern period, Richard Popkin identified a tendency he dubbed "liberal racism." He defined this as "making the best of the European experience the model for everyone, and the eventual perfection of mankind consisting in everyone becoming creative Europeans."[15] In Popkin's own view, this variety of racism is not racism in the fullest sense, since in order to be so qualified a view of human diversity must hold that one group is necessarily, irreparably inferior to another group, rather than simply inferior due to contingent, cultural disadvantages.

Liberal racism was an effective ideological complement to the project of colonial expansion, Popkin observes, since, obviously, colonizing and dominating an indigenous group of people is more easily justified if it is presumed that this group is improvable, or, which is the same thing, that colonization by Europeans will lead to an eventual Europeanization. If a group of people is thought to be naturally or necessarily inferior, then the project of colonial domination becomes much harder to explain. On the supposition of natural inferiority, the only sort of domination that

[15]Richard H. Popkin, "The Philosophical Basis of Modern Racism," in Richard A. Watson and James E. Force (eds.), *The High Road to Pyrrhonism*, San Diego: Austin Hill Press, 1980, 89.

makes sense is *total* domination, and it is not surprising that for this reason, at the high point of the era of slavery, some racial theories, such as that of Christoph Meiners, held that while Europeans, Asians, and Native Americans differed only as a result of contingent and reversible environmental circumstances, Africans must be supposed to have an entirely separate origin. We see a similar polygenetic account of the origins of Native Americans, but not of Africans, at the high point of the Spanish conquest of Latin America some two centuries earlier. We will be discussing the theories of monogenesis and polygenesis in later chapters, but for now they are worth mentioning in passing simply in connection with the suggestion that liberal racism may well be the order of the day wherever the aim is colonial domination of a people and the maintenance of the dominated in a second-class social position, yet still with the possibility of aspiration toward the social advantages that come with the adoption of the values and habits of the colonial power. Outright or overt racism by contrast makes sense only when it is politically expedient to completely dehumanize the target group, as in the case of the genocide against the Native Americans, or the transatlantic slave trade.

Of course, we seldom find people self-identifying as liberal racists; indeed they may well be less likely to do so than racial essentialists would be prepared to identify themselves as racial essentialists. Liberal racism involves nearly of necessity a practical contradiction, a failure to treat groups of people as equal to one's own, even as one insists, theoretically, on the unity of the human species. Liberal racism, moreover, has generally served as the ideological background on which the practices that support racial essentialism have been developed. In other words, thinkers like Kant and Blumenbach (as we will see in detail in the final chapter) have explicitly supported some version of liberal racism, without of course calling it by its name, while at the same time offering the sort of racial taxonomies, and, in Blumenbach's case, the quantitative data about cranial and skeletal measurements, that would serve to support overt illiberal racism.

Thus, it will not do, in reconstructing the origins of modern racism out of early modern philosophy, to simply identify authors who expound the view that one group is inherently inferior to another, or that there are separate creations or natural origins for different groups of people. The authors who defended such views are in a minority in modern history, and in any case focusing on them causes us to overlook the ways in which practices and discourses (to use a loaded term, but one that is not entirely unhelpful) themselves served to constitute the categories that would abet or underlie modern racism, a racism that needed no explicit solid philosophical or scientific theory to support it. It was the practice of taxonomizing the human species and of elaborating differences between groups, even if these differences were not explicitly held to be essential, or indeed even if it was explicitly denied that they were essential, that served as the principal support of modern racism. To put this another way, the availability of categories such as "black" and "white" as social kinds leads willy-nilly to their being treated as natural kinds. Thus we may speak not only, with Hacking, of the "looping" of natural kinds, where they come to appear particularly real or ineliminable as a result of the way the people to whom they are applied adopt, appropriate, or otherwise respond to them.[16] We may also speak, perhaps, of "slippage," where distinctions between kinds of people that are explicitly identified by the people making them as "artificial" (for example, Kant's *On the Different Races of Men*, treated at length in chapter 9) nonetheless come to appear as natural once they are made.

Popkin's account of liberal racism, as the name implies, characterizes a tendency of the secular Enlightenment or of high modernity, one that takes shape in connection with the liberal political philosophy of the age of antimonarchism, opposition to state religion, and so on. But what this misses is that liberal racism might also plausibly be seen as a secularized descendant of Christian missionary universalism, which holds that all people throughout the world are at least potentially improvable, to the extent that they are prepared to become like us. In the Christian version, the improvement is instantaneous, occurring in the moment of baptism (even if this must be followed by observable behavioral changes), while in the secularized form the criteria for becoming "like us" grow rather more complicated, and come to be rooted in manners and forms of life rather than in a moment of conversion.

Naturally, this is an oversimplification of the differences. We may in fact see the early modern "rites controversy,"[17] particularly as it played out between Jesuits in China and their superiors in Rome, as a conflict in large part over the question as to what, besides the speech act of conversion, a non-European must subsequently do to convince the Europeans of his or her improved state. In any case, the racism associated with Enlightenment-era universalism did not appear out of nowhere. Like Christian missionary universalism, moreover, it does not require explicit claims about inferiority, circumstantial or essential, in order for the categories it deploys to appear to those who use them as natural. This is the case for "Christian" and "barbarian" as much as for "black" and

[16]For the clearest statement of the thesis concerning looping in the creation of social kinds, see Ian Hacking, "The Looping of Human Kinds," in Dan Sperber and A. J. Premack (eds.), *Causal Cognition*, Oxford: Oxford University Press, 1995, 351–83.

[17]See in particular David E. Mungello (ed.), *The Chinese Rites Controversy: Its History and Meaning*, Monumenta Serica Monograph Series XXXIII, Nettetal: Steyler Verlag, 1994.

"white."[18] As Philippe Descola has sharply emphasized, frequent inversions of the value system implied by the distinction between, for example, the civilized and the savage, are not for that a rejection of the terms of the value system. Observing the way in which classical ideas about the Barbarian Germans or Bretons in Tacitus would be transplanted in the modern period onto the Tupinamba or the Hurons, Descola notes,

> One will say perhaps that this running together of sense and of epoch opens the possibility of an inversion that Montaigne or Rousseau will know to exploit: the savage can henceforth be good and the civilized man bad, with the first incarnating the virtues of ancient simplicity that the corruption of morals has caused the other to lose. This is to forget that such a rhetorical artifice is not entirely new—Tacitus himself gave in to it—and that it in no way calls into question the game of reciprocal determinations that make the savage and the domestic constitutive of one another.[19]

The transposition of the ancient figure of the barbarian into the new American context fits within the broader pattern both of accounting for the unfamiliar in terms of the familiar, as well as of attempting to account for human dispersion across the globe in monogenetic terms. The conceptualization of Americans as Scythians or other sorts of barbarians familiar to the Greeks may seem a stretch to us, but it is important to understand the central, and substantially correct, concern to maintain the unity of the human species by subsuming Native Americans into known history. In any case here Descola's interest is in arguing that for early modern thinkers the Greeks not only provide a model of the barbarian that may now be adopted, for example, by Jesuits in their ethnography of the Huron or the Iroquois;[20] in addition, the Greeks already provide an example of alternating exaltation and debasement of the barbarians in question, and thus show already that an inversion of an essentialist and dichotomous scheme is something quite different from its elimination. It does not matter whether one denies any inherent difference between the savage and the civilized, or even exalts the savage over the civilized: to re-

[18]For an important anthropological investigation of the widespread classificatory distinction between "us" and "them," see Eduardo Viveiros de Castro, "Cosmological Deixis and Amerindian Perspectivism," *Journal of the Royal Anthropological Institute* NS 4 (1998): 469–88; Viveiros de Castro, "Os pronomes cosmólogicos e o perspectivismo ameríndio," *Mana* 2, 2 (1996): 115–44.

[19]Philippe Descola, *Par-delà nature et culture*, Paris: Gallimard, 2005, 80–81.

[20]For perhaps the most compelling and insightful account of the role of classical Greek sources in early modern ethnography of the New World, as examined through the case study of the Jesuit missionary to New France, Joseph-François Lafitau, see Jean-Pierre Vernant and Pierre Vidal-Naquet, *La Grèce ancienne 3: Rites de passage et transgressions*, Paris: Seuil, 1992, chapter 1.

main constrained to thinking of the human species as consisting in these hierarchalized subtypes is to discursively maintain these subtypes in existence, to keep them appearing as natural, again, even if one's goal, as with Montaigne and Rousseau, is to overturn them.

Now there is some ground for thinking that the case is different in Enlightenment-era distinctions between races than in missionary distinctions between Christian and heathen, or between Descola's preferred categories of the civilized and the savage. But all of these are alike in that—apart from a few full-fledged racial realists such as Christoph Meiners, and various freethinkers, who adopted polygenesis theory for its shock value—(1) they are categories that are applied in view of a population's current circumstances, and are not held to be based on fundamental divisions between real human subtypes, even if races are somewhat more durable than categories like "heathen" to the extent that they cannot change over the course of a single lifetime (though even here there was some debate); (2) the hierarchy of supposedly superior and inferior races is often inverted, just as Montaigne and Rousseau invert the relationship between civilized and savage, without, for that, being overcome. Thus, Blumenbach, commonly cited as the very founder of modern racial science, writes, in an unmistakably Rousseauian vein,

> I am of the opinion that after all these numerous instances I have brought together of Negroes of capacity, it would not be difficult to mention entire, well-known provinces of Europe, from out of which you would not easily expect to obtain off-hand such good authors, poets, philosophers, and correspondents of the Paris Academy. And on the other hand, there is no so-called savage nation known under the sun which has so much distinguished itself by such examples of perfectibility and original capacity for scientific culture, and thereby attached itself so closely to the most civilized nations of the earth, as the Negro.[21]

This is, transparently, a straightforward inversion of the more familiar statements of the era about the improvability of Africans. David Hume, for example, notoriously wrote in 1748,

> I am apt to suspect the negroes, and in general all the other species of men . . . to be naturally inferior to the whites. There never was a civilized nation of any other complexion than white, nor even any individual eminent either in action or speculation. . . . Not to mention our colonies, there are Negro slaves dispersed all over Europe, of whom none ever discovered

[21] Johann Friedrich Blumenbach, *De generis humani varietate nativa*, 3rd ed., Göttingen: Vandenhoek and Ruprecht, 1795. Here and in most subsequent discussion citations of Blumenbach are from *The Anthropological Treatises of Johann Friedrich Blumenbach*, trans. and ed. Thomas Bendyshe, London: Anthropological Society, 1865, 312.

> the symptoms of ingenuity; though low people, without education, will start up amongst us, and distinguish themselves in every profession. In Jamaica, indeed, they talk of one Negro as a man of parts and learning; but it is likely he is admired for slender accomplishments, like a parrot who speaks a few words plainly.[22]

Blumenbach's claim is an inversion of claims such as Hume's; but it is not a rejection of the categories Hume deploys. Moreover, while Hume is infamous for his baseless claims about the inferiority of Africans, these claims were more an echo of already existing commonplaces than a significant contribution to the production of early modern racial realism; Blumenbach's quantitative and methodical study, by contrast, played a very important role in the emergence of nineteenth-century racial science, even if Blumenbach himself was far more circumspect than Hume about the significance of the differences he was measuring for marking out real, essential divisions within the human species. Yet, as has already been suggested, liberal racism does not need the likes of Hume in order to thrive; it is enough that there be Blumenbachs.

1.4. Constructionism and Eliminativism

There has been a common presumption in debates surrounding social construction that to catch out some entity or category as so constructed is at the same time to condemn it. Thus Hacking notes that "a primary use of 'social construction' is for consciousness raising"; it is "critical of the status quo." Social constructionists generally move, Hacking argues, from the argument that a given entity or category *X* "need not have existed," to the view that "[w]e would be much better off if *X* were done away with, or at least radically transformed."[23] On this line of thinking, every entity or category is expected either to be a real feature of the world, something left over when the world is carved at its joints, or it is to be exposed as constructed and by the same measure be relegated to the scrap heap along with phlogiston, the ether, and so on.

P. J. Taylor has argued, by contrast, against the view that, typically, categories tend to go away when they are shown to be constructed.[24] The category of "race" seems to confirm Taylor's analysis. Since the mid-

[22]David Hume, footnote to "Of National Character" (1748), in *The Philosophical Works of David Hume*, Bristol: Thoemmes Press, 1996, 3:228.

[23]Hacking, *Social Construction of What?*, 6.

[24]See P. J. Taylor, "Building on Construction: An Exploration of Heterogeneous Constructionism, Using an Analogy from Psychology and a Sketch from Socio-economic Modeling," *Perspectives on Science* 3, 1 (1995): 66–98.

twentieth century no mainstream scientist has considered race a biologically significant category; no scientist believes any longer that "Negroid," "Caucasoid," and so on represent real natural kinds, carve nature at its joints, or whatever the preferred metaphor may be. For several decades, particularly since the work of Richard Lewontin in the early 1970s, it has been well established that there is as much genetic variation between two members of any supposed race as between two members of supposedly distinct races.[25] This is not to say that there are no real differences, some of which are externally observable, between different human populations; it is only to say, in Lawrence Hirschfeld's words, that "races as socially defined do not (even loosely) capture *interesting clusters* of these differences."[26]

In very recent years, there has been a return of what is sometimes called "new racial naturalism," which argues on the basis of new genetic research that Lewontin and others had moved somewhat too hastily to their antirealist conclusions. As early as 1977,[27] Jeffry Mitton published a significant article questioning Lewontin's methodology. This was followed by a more direct attack on "Lewontin's fallacy" by A.W.F. Edwards in 2003.[28] The fallacy in question is precisely Lewontin's view that the analysis of genetic data in a locus-by-locus way is sufficient for the establishment of conclusions about racial categories. As Edwards writes, "There is nothing wrong with Lewontin's statistical analysis of variation, only with the belief that it is relevant to classification."[29]

Philosophers by and large have not been very impressed with the opposition to Lewontinian antirealism that has been gaining some momentum among geneticists and natural scientists in related fields. One particularly forceful argument against the new realism is the recent article

[25] See in particular Richard Lewontin, "The Apportionment of Human Diversity," *Evolutionary Biology* 6 (1972): 391–98.

[26] Lawrence Hirschfeld, *Race in the Making: Cognition, Culture, and the Child's Construction of Human Kinds*, Cambridge, MA: MIT Press, 1998, 4.

[27] Jeffry Mitton, "Genetic Differentiation of Races of Man as Judged by Single-Locus and Multilocus Analyses," *American Naturalist* 111, 978 (1977): 203–12.

[28] A. W. F. Edwards, "Human Genetic Diversity: Lewontin's Fallacy," *Bioessays* 25, 8 (2003): 798–801. Another important volley from the same period, which helped to reopen the debate about the precise extent to which human races are genetically determined, was Noah A. Rosenberg, et al., "Genetic Structure of Human Populations," in *Science* 298 (2002): 2381–85. For an important philosophical engagement with both the old and the new arguments about racial realism, which makes a strong case for the need among the antirealists to come up with more compelling, and scientifically better grounded, arguments, see Neven Sesardic, "Race: A Social Destruction of a Biological Concept," *Biological Philosophy* 25 (2010): 143–62.

[29] Edwards, "Human Genetic Diversity," 800.

by Adam Hochman, "Against the New Racial Naturalism."[30] Hochman argues, first of all, that the new realism applies different criteria for the distinction of human races than are standardly employed by biologists for the identification of subspecies in other animals, and that in this respect the enumeration of human races departs without explanation from the best scientific standards, and indeed from what one would expect from any inquiry that purports to be treating human beings as products of nature alongside finches and macaques.[31] Second, he argues that while genetics has gone beyond Lewontin's facile dismissal of the genetic clustering in human populations in different geographical regions, the new realists have yet to show that this genetic clustering yields anything that is not biologically superficial. Genetic data can with some degree of accuracy tell us a person's continent of origin, but this, in Hochman's view, supports only a notion of "biogeographical ancestry," rather than of robust racial naturalism.

Philosophers are for the most part unimpressed with the sort of evidence adduced by Mitton and Edwards, because they believe that in the end the philosophical question of race cannot be answered by the geneticists, whatever evidence they are able to pull up for or against race as a natural kind. As Lisa Gannett explains, "race is socially constructed by enlisting biological differences and investing these with socio-cultural meanings."[32] This account seems plausible for many cases of the perception of racial difference, but not necessarily all of them. In fact, Gannett's claim may well be an extrapolation from a distinctly American, or perhaps North Atlantic, context, in which the vastly most salient racial distinction over the past several centuries is one that happens to have emerged between populations that tend to exhibit visible phenotypic differences. But it is by no means clear that this division, as a social phenomenon, is different in any important way from the racialized distinction

[30]Adam Hochman, "Against the New Racial Naturalism," *Journal of Philosophy* 110 (2013): 331–51. For a compelling refutation of some of the central claims of Hochman's argument, see Quayshawn Spencer, "The Unnatural Racial Naturalism," in *Studies in History and Philosophy of Science Part C: Biological and Biomedical Sciences* 46 (June 2014): 38–43. See also Spencer, "What 'Biological Racial Realism' Should Mean," *Philosophical Studies* 159 (2012): 181–204. For other important recent work questioning the inference from the study of genetic variation to a viable classification of races, see Jonathan Michael Kaplan and Rasmus Grønfeldt Winther, "Prisoners of Abstraction? The Theory and Measure of Genetic Variation, and the Very Concept of 'Race,'" *Biological Theory* 7 (2013): 401–12.

[31]For further important investigation of the problem of which sort of groupings below the species rank warrant being called "races," see Matthew Kopec, "Clines, Clusters, and Clades in the Race Debate," *Philosophy of Science* (forthcoming).

[32]Lisa Gannett, "Questions Asked and Unasked: How by Worrying Less about the 'Really Real' Philosophers of Science Might Better Contribute to Debates about Genetics and Race," *Synthese* 177 (2010): 363–85, 375.

between castes in India, or that between the majority population and the Burakumin in Japan, who are the product of a vocation-based process of ethnogenesis. In these cases, the same attribution of essentialized or biologically grounded differences occurs as in the case of the perception of "black" and "white" in the United States, but there simply is no genetic difference on which this attribution can be anchored, the lower castes having emerged historically as a result of the stigmatization of certain professions.

In any case, if we return to the local or regional context of the Atlantic world, the possible presence of real biological differences, for which the new realism provides some genetic evidence, need not be seen as a threat to the constructionist view that in any case what is important in discussions of race is the way in which social categories build on top of biological ones, and, moreover, there is nothing in the biological differences that predetermines the different ways in which the social categories might be built. Philosophy of race is by now overwhelmingly constructionist in the sense Gannett describes. Most contributors to this field agree that the perception of racial difference has little, perhaps nothing, to do with pregiven physiological features, but rather with the way a person is "racialized." To be racialized in a certain way is to be fitted into a complex social scheme in view of such factors as one's name, one's neighborhood, one's accent, and so on. Clusters of such features also sometimes map onto groups of people who tend to share phenotypic traits with one another, but it is never in virtue of these phenotypic traits that a group of people are racialized in the first place. Thus when we perceive someone, say, as "black," we are never simply reading a racial fact off of some pregiven observable traits. Rather, we are making a social judgment, and then correlating it, however loosely, with these traits. And the traits thereby come to take on an apparent significance that cannot be explained on subsequent scrutiny.

Consider the well-known historical example of the description of the hair of sub-Saharan Africans as "wool": not just as being *like* wool, but literally as wool. But what then is wool? To the early modern European naturalist, such a description would have been presumed to be a straightforward observation of a fact about the physical diversity of the human species. Yet it relies on a prior decision to deploy a definition of "wool" that extends it to the human species, and in so doing treats the "wool" of Africans differently from, say, the short, straight hair on the head of a European man (which by parallel reasoning we might just as rightly call a "pelt"). To call an African man's hair "wool" because it is curly is to implicitly (or perhaps not just implicitly) insert the African into the order of animal nature, alongside other objects of study such as goats and ibexes, in a way that the European naturalist had traditionally declined to do for his own physical attributes. And this inconsistent approach to the description of physical differences between different groups has the result that the difference it picks out comes to appear as being of much greater significance than it in fact is: the African shares a trait with sheep, it says. And this is no small claim. Again, this significance arises from a definitional choice, prior to the observation of physical differences. It does not arise from a bare description of physical traits.

And yet the category of race continues to be deployed in a vast number of contexts. The history of race, then, is not like the history of phlogiston: race is not an entity that is shown not to exist and that accordingly proceeds to go away. How are we to explain this difference? One striking feature of the concept of race in current academic parlance is the evident ambivalence with which it is deployed. Thus for example Anthony Appiah identifies himself as a racial skeptic to the extent that the biological categories to which racial terms refer have been shown, he agrees, not to exist.[33] Yet at the same time he acknowledges that the adoption of "racial identities" may often be socially expedient, and even unavoidable, for members of perceived racial minorities.[34] Along the same lines, Naomi Zack forcefully writes,

> Tigers have to be dismounted with care. It's one thing to understand within a safe forum that race is a biological fiction. In American culture at large, the fiction of race continues to operate as fact, and in situations of backlash against emancipatory progress, the victims of racial oppression, nonwhites, are insulted and injured further for their progress against oppression. If those who practice such second-order oppression begin to employ the truth that race is a fiction, gains already secured against first-order oppression (or in redress of it) could be jeopardized.[35]

[33]K. Anthony Appiah, "The Uncompleted Argument: DuBois and the Illusion of Race," in L. Bell and D. Blumenfeld (eds.), *Overcoming Racism and Sexism*, Lanham, MD: Rowman & Littlefield, 1995. Particularly important scholarship has been published in recent years on the utility of racial data in medical statistics for diagnosis and for the improvement of health care for disadvantaged populations. For a particularly compelling argument for the epidemiological usefulness of racial categories, quite apart from their context-independent "reality," see Neil Risch, Esteban Burchard, Elad Ziv, and Hua Tang, "Categorization of Humans in Biomedical Research: Genes, Race and Disease," *Genome Biology* 3, 7 (2002): 1–12. For a particularly useful introduction to this scholarship, see also Eric J. Bailey, *Medical Anthropology and African American Health*, Westport, CT: Greenwood, 2000.

[34]Anthony Appiah, "Race, Culture, Identity: Misunderstood Connections," in Anthony Appiah and Amy Gutmann (eds.), *Color Conscious*, Princeton: Princeton University Press, 1996.

[35]Naomi Zack, "Mixed Black and White Race and Public Policy," *Hypatia* 10, 1 (1995): 120–32, 130. Toni Morrison makes a similar point as follows: "The act of enforcing racelessness . . . is itself a racial act" (*Playing in the Dark: Whiteness and the Literary Imagination*, Cambridge, MA: Harvard University Press, 1992, 46).

Zack is surely right: it is not up to any single philosopher to make the eliminativist case; if race lingers as a category for making sense of the world, even or especially among the people who have been historically disadvantaged by it, and who are well aware of its lack of grounding in biology, then we will do far better, in the pursuit of both truth and justice, to analyze the concept, to lay bare its complex history, rather than to insist that it is a nondenoting term and that anyone who makes appeal to it is mistaken.

Along similar lines to those laid out by Zack, Ron Mallon has distinguished between metaphysical views of race on the one hand and normative views on the other, dividing the latter into "eliminativist" and "conservationist" camps.[36] On his scheme, one may very well coherently remain metaphysically antirealist about race but still defend the conservation of the concept on normative grounds. Here we are principally interested in what Mallon would call the metaphysical question of race, yet there is cause for some passing concern about the distinction: under any circumstances where we know our concepts do not match up with the world, we must have strong grounds for wishing to hold onto them anyway; and this is so a fortiori when, as in the case of race, there appear to be at least a few perfectly adequate, coextensive terms—"ethnie," "culture," "ethnocultural group"—that unlike "race" could not be taken to imply any false metaphysical or scientific commitments on the part of the person who deploys them. Why not then trade the social expediency of "racial identities" for cultural ones?

There is indeed a very simple, logical problem involved in classifying human beings in terms of a supposed fact of the matter about their racial identity, rather than in terms of self-understanding rooted in cultural identity. The problem is that the global categories must be supposed to exist in order to categorize individual people or populations in terms of them; but it is only through the physical-anthropological description of such individuals and local groups that one could possibly arrive at an idea of what the global groupings are. For this reason, one might justly worry that people who continue to use scientific research as a way of strengthening their case for the reality of race are begging the question: they are assuming at the outset the reality of the very thing they are supposed to be trying to prove. Yes, traits cluster in populations, but it is only if you have already presupposed that the human species breaks down into a fixed set of real, biologically significant subdivisions that you will find it meaningful to subsume new information about such clustering into a racial schema.

[36]Ron Mallon, "Race: Normative, Not Metaphysical or Semantic," *Ethics* 116, 3 (2006): 525–51.

Otherwise, what you will notice are all the salient respects in which the population that is the locus of such clustering does not amount to a discrete kind. For one thing, it is entirely permeable at its boundaries, and thus has nothing in common with the isolated reproductive communities that constitute biological species on analogy to which races are, consciously or unconsciously, modeled. For another, trait clusters tend to be noticed in populations that were already of interest to us as purported races for initially nonscientific reasons. Take the example of media reports on the recent discovery of the mutation that led to thicker hair shafts, more sweat glands, and smaller breasts in many East Asian populations.[37] This discovery purportedly provided new empirical evidence for the distinctness of the Asian race, but one might suppose that the production of the evidence unfolded somewhat as follows: we assumed at the outset that there were such people, constituting a real subdivision of the human species, and then we went in search of their distinctive features. We found some in the sweat glands, hair follicles, and breast size of females. But would we find the same traits clustering in, say, Tungusic peoples, or the Chukchi? They are East Asian too, after all, Eastern Siberian, to be precise, and it is a contingent fact about human history that they are outnumbered by, say, Han Chinese. If we sample all of the peoples of the world, rather than the ones that are salient to us on prescientific cultural and historical grounds, we will notice that our conception of where the racial boundaries lie is rooted in our prescientific interests, and only subsequently filled out, as best it can be, by new genetic research.

This is particularly evident in the local U.S. context, in which genetic and medical information about African Americans becomes naturalized and universalized in such a way as to purportedly tell us something about a significant subdivision of the entire human species, one that was formerly called "Negroid." But of course such information tells us nothing of the sort: it is useful for diagnostic and therapeutic purposes for doctors in the United States to know that someone is "black," but this in no way implies that the same information about a person in the Kalahari, Ethiopia, or even (depending on the culture and period of history you inhabit) New Guinea, Sri Lanka, or Australia could be used in the same way. The information is of strictly local interest, and yet it calls upon a global system for subdividing the human species, one that places Khoi-San, Ethiopians, and African Americans, at least, in the same quasi-natural kind.

And this point is also useful for understanding the simple and obvious, but too-often neglected, response to the racists who invoke supposedly scientific evidence concerning the superior performance of black athletes

[37] See for example Nicholas Wade, "East Asian Physical Traits Linked to 35,000-Year-Old Mutation," *New York Times*, February 14, 2013.

in track events, or the inferior performance of black students on standardized tests. It is seriously unlikely that a mass-scale standardized test of everyone who is placed for historical and nonscientific reasons in the folk category "black" could ever be carried out in a sufficiently rigorous way to warrant a conclusion of the sort: "'Blacks' perform worse on standardized tests than 'whites.'" Again, what this would involve is devising a test that could be given to Namibians, Ethiopians, Haitians, and so on, and whose results could then be compared with those of the same test as given to Norwegians, Circassians, Scotch-Irish West Virginians. This will never happen, but that does not matter to the racial realists, because anyway what they *really* mean when they invoke such tests to ground claims of racial difference is that *here*, in our local context, there is such a difference. But race is supposed to be global, natural, a result of evolution, and so forth, while local differences are obviously the result of only civil history and culture. And this is the great inconsistency of the pseudoscientific concept of race: that it is reaching too far too fast, invoking a global natural order to which claims about local racial difference never accurately apply, and failing to notice that the local differences admit of a much more parsimonious explanation than the one that has to move all the way down to the level of biology.[38]

Let us, with these considerations in mind, attempt to get out of our own local situation, to move away from the familiar context of racial distinctions in American society, and return again to the classical period, in the hope that the very antiquity of the examples under consideration can help us to understand the nature of racial and ethnic divisions in general. It has already been suggested that "race" might be suitably replaced by "ethnie," a term whose denotation is coextensive with the term it replaces, yet without implying any essentialistic or biological theory of the groups it picks out. But what is an ethnie? One way to go about knowing a thing, as Aristotle noted, is by considering how it comes to be, so let us try that route. Recent scholarship has suggested that ethnogenesis almost as a rule occurs through elective confederation, generally for military purposes; even the purest nations are creole in their roots. To cite a well-known example: the alliance of Germanic tribes known as the Allamani (whence the current French word for "German": *allemand*) is evidently the transformation into an ethnonym of what was initially a description; to put it in modern German: *alle Männer*. As Herwig Wolfram argues, our modern conception of an ethnie, a *gens*, or a nation is

[38]Recent scholarship in anthropology has been particularly useful in revealing the fluid boundary between the discrimination of races based on supposedly natural features, on the one hand, and, on the other, cultural differences such as religion. See in particular Peter Wade, *Race, Nature and Culture: An Anthropological Perspective*, London, Pluto, 2002; Wade, *Race and Ethnicity in Latin America*, 2nd ed., London: Pluto, 2010.

one that derives entirely from the concept of nationhood that was created at the time of the French Revolution.[39] These terms tend to refer to a community that is at least supposed to be of shared biological descent. Wolfram notes however that in the tribal sagas of the peoples he studies, the Goths, *people* is equated with *army*, and to this extent "the sources attest the polyethnic character of the *gentes*. These *gentes* never comprise all potential members of a *gens* but are instead always mixed. Therefore their formation is not a matter of common descent but one of political decision. . . . Whoever acknowledges the tribal tradition, either by being born into it or by being 'admitted' to it, is part of the *gens* and as such a member of a community of 'descent through tradition.' "[40] This account of the ethnogenesis of the ancient Germans could with some adjustments be compared, for example, to the formation of the identity of dustbowl "Okie" migrants to the West of the United States, who tended to come from a similar background (Scotch-Irish) and shared many of the same traits, both phenotypic and cultural, yet could also easily absorb people of different ethnic backgrounds (as long as they were not so far as to be perceived as nonwhite). Another related example are the so-called Travellers of the United Kingdom and Ireland, who evidently started out as a social class involved in certain professions, such as blacksmith or cobbler, and since 2000 have been recognized as a legal ethnic minority in the United Kingdom under the Race Relations Act.[41] We also know of at least one Amazonian ethnie that reproduces itself from one generation to the next not principally through biological reproduction but rather through adoption (or, to be more precise, kidnapping).[42]

All this is to say that not only ethnogenesis but even, so to speak, ethnostasis need not necessarily be undergirded by mutual genetic proximity. Ethnies are plainly not "races" in a biological sense, even if what people often have in mind when they speak of "race" is in fact an ethnie. In many instances, ethnic groups come to be perceived in racial terms, and indeed seek to present themselves as racially distinct, as a maneuver within a larger struggle for political autonomy. To cite one very clear example of this, a common trope of the Québécois sovereignty movement

[39]Herwig Wolfram, *Geschichte der Goten: von den Anfängen bis zur Mitte des 6. Jahrhunderts*, Munich: Beck, 1979, translated as *History of the Goths*, by Thomas J. Dunlap, Berkeley: University of California Press, 1988. See also Herwig Wolfram and Walter Pohl (eds.), *Typen der Ethnogenese unter besonderer Berücksichtung der Bayern*, 2 vols., Vienna: Österreichische Akademie der Wissenschaften, 1990.

[40]Wolfram, *History of the Goths*, 5–6.

[41]See Chris Gray, "Irish Travellers Gain Legal Status of Ethnic Minority," *Independent*, August 30, 2000, http://www.independent.co.uk/news/uk/this-britain/irish-travellers-gain-legal-status-of-ethnic-minority-710768.html.

[42]See Maurice Godelier, *Métamorphoses de la parenté*, Paris: Fayard, 2004.

(though also one rejected by most within the movement) has been to argue for political self-determination on the grounds of racial distinctness, even though elsewhere in North America descendants of French immigrants are uncontroversially classified along with English immigrants as "white."[43] Such variability and fluidity is precisely what one might expect as a consequence of the legacy of liberal racism.

1.5. Natural Construction

Given, then, that we now know that the identity groups in modern multicultural states are plainly constituted on ethno-linguistic and cultural grounds, rather than on biological-essential grounds, why, again, do so many people remain normatively committed to racial identities? Sometimes it does happen that the removal of a social kind's natural undergirding in turn causes that category to largely wither away. Thus at least in the European cultural sphere, the category of "witch," surely a social kind and not a natural one, has, like phlogiston, largely withered away as a result of broad convergence by sometime in the eighteenth century upon the view that the term has no referent in the world. So, again, if witches can go the same way as phlogiston, why not race? Part of the answer to this question, as we have seen, may come from recent empirical work in the cognitive anthropology of race. Francisco Gil-White, Edouard Machery and Luc Faucher,[44] and others have argued that the categorization of human subgroups is grounded in a natural disposition of the human mind.[45] On this account, we are cognitively predisposed to perceive differences between biological kinds as rooted in something essential.

In many respects, the way was paved for this approach to race by the earlier work of Brent Berlin,[46] Scott Atran, and others, who sought a cognitive basis for prescientific folk taxonomies of the biological world.

[43] See for example Claude Jasmin, "Je suis fier de ma race," *Le Devoir*, May 30, 2013.

[44] In addition to the article cited above, see Edouard Machery and Luc Faucher, "Social Construction and the Concept of Race," *Philosophy of Science* 72 (December 2005): 1208–19.

[45] For other influential work on the question of racial perception from a cognitive or evolutionary point of view, see "Ethnicity and Ethnocentrism; Are They Natural?," in Raymond Scupin (ed.), *Race and Ethnicity: An Anthropological Focus on the United States and the World*, Upper Saddle River, NJ: Prentice Hall, 2003; R. A. Brown and George J. Armelago, "Apportionment of Racial Diversity: A Review," *Evolutionary Anthropology* 10 (2001): 34–40; Leda Cosmides, John Tooby, and Robert Kurzban, "Perceptions of Race," *Trends in Cognitive Science* 7 (2003): 173–79.

[46] See in particular Brent Berlin, *Ethnobiological Classification: Principles of Categorization of Plants and Animals in Traditional Societies*, Princeton: Princeton University Press, 1992.

Atran in particular has argued that what would eventually come to be called "biology" emerged out of a conception of the natural world fundamentally based on the attributions of essences to different sorts of natural being. Correlatively, the emergence of "racial science," more or less on the heels of modern biological taxonomy in the seventeenth and eighteenth centuries, may be seen as an outgrowth of a related cognitive disposition—that is, a presumably innate propensity to interpret salient features of the natural world in a certain way—to subdivide the human world into distinct groups with essential or species-like properties. There may be some concern that the historical account here pulls in a different direction than the argument about the cognitive parameters of our social categories: after all, the historical part highlights change and difference in the way people think about race from one epoch to another, while the cognitive part emphasizes stability. However, this disparity in fact serves our analysis very well, to the extent that the cognitive considerations reveal, as we will see, a stable background against which historical variations can occur and in reference to which these variations may be interpreted.

One of the broad implications of the recent research in cognitive science and related fields has been that, even if something might not be real, there might still be good reasons why we tend to act as though it is. The study of these good reasons amounts to a study of the foundations of what Atran has called "the anthropology of science." What he has in mind is this: assuming the correctness (or at least fruitfulness) of Chomsky's view that "each fundamental type of human knowledge arises from a specialized cognitive aptitude," then science, "which is patently different from other forms of human knowledge, should also be innately grounded in some special 'science forming faculty.' "[47] Atran proposes to illustrate this claim by studying folk knowledge of living kinds, in order to determine the extent to which such knowledge is continuous with the systems of scientific taxonomy that would emerge over the course of the fifteenth to eighteenth centuries. His conclusion is that while there is in fact no specific "science-forming faculty" of the human mind, nonetheless "certain sciences seem fitted to specific common-sense domains."[48] The science of biological classification, in particular, "emerged as an elaboration of universal cognitive schema common to all and only folkbiological taxonomies."[49] The argument here is that certain domains of scientific inquiry, particularly the ones centered on classification, are given shape by folk-taxonomical dispositions that precede and underlie the elabora-

[47] Atran, *Cognitive Foundations of Natural History*, x.
[48] Atran, *Cognitive Foundations of Natural History*, x.
[49] Atran, *Cognitive Foundations of Natural History*, x.

tion of properly scientific taxonomic systems. These dispositions, in turn, are believed to rest on a universal cognitive disposition in human beings to organize the variety of things in the natural world in a certain way. As Nancy Nersessian has described the project of cognitive history of science, this endeavor assumes at the outset that "science is one product of the interaction of the human mind with the world and with other humans. It presupposes that the cognitive practices scientists have invented and developed over the course of the history of science are sophisticated outgrowths of ordinary thinking."[50]

Working in this vein, as we have seen, recent scholars have turned their attention from the folk science of biological taxonomy to that of racial science. Some have asked whether the innate cognitive schema responsible for making racial distinctions is not the same as, or at least does not work according to the same assumptions as, the one for biological species. Gil-White, again, to cite one of the more controversial representatives of this approach, has argued that the tendency to essentialize ethnic difference, that is, to perceive different human groups as if they were divided up into real natural kinds, not only is innate, but indeed piggybacks on the innate cognitive tools available to the human mind for classifying biological species. Whatever controversy his work has generated, Gil-White has at least convincingly demonstrated that "categorization into two different ethnies biases people's expectations of phenotypic differences, and also that 'racial' phenotypic differences may not be necessary for essentialism."[51] To put this somewhat differently, the perception of "racial" as opposed to ethnic difference is not necessarily a bare response to visible properties of another person; it is sooner a *consequence* of the prior supposition of essential difference. This conclusion has also been one of the core lessons of recent scholarship in the noncognitivist (and even anticognitivist) scholarship in the philosophy of race, as well as in the key authors writing from a postcolonial perspective.[52]

[50]Nancy Nersessian, "Opening the Black Box: Cognitive Science and History of Science," *Osiris* 10 (1995): 194–211, 195. See also Nancy Nersessian, *Creating Scientific Concepts*, Cambridge, MA: MIT Press, 2008. For an attempt to place the problem of race at the intersection between cognitive science and history of science, see Justin E. H. Smith, "'Curious Kinks of the Human Mind': Cognition, Natural History, and the Concept of Race," *Perspectives on Science* 20, 4 (Winter 2012): 504–29.

[51]Gil-White, "Cognition of Ethnicity," 183.

[52]Some of the recent work in the philosophy of race has already been cited above. For postcolonial sources that argue for a related thesis, see in particular Frantz Fanon, *Peau noire, masques blancs*, Paris: Seuil, 1952; Albert Memmi, *Le racisme: description, définition, traitement*, Paris: Gallimard, 1982. For a compelling Marxist account of the function of the category of race in the articulation and preservation of class structure, see Stuart Hall, "Race, Articulation, and Societies Structured in Dominance," in Philomena Essed and David Theo Goldberg (eds.), *Race Critical Theories: Text and Context*, Oxford: Blackwell,

In most of the social-constructionist literature, as we have seen, to identify an entity or phenomenon as constructed is very often also, as if ipso facto, to disapprove of it, and to initiate a process of overcoming or discarding it. Related to this, there has also been a tendency to move from the idea of construction to that of invention: that is, to suppose that because race does not really exist, it must have been actively brought into existence by human design. On the cognitive view, however, this tendency is misguided, since, to speak with Machery and Faucher, *bad* folk science may in fact be *good* epistemology.[53] There may indeed be compelling reasons why in traditional social environments human beings have tended to essentialize their own group's distinctness from other groups. In fact, Machery and Faucher have identified a number of salient properties that ethnies do in fact have in common with species. In particular, "coethnics have a distinctive morphology (dress etc.), coethnics behave in a characteristic way, ethnic membership is based on descent, and reproduction is endogamous."[54] Machery and Faucher thus speculate that there may be "an evolved, canalized disposition to think about ethnies in a biological way."[55]

With race, then, we are dealing with the seemingly paradoxical case of what might be called a *natural construction*. Naturally constructed, we might say, are those entities or categories that do not fade away when human inquiry finds that they are not in fact robust features of the natural landscape, and that linger as a result of a natural propensity of the human mind to organize the world by means of them. While it is certainly difficult to establish unambiguously the existence of such a thing, we may nonetheless cautiously hypothesize that race is just such a natural construction, to the extent that it enables us to understand why racial thought has had such a long history, in spite of the continual variation in the way racial classifications are made, and in spite of the fact that, today, such classifications continue to have normative significance even for many who are fully aware that they have no firm ground in nature.

2002, 38–68. See also Étienne Balibar and Immanuel Wallerstein, *Race, Nation, Class: Ambiguous Identities*, London: Verso, 1991.

[53] Edouard Machery and Luc Faucher, "Why Do We Think Racially? Culture, Evolution and Cognition," in Henri Cohen and Claire Lefebvre (eds.), *Categorization in Cognitive Science*, Amsterdam: Elsevier, 2005, 1009–33.

[54] Machery and Faucher, "Social Construction and the Concept of Race," 1212. The authors neglect, however, significant cases from the anthropological literature revealing that in many cases the reproduction and maintenance of an ethnic group occur through incorporation of unrelated members from outside, not only through affine kinship, but also, very commonly, through adoption and abduction. See Descola, *Par-delà nature et culture*; see also John K. Thornton, *A Cultural History of the Atlantic World, 1250–1820*, Cambridge: Cambridge University Press, 2012, chap. 4, "The American World, 1450–1700."

[55] Machery and Faucher, "Social Construction and the Concept of Race," 1210–11.

If the evidence that race is naturally constructed is only now beginning to accumulate, it is worth bearing in mind here that we already have a tremendous amount of evidence, and a somewhat longer trail of scientific literature, suggesting that a number of the familiar categories of human perception of the natural world are indeed constructed in a similar way. Species, in particular, have been shown by science not to be at all what they seem in our ordinary way of thinking about them. To be sure, within the philosophy of biology there is a tremendous diversity of viewpoints as to their ontological status. Many today believe that species are not so much classes or sets as they are scattered individuals or homeostatic property clusters, or some such thing that is much more concrete than abstract, much more like an individual than like a kind.[56] Notwithstanding the fact that the debate about species continues, it is now most certainly correct to say, in light of evolution, that species are not *fixed* natural kinds, but rather something more like momentary representatives of lineages that include, if one goes far enough back, organisms that are not members of the same species. Nonetheless, those of us who have studied biology, no less than the "folk," appear universally to suppose in our ordinary lives that a species is a really existing kind of thing, with multiple instantiations. These individual time slices of lineages are in fact biologically significant, unlike race, but this does nothing to diminish the force of the observation that, much like race, species, even if we *know* what they are, aren't what we (ordinarily) *think* they are. As Atran has compellingly argued, human cognition of the natural world proceeds through its most "phenomenally salient" elements, particularly plants and animals. "[O]ur universally held conception of the living world," he writes, "is both historically prior to, and psychologically necessary for, any scientific—or symbolic—elaboration of that world."[57]

Now cognitive science is often faulted for playing fast and loose with speculations about the past, particularly about human prehistory, and indeed the trained historian is right to ask what is really at stake in claims

[56]The "species as individuals" thesis has been most forcefully defended, separately, by David Hull and Michael Ghiselin. See in particular David Hull, "A Matter of Individuality," *Philosophy of Science* 45 (1978): 335–60; Michael Ghiselin, "A Radical Solution to the Species Problem," *Systematic Zoology* 23 (1974): 536–44. Serious doubts have been raised about this approach, however. Philip Kitcher, in particular, provides a modified defense of the natural-kinds view of species. See Kitcher, "Species," *Philosophy of Science* 51 (1984): 308–33. For more recent contributions to the debate, see in particular Ruth Millikan, "Historical Kinds and the 'Special Sciences,'" *Philosophical Studies* 95 (1999): 45–65; Robert Wilson, "Realism, Essence, and Kind: Resuscitating Species Essentialism?," in Robert Wilson (ed.), *Species: New Interdisciplinary Studies*, Cambridge, MA: MIT Press, 1999; Marc Ereshefsky and Mohan Matthen, "Taxonomy, Polymorphism, and History: An Introduction to Population Structure Theory," *Philosophy of Science* 72 (2005): 1–21.

[57]Atran, *Cognitive Foundations of Natural History*, 13.

of historical priority of the sort Atran is making. Historians, particularly wary of anachronism, will want to know whether Atran's claim amounts to saying that Aristotle's distinction between a jellyfish and a cuttlefish, say, was a *biological* distinction, even if the Greek natural philosopher himself lacked such a term. If we see Aristotle's study of animals as based on the same cognitive aptitudes out of which Linnean taxonomy would later emerge, then the answer would seem to be "yes" and "no" at once. "No," to the extent that there was no such thing as biology, in the sense of an explicit research program with well-defined aims; but "yes," to the extent that the entities he was picking out, *and* the features of these entities that were of interest to him, are some of the same as those that we today call "biological."

And so, perhaps, with race: the features that were picked out in antiquity, or indeed in prehistory, as phenomenally salient in other groups were at least partially the same as those we would pick out as "racial": a mixture of apparently shared physical features, together with invisible, "essential" features of which the physical ones are thought to be signs. Again, and of course, race unlike species has turned out to be biologically insignificant. And yet this point of disanalogy should not cause us to abandon altogether the thought of a parallel history of the biological and the racial. For again, distinctions that are not about something real are not for that reason not real distinctions, and if we take the prospect of a cognitive turn in the study of the concept of race seriously, then whether the distinctions pick out something real is of little concern.

Atran believes that by the nineteenth-century scientific taxonomy properly speaking had broken away from the folk-rooted tradition of natural history, and that, as a result, "natural history's common-sense preoccupation with comprehending phenomenal reality gave way to biology's quest to explain the unforeseen."[58] Yet if the extension of the cognitive turn from living kinds to human kinds is justified, it appears, as we will see, that natural-historical thinking in the latter domain held out somewhat longer. Even as new, more sophisticated questions came to be asked about biological taxa, the older natural-historical project was still well under way with regard to human subtypes, and this *even though* the majority of the practitioners of this late branch of natural history sensed that the phenomenal reality they were describing with their racial typologies had no place in the emerging, speculative science of biology, which was, as Atran puts it, centered around "a dynamic reassessment of the relation between species and higher-order taxa in terms of biological functions, anatomical structures and historical processes."[59]

[58] Atran, *Cognitive Foundations of Natural History*, 10.
[59] Atran, *Cognitive Foundations of Natural History*, 10.

Machery and Faucher have emphasized the importance of bringing together the literature on social construction with the new cognitive-evolutionary approach they are helping to develop, and this synthesis will surely be fruitful. But one might add that another approach that might supplement these newly unified strands is the historical one, which might either falsify or support the hypothesis of an evolved disposition to carve up the human species into essentialized ethnies or "races." Like data from psychological studies of infants,[60] or from cross-cultural fieldwork such as Gil-White's, the study of different historical periods in the aim of determining what remains stable and what changes in the way the concept of race is deployed, can itself play an important role in the discussion. This much has already been shown by Atran in the parallel but related area of biological taxonomy, where his twofold approach treats the problem both historically (based on Western sources) and ethnographically (based on comparative fieldwork).

Such a contribution is important, since one widespread view in recent years, as we have already seen, has held that race is not just a construction, but indeed a distinctly *modern* construction, and if this is in fact so then the case for a cognitive-evolutionary approach to race is significantly weakened. Some suppose that "race" as a social category is simply coterminous with the spread of racial typologies in the eighteenth and nineteenth centuries.[61] Other scholars date the invention of race to the early modern period, roughly contemporaneously with the rise of the transatlantic slave trade.[62] In general, scholarly work that is placed under the banner of postcolonial theory supposes that in some way or other racial thinking is a part of the legacy of modern European colonialism.

Most philosophers of race, whatever their normative commitments, agree at least that there are certain respects in which the concept of race exists prior to the development of modern scientific racial theory, which is to say modern scientific racism, and other respects in which what we mean by "race" today is conditioned by our era and culture. One point that is often emphasized is that even if there is some sense in which human beings cannot but think of the human species as falling into different subtypes, there is no obvious way in which the various races we recognize today dictate to us what these subtypes are. Correlatively, again, there is no respect in which we simply read a person's racial identity off of their appearance; to identify someone as belonging to this or that race is not to

[60] See in particular Hirschfeld, *Race in the Making.*

[61] See, e.g., Michael Banton, "The Concept of Racism," in Sami Zubaida (ed.), *Race and Racialism*, London: Tavistock, 1970.

[62] See in particular Ivan Hannaford, *Race: The History of an Idea in the West*, Baltimore: Johns Hopkins University Press, 1996.

give a report of the visible features of that person that are given in bare perception.

Recent history shows, in fact, that ethnic groups can easily be shifted from one racial category to another,[63] but it would be absurd to suppose that a change in racial designation ever results from a closer inspection of the physical features of members of these groups. Instead, plainly, to call someone "white" is not to give a bare report of the perception of a white-colored entity, but rather to offer a sort of color-coded value judgment. Thus the recent fear that the United States is on its way to becoming a minority "white" country is entirely contrived:[64] if "whites" wish to remain a majority, they are perfectly free to redefine the category in such a way as to ensure that by definition there are more whites than nonwhites. Indeed, one cannot help but conclude that the sort of journalism that follows the supposed decline of the white race is responsible for maintaining in existence the very categories on which it purports to be reporting.

1.6. Conclusion

Race, it seems, is in the end really just an evaluative notion masquerading as a natural kind. The abolitionist Henri Grégoire well understood this in his *De la littérature des Nègres* of 1808 when he wrote: "Those who have wanted to disinherit the Negroes have used anatomy to their advantage, and the difference of color gave rise to their first observations."[65] In other words, the desire to disinherit is prior to the observation, and skin color is simply the first feature given in observation on which the defender of slavery hopes to be able to base his argument for disinheritance: a hope based on a profound non sequitur, yet one that has generally gone unno-

[63] See Noel Ignatiev, *How the Irish Became White*, New York: Routledge, 1995; Karen Brodkin, *How the Jews Became White Folks*, New Brunswick, NJ: Rutgers University Press, 1998. Francisco Bethencourt has convincingly argued, on the historical example of Jews as well as Armenians, that there can be no clear separation between "racial" discrimination based on supposedly natural hierarchies and racial discrimination based on religious hierarchies. See Bethencourt, *Racisms: From the Crusades to the Twentieth Century*, Princeton: Princeton University Press, 2013.

[64] See for example Sam Roberts, "Census Benchmark for White Americans: More Deaths than Births," *New York Times*, June 13, 2013, http://www.nytimes.com/2013/06/13/us/census-benchmark-for-white-americans-more-deaths-than-births.html?_r=0; James Hamblin, "'Rise of the Colored Empires.' White Babies Are No Longer the Majority in the US," *Atlantic*, June 13, 2013, http://www.theatlantic.com/health/archive/2013/06/rise-of-the-colored-empires/276844/.

[65] Henri Grégoire, *De la littérature des Nègres, ou, recherches sur leurs facultés intellectuelles, leurs qualités morales et leur littérature: suivies des notices sur la vie et les ouvrages des Nègres qui se sont distingués dans les sciences, les lettres et les arts*, Paris, 1808, 14.

ticed in the history of racial thinking. That this is how racial distinctions work is clear after only a modicum of reflection, and yet many who have undertaken this reflection continue to consider the notion indispensable for making sense of the diversity of the human species. Understanding why this is the case requires an investigation of the historical development of the concept.

Chapter 2

Toward a Historical Ontology of Race

2.1. False Positives in the History of Race

It has been argued, so far, that an historical examination of the emergence of our current racial categories is necessary in order to gain a clearer picture of what is recent and contingent in racial thinking, and what, by contrast, is deep-seated. Yet a methodological problem presents itself straightaway: if we go looking in old texts for what people had to say about race in the past, does this not presuppose that there is a fixed thing, "race," about which people have spoken in the same easily identifiable way across different historical eras? But if we do not suppose there is such a fixed thing, then how can we know, at the outset, what we are looking for when we set out to study the various alternative ways people had in the past of accounting for the diversity of the human species?

There are no easy answers to these questions, other than the commonsense advice that one should not presume too quickly that one has understood the relevance of a historical text to understanding issues related to the problem of race as it is understood today. Often it is not at all clear whether some historical account of a person or a group of people can properly be said to have to do with race or not. In many cases, what we find instead is a rather haphazard mixture of racial or quasi-racial description, together with a characterization of national customs, religion, and other features that we today clearly mark off as cultural rather than racial. Thus for example Pierre Gassendi, the great materialist philosopher and influential figure for François Bernier, makes a revealing comment, if only in passing, about the variety of the human species and about the problem of determining its boundaries. In his *Vie de Peiresc*, published in English translation in 1657, Gassendi describes a report from a certain Thomas Arcosius to Nicole-Claude Fabri de Peiresc, in which he relates

> what had happened to one of *Ferrara*, when he was in a Country of *Marmarica*, called *Angela* [i.e., Angola]. For he happened one day upon a Negro, who hunted with Dogs certain wild men, as it seemed. One of which

> being taken and killed, he blamed the Negro for being so cruel to his own kind. To which he answered, you are deceived; for this is no man, but a Beast very like a man. For he lives only upon Grasse and has guts and entrals like a Sheep, which that you may believe, you shall see with your eyes; whereupon he opened up his belly. The day following, he went to hunting again, and caught a male and a female. The female had Dugs a foot long; in all other things very like a Woman. . . . Both their Bodies were hairy all over, but the hair was short and soft enough.[1]

Now of course this anecdote is mediated through several parties by the time it reaches Gassendi, yet it reveals something significant about the way variety within the human species was understood by Gassendi's contemporaries, a way that Gassendi himself transmits without opposition: there is a fact of the matter as to who is a human being and who is not, even if there is great variety in physical appearance. If beasts, such as those hunted by the "Negro" in this tale, happen to have humanoid features, there is nonetheless a way of determining absolutely whether they are in fact human or not; in this case, cutting them open and looking at internal anatomy suffices. If a creature is not of one's own kind, but only a beast, then there is no moral concern about cruelty toward it. But while the European supposes that the kind in question here is "Negro," the "Negro" in turn denies that the beast belongs to the *human* kind: he does not maintain that it is "very like a Negro," but that it is "very like a man," and "Negroes," as all parties agree, are men. When "kind" is used here in reference to "Negroes," this does not isolate this group in any natural or essential way from the human kind of which it is a part.

In general one must be careful not to suppose that every historical occurrence of a description of what appear to be physical racial traits in fact has to do with race in any robust sense, since there are very many features of human beings that could in fact be picked out by, for instance, color terms, only a few of which involve what we today would think of as racial considerations. Indeed, many color terms are used to describe people that have no parallel racial connotation at all: someone is blue when sad, for example, or green with envy. The association of envy with greenness is obviously very arbitrary, and unlikely to extend beyond a fairly small linguistic community. If a foreign interpreter were attempting to make sense of an English text in which someone is described as "green" in this respect, he or she would be fairly naïve to suppose that actual, visible greenness is implied. Mutatis mutandis, the same point may be made about our reading of authors from several centuries ago who describe

[1] Pierre Gassendi, *The Mirrour of True Nobility and Gentility: Being the Life of the Renowned Nicolaus Claudius Fabricius, Lord of Pieresk, Senator of the Parliament at Aix*, trans. William Rand, London: Humphrey Moseley, 1657, book 5, 92–93.

people as "black": not every occurrence of this term is racial, and in fact, like "red" for Native Americans, even when it is meant as racial, the term often functions more as a symbolic color coding of people with a wide range of skin tones than as an actual description of a person's physical appearance. There is no pregiven, bare fact about the way different human groups look that determined in advance who would end up being called "black," who "white," who "red," and so on. To focus on just one, particularly important example, how might the term "black" be used, other than racially? Let us consider a few instances.

2.2. "Erst Spruce, Now Rusty and Squalid"

To the extent that the temperament and conformation of the body were long seen as flowing simply from the way one kept one's body, it is not surprising to find premodern and early modern authors describing social class, and in particular the laboring and poor classes, in a way that elides class distinctions with what look to us like racial ones. Thus Robert Burton, in a subsection of *The Anatomy of Melancholy* on "Poverty and Want," describes "African negroes, or poor Indian drudges," who "are ugly to behold, and though erst spruce, now rusty and squalid, because poor."[2] In other words, it is not that they are held to be suitable for a life of labor because they are already dark; rather, their complexion results from the fact that they live their lives as laborers. Burton says of laborers in general that they "are commonly such people, rude, silly, superstitious idiots, nasty, unclean, lousy, poor, dejected, slavishly humble, and, as Leo Afer observes of the commonalty of Africa, their life is full of misery, toil and suffering, want and misfortune; they are more ignorant than asses, and you would say they were the offspring of brutes."[3] This is hardly a flattering portrayal, but here what interests Burton are the habits of the lower classes, and he takes Leo the African's description of the "commonalty" of the continent, from which Leo himself has his moniker, as an example of the effects of poverty, not as an account of the innate features of a subtype of humanity. The reference to Leo's description of Africans occurs right alongside numerous other examples of "base villains, hunger-starved beggars, wandering rogues," and so on, who are indigenous to Europe. They as well are held to be rusty or dark as a result of their condition.

[2] Robert Burton, *The Anatomy of Melancholy*, New York: New York Review Books Classics, 2001, 351.

[3] Burton, *Anatomy of Melancholy*, 351.

To cite another, related example, in his 1661 *Fumifugium* John Evelyn expounds at great length on the relationship among lifestyle, modes of labor, and what we would think of as racial attributes:

> The more hot promotes indeede the Witt, but is weak and trifling; and therefore Hippocrates speaks the Asiatique people Imbelles and Effeminate, though of a more artifical and ingenious Spirit: If over cold and keen, it too much abates the heat, but renders the body robust and hardy, as those who are born under the Northern Bears, are more fierce & stupid, caused by a certain internal Antiperistasis and universal Impulsion (that is, the heat of their bodies is condensed and exercised by the coldness of the atmosphere surrounds them). The drier Aer is generally the more salutary and healthy, so it be not too sweltery and infested with heat or fuliginous vapours, which is by no means a friend to health and Longaevity, as Avicen notes of the Aethiops who seldome arived to any considerable old Age.[4]

Now this might appear to be a partitioning of the human species into races, but Evelyn immediately follows his account of human variety with a revealing proverb: *Mores Hominum do corporis temperamentum Sequi*: "I give the habits of men, [and from this] the temperament of the body follows."[5] In other words, the air you breathe, and the water you drink, and the practices that together constitute your hygiene in a given natural environment are what cause your body to be the way it is. A human body that looks very different from another human body is not to be explained by postulating different subtypes of the human species, but only different hygienic practices and different "lifestyles."[6]

From the time of the Hippocratic treatise *Airs, Waters, Places* in the fourth century BCE, the varieties of human appearance and behavior were commonly understood to result from environmentally induced variations in the balance between the humors. How, now, were these variations different from modern races? For one thing, they were held to be relatively reversible and contingent features of the populations they characterized. Moreover, it would be enough to subject a single individual to the same environmental conditions as the members of another population face in order to bring about in that person the same temperament and constitution as the members of the other population. Thus if there

[4]John Evelyn, *Fumifugium, or The Inconveniencie of the Aer and Smoak of London Dissipated. Together with some Remedies Humbly Proposed by J. E. Esq*, London: W. Godbid for Gabriel Bedel, and Thomas Collins, 1661, 13–14.

[5]Evelyn, *Fumifugium*, 13–14.

[6]For a lucid account of the particular importance of environmental explanations of "racial features," and of their utility for the formation of class hierarchies in the colonial American context, see Carlos López Beltrán, "Hippocratic Bodies, Temperament and Castas in Spanish America (1570–1820)," *Journal of Spanish Cultural Studies* 8 (2007): 253–89.

were, say, a microregion of Germany where "Asiatic" environmental conditions prevailed, a person who settled in that microregion would end up with Asian attributes. Thus humoral accounts of human diversity focused on the way environments shape individuals, rather than the way populations share traits.

The humoral usage of the adjective "black" is a very common source of misunderstanding in the interpretation of historical texts. Consider the following passage from Baruch Spinoza's letter to Pieter Balling of July 1664:

> One morning, as the sky was already growing light, I woke from a very deep dream to find that the images which had come to me in my dream remained before my eyes as vividly as if the things had been true—especially the image of a certain black, scabby Brazilian [*cujusdam nigri, & scabiosi Brasiliani*] whom I had never seen before. For the most part this image disappeared when, to divert myself with something else, I fixed my eyes on a book or some other object. But as soon as I turned my eyes back away from such an object without fixing my eyes attentively on anything, the same image of the same Ethiopian [*Aethiopis*] appeared to me with the same vividness, alternately, until it gradually disappeared from my visual field.[7]

This letter has been treated as a probable example of antiblack racism in early modern philosophy. Michael Rosenthal notes the critics who "argue that such apparently ad hoc examples are really symptomatic of a larger pattern of racial discrimination."[8] Rosenthal maintains that there is "an obvious racial aspect of the dream," adding that Spinoza "makes a point of adding the adjective 'black' to his description of the man and, lest the reader think he was indicating nationality with the term 'Brazilian,' the second reference to 'Ethiopian' makes it clear that he is not."[9]

It is indeed impossible to deny that the term "Ethiopian" is being used in a racial sense here. This term along with "Moor" were conventional tags for people with the phenotype that would later be called "black" (though "Moor" would also often be used as a synonym of "Muslim"). Thus we see for example John Evelyn's 1661 description of the difference between "Asiatique people" and "Aethiops," and several decades later Anton Wilhelm Amo (who will be the focus of chapter 8), from West Africa, is identified in the enrollment records of the University of Halle as an

[7]Benedict de Spinoza, *Opera*, ed. Carl Gebhardt, 4 vols., Heidelberg: C. Winter, 1925, 4:76.25–77.6.

[8]Michael A. Rosenthal, "'The Black, Scabby Brazilian': Some Thoughts on Race and Early Modern Philosophy," *Philosophy and Social Criticism* 31, 211 (March 2005): 211–21, 212.

[9]Rosenthal, "'Black, Scabby Brazilian,'" 212.

"Aethiops." But is there anything else happening in this letter? Why, for example, is the apparition not just black, but indeed scabby and black?

There appears to be a medical, and in particular a humoral, background for Spinoza's dream. The apparition is not only a stock racial caricature: to be "black" in a humoral sense is to have a surfeit of black bile or melancholy, and thus to be earthy, and thus, in turn, to be associated with corpses and other supposedly dark things. In extreme cases, a melancholic person, in addition to being dark or "muddy" himself or herself, can also hallucinate corpses or, alternatively, "black" men. In *The Anatomy of Melancholy*, Burton attributes to Jovianus Pontanus the view that those born under the influence of Saturn are "very austere, sullen, churlish black of colour."[10] Hercules de Saxonia is cited as holding that "these that are naturally melancholy" are "of a leaden colour or black."[11] Burton himself adds that a certain subspecies of melancholics are "of a muddy complexion."[12] It is precisely such people, according to a certain Guianerius, who "think themselves dead many times, or that they see, talk with black men, dead men, spirits and goblins frequently."[13] Gordonius writes that extreme melancholics see and talk "with black men, and converse familiarly with devils."[14] Gentilus Fulgosus is cited as relating the story of a melancholy friend, who "'had a black man in the likeness of a soldier' who followed him wherever he was."[15]

In a footnote to the citation from Guianerius, Burton distinguishes two different sorts of reaction to sudden, unexpected visions: there are those who *blandiuntur*, and those who *territant*, that is, who are "terrified," which is to say, etymologically, those who are *rendered earthy*, of which a sign is the sudden darkening of the face. The extreme consequence of this, for people of a certain humoral disposition, is to come to believe that they themselves are corpses. In Burton, this particular sort of hallucination has no racial connotation whatever, but is rather the straightforward consequence of the excessive production of black bile. That blackness is associated with earth in humoral medicine enables in turn the association of the melancholic condition with corpses, and the interchangeability of the figure of the black man with that of the corpse.

It is likely that between the time of Burton's work and Spinoza's letter, there is a superimposition of a racial or quasi-racial connotation upon the figure of the black man, and that the term is gradually deprived of its original humoral significance. In Spinoza's case the figure not only is

[10] Burton, *Anatomy of Melancholy*, 397.
[11] Burton, *Anatomy of Melancholy*, 399.
[12] Burton, *Anatomy of Melancholy*, 400.
[13] Burton, *Anatomy of Melancholy*, 400.
[14] Burton, *Anatomy of Melancholy*, 402.
[15] Burton, *Anatomy of Melancholy*, 402.

racialized, but is explicitly associated with a particular geographical region. And yet, at the same time he retains the "scabbiness" of the old visions of corpses that interested Burton. The co-occurrence in Spinoza of the humoral and the racial connotations of "black" likely indicates that the philosopher is writing at a moment when the racial sense of "black" is displacing the humoral one. Spinoza himself probably does not give much thought to the sources of his dream, and does not seem concerned to explain how someone could be Brazilian and Ethiopian at the same time. Rosenthal is certainly correct to identify this passage as involving a negative portrayal of a person who is black in a racialized sense. But race doesn't tell the full story. The apparition is scabby not because Spinoza conceptualizes racially black people in general as ghoulish or sickly; the scabbiness is in fact associated with a different, nonracial sense of "black."

This clarification of what is going on in Spinoza's letter is not intended as special pleading on his behalf. We have dwelt on this example at some length only because it clearly illustrates how, in trying to make sense of ideas about race in the early modern period, very often there is more than meets the eye in texts that are concerned with, or that appear to be concerned with, race.

While historical texts can conceal or obscure racialized perceptions, or can cause us to mistakenly believe we are looking at a treatment of racial difference when in fact we are not, these texts can also, sometimes, vividly reveal the contingency of current ideas about race. Consider the strange history of the dreadlock, a style of wearing one's hair that, since at least the mid-twentieth century, has been racialized as black, indeed to such an extent that a racialized white person will often be met with surprise when he or she succeeds in growing dreadlocks. But little do most know today that throughout the early modern period dreadlocks were coded not as black, but as Polish. In François Bernier's 1669 letter to Jean Chapelain, for example, written from the Persian city of Shiraz, the French traveler describes the appearance and practices of the yogis he had seen during his sojourn in India. "There are some [yogis] who have hair that falls to half-way down their thigh," he writes, "and that is matted into branches like the massive hairs of our water-dogs, or rather like the hair of those who suffer from the illness from Poland that we call the 'Polish plait.' I saw in many places ones who held an arm, and sometimes both arms, elevated and stretched continually above their heads."[16] Bernier's account of dreadlocks as a medical condition, and one associated, specifically, with Poland, is by no means unusual. His near-contemporary

[16]François Bernier to Jean Chapelain, October 4, 1669, in *Un libertin dans l'Inde Moghole. Les voyages de François Bernier (1656–1669)*, ed. Frédéric Tinguely, Paris: Éditions Chandeigne, 2008, 316–18.

Thomas Hall would similarly see this supposed "hair condition" as a distinctly Polish affliction. In a scolding and moralistic manner very akin to the work of John Bulwer, to which we will turn later, Hall writes in his 1654 book, *The Loathsomnesse of Long Haire*, that dreadlocks are a sort of modern plague, and a sign of the wrath of God against the entire nation of Poland: "Feare him that did that Plica send, / And those sad Crawlers: and hath more / Unheard of Judgments still in store."[17]

A number of medical reference works, well into the nineteenth century, describe this hairstyle as a sort of trichoma, a "strange illness," indeed a "hair disorder."[18] But why do they continue to associate it with Poland? An 1834 edition of the *Cyclopaedia of Practical Medicine* suggests both environmental causes, as well as the influence of foreign invasion. This standard reference work informs us that the frequency of the condition in Poland results from a long tradition, going back to the eleventh century, of wearing the hair "after the manner of the Tartar and some Indian tribes," namely, in "a single tuft of hair being left to grow from the top of the scalp."[19] The Poles themselves, we learn, "believe that it was carried into their country by the Tartars in the twelfth or thirteenth century."[20] The inhabitants of the country are described as "being wretchedly lodged and clothed, and exposed to the combined injurious influence of a marshy soil and a damp variable climate," all of which contributes to the "greatly increased growth of this portion of the hair, and the unnatural quantity of viscid secretion which is at the same time thrown out."[21]

But how, in turn, can a hairstyle be thought to be an illness? If the dreadlocks become problematic, they can simply be cut off. Or can they? The authors of the *Cyclopaedia* entry tell of a patient, a boy of about fifteen years of age, who "lay in a most filthy state, and his black hair, knotted long and matted together, gave out an intolerable stench."[22] It is related that none other than the Polish physician in attendance strongly opposed the option of "cutting off the hair, on the ground that the humour exuded on the hair might turn in on the brain and cause apoplexy."[23]

This is curious material, to be sure, but there is a serious lesson in it: today dreadlocks are not seen as a medical condition, but they are racially

[17]Thomas Hall, *The Loathsomnesse of Long Haire*, London: J. G. for Nathanael Webb and William Grantham, 1654, 4.

[18]John Forbes, Alexander Tweedie, and John Conolly (eds.), *Cyclopaedia of Practical Medicine, comprising Treatises on the Nature and Treatment of Diseases, Materia Medica and Therapeutics, Medical Jurisprudence, Etc.*, vol. 3, London: Sherwood, Gilbert, and Piper, and Baldwen and Craddock, 1834, 400.

[19]Forbes, Tweedie, and Conolly, *Cyclopaedia*, 400.

[20]Forbes, Tweedie, and Conolly, *Cyclopaedia*, 400.

[21]Forbes, Tweedie, and Conolly, *Cyclopaedia*, 400.

[22]Forbes, Tweedie, and Conolly, *Cyclopaedia*, 401.

[23]Forbes, Tweedie, and Conolly, *Cyclopaedia*, 401.

coded, and when spotted on European heads are held to be a transposition of a hairstyle "naturally" suited to African heads. But what we learn in fact is that in the European imagination the dreadlock originally came from a country on the eastern frontier of Europe that, in the early modern period, was easily associated in the imaginations of French and English authors with the exotic and terrifying world of the Tatars (large numbers of whom had settled in the Polish-Lithuanian Duchy and its antecedent states from the fourteenth to the seventeenth centuries). One easily suspects that the association of matted hair with the Afro-Caribbean and African American cultural sphere is itself the transposition of prejudices that are much more deeply rooted than we ordinarily perceive, and that have to do as much with perceptions about the boundaries between civilization and savagery as with informed judgments about the physical properties of different varieties of human hair.

2.3. Race and Dualism

We have already noted that often questions of race come up in early modern philosophy texts not so much because the author is interested in race for its own sake, but simply because an example or anecdote involving it helps the author to illustrate a point about some other philosophical problem. It was noted, in reference to Charles W. Mills, that in order to adequately investigate the history of the concept of race, one must be prepared to dispense with the common view among philosophers according to which examples chosen by philosophers are not in themselves significant, but only the point the philosopher seeks to illustrate by means of them. One remarkable and illustrative case of race being deployed in such a way can be found in the French Jesuit Gabriel Daniel's 1691 work, *Voiage du monde de Descartes*. This is a satirical work, a fictional romp whose purpose is as much to ridicule Descartes, and to make light of as many other people and ideas as the author can touch upon along the way. The work's plot, such as it is, involves an encounter with the disembodied spirit of Descartes, who takes the narrator on a tour of the "world" described in the philosopher's 1633 *Le Monde*. Daniel's principal aim is to deflate and satirize Descartes's vortex theory, and in order to do this he imagines a trip to the moon, in turn inspired by Cyrano de Bergerac's *Histoire comique des états et empires de la Lune*, first published in 1655.[24] In the course of the narrator's space travels he encounters the

[24]Hercule Savinien Cyrano de Bergerac, *Histoire comique des états et empires de la Lune et du Soleil*, ed. P. L. Jacob, Paris: Adolphe Delahays, 1858 [1655],

disembodied spirits of many dead philosophers, who have, it turns out, taken up residence on the lunar surface.

In Daniel's parallel reality, Descartes's discovery of the strict separation between body and soul has brought with it a practical application: the ability of the soul to leave the bodily machine at will. In the story, this is an ability that Descartes guards as a sort of secret, and in this way Daniel effectively portrays Cartesian mechanism and dualism as a sort of magical sect; at one point the narrator proclaims, "I was very worried that this was some sort of sorcery and magic, and that under the pretext of taking me to Descartes's World, they were in fact taking me to a sabbath."[25] Here it is important to note that the power of a soul to intentionally leave the body behind and "go wandering" was in the seventeenth century commonly invoked in connection with ideas about witchcraft in popular culture: Carlo Ginzburg, to mention one influential scholar, has described in great detail the phenomenon of "night wandering" among early modern Friulian peasants.[26] It is also important to note that the idea of night wandering was not entirely rejected in learned circles. Henry More, for example, an important correspondent of Descartes, writes in his 1653 *Antidote against Atheism* that it is not entirely unreasonable to believe that spirits leave their bodies at times, "and see and do such & such things, meet one another, discover secrets and the like."[27] Curiously, More argues that it is at least no more absurd to believe this really happens than it is to agree with "Cartesius, that stupendious Mechanicall wit," in his claim "that Animalls themselves were mere Machina's."[28] But in Daniel's distorted world it is Descartes himself who believes that souls go wandering (even if Daniel's Descartes remains sufficiently anchored in the real Descartes as to deny, true to character, that animals have souls to begin with that might leave the body), and, again, the ability to do so is described as a sort of closely guarded secret, and as a practical application of the doctrine of mind-body dualism.

It is within the context of this farce that Daniel introduces the figure of the African servant who inadvertently gets initiated into this secret, with tragic results. It is worth citing the long excursus on the servant's fate in full:

[25] Gabriel Daniel, *Le voiage du monde de Descartes*, Paris: Simon Bénard, 1691, 39. "J'apprehendois fort qu'il n'y eût ici de la sorcellerie & de la magie; & que sous pretexte de me mener au Monde de M. Descartes, on ne voulût me mener au sabat."

[26] See Carlo Ginzburg, *I benandanti. Stregoneria e culti agrari tra Cinquecento e Seicento*, Turin: Einaudi, 2002.

[27] Henry More, *An Antidote against Atheism, or, An Appeal to the Naturall Faculties of the Minde of Man, Whether There Be Not a God*, London: J. Flesher, 1655 [1653], book 3, chap. 11, 237.

[28] More, *Antidote against Atheism*, book 3, chap. 11, 239.

"Oh my God! Sir!," I cried out, terrified. "What is this you are causing me to see?"

"Gentle now," he said to me, "gentle. Do not be alarmed. What you are seeing is not so devilish, even if it is black. This is not at all a devil; it is the soul of a little Negro, who is in the service of M. Descartes, and whose adventure I will tell you in two words, to relieve you of any doubt or worry.

"This little Negro was once the valet of M. Regius, the famous professor of medicine at the University of Utrecht, who, as you know, was previously a close friend, disciple, and admirer of Descartes. In view of these qualities, he earned from him knowledge of the secret of the separation of the body and soul. But afterward they had a falling out, to the point that Descartes thought himself obliged to write against him, since he corrupted the doctrine, and even made it scandalous. Regius, whose manners were not always those of the most gallant man in the world, at least as Descartes described him, took his vengeance against him, and displayed his contempt for the thing that Descartes valued the most, by teaching the secret to this little Negro, who took it into his head more than once to take advantage of it.

"As he was coming back one day from the countryside, where his master had sent him, finding himself tired, and having sat down in the shadow of an oak, his soul left his body there to rest and sleep, and went to amuse itself I don't know where. But some robbers killed a man nearby. The officer who was in the area, having been notified, came without delay with his soldiers: the noise was so great, that he awoke the body of the little Negro. . . . The machine [of his body], determined [into motion] by this noise, and by the strong impression that the presence of these armed men made on its organs, began to flee. They ran after him, caught him, and interrogated him. He halted at every word in his responses, which, in the absence of a soul, could not be very coherent. The officer, who was a bit hurried, took his fleeing, and the fright that appeared in his face and in his speech, as decisive proof of his crime, and ordered him to be hanged from a tree then and there as an accomplice to the murder that had just occurred. The soul, returning a moment later, found its body striking the horrific figure of a hanged man. Obliged as it was to stay away, it found itself in a difficult position. Most separated souls that move about throughout the expanse of the world, being philosophers' souls, and souls of importance, who, in an assembly that the most formidable among them had summoned, had declared to be true that philosophical opinion according to which not all souls are of the same species, did not wish to accept that the soul of an ignorant Negro should have the same privileges as they themselves, and chased him away wherever he appeared. In the end, for the sake of his happiness, he dared to leave our vortex, and passed to the place where the spirit of M. Descartes was meditating. The spirit had compassion for it, and permitted it to stay with him."[29]

There are a number of noteworthy elements in this passage, simultaneously rich and problematic as it is. One thing to note is that the author falls back on the same association between blackness and ghoulishness that we have already seen in Spinoza's dream about the scabby Brazilian. This, we may presume, is likewise a vestige of the medical tradition that associated blackness with a surfeit of black bile, as well as with earth and ash, and thus with cadavers. In this case, however, the association is quickly identified as an error, and the apparent color of the spirit arises from the fact that during its lifetime it had been associated with a person of African descent. But the more important questions here are these: Why does Daniel invent this character at all? Which figures or ideas exactly is Daniel lampooning? While Descartes is the ostensible target of Daniel's farce, in the end it is Descartes who comes out looking less ridiculous than the disembodied philosophers who have taken refuge on the moon. The servant falls into trouble in the first place only because he learns the "secret" and begins to dabble in Cartesian magic. But once he finds himself in trouble it is only Descartes's ghost, among all the dead philosophers, who is willing to have him. The other dead philosophers, in apparent bad faith, invoke the medieval doctrine according to which disembodied souls must be individuated with respect to species—that is, any immaterial being must occupy its own species—as a way of refusing to acknowledge any community with the soul of the African. For Descartes, by contrast, there is only one species of created substances that has a soul at all, and any disembodied soul from this species must of necessity be of the same nature as any other. The fact that the African has a soul in the first place that is capable of going wandering from the body means that he is not a mere machine, for in going wandering the soul leaves its machine behind. And if one is not a mere machine on the Cartesian ontology, there is only one other thing one can be, and that is a thinking thing. A thinking thing, finally, is simply not the sort of thing that can be of a certain race or other, let alone of a certain species or other.

In seeking to mock Descartes, Daniel inadvertently—and accurately—draws attention to an important feature of Cartesian mind-body dualism and its corollary, Cartesian human-animal dualism: that in such an ontology there simply is no space for finer gradations of race, where race is understood as marking out physical differences between different human groups that in turn correspond to differences of mental capacity. Descartes

[29]Daniel, *Le voiage du monde de Descartes*, 44–45.

is not a programmatically antiracist thinker, by any means; but his ontology makes racial thinking strictly speaking incoherent, as Daniel appears to detect. And so, in the tale, the African suffers a terrible fate, and is the victim of what must rightly be called a racist lynching. But in the end his wandering soul ends up in the Cartesian camp, and notwithstanding the author's intentions, in departing the vortex of the moon, takes at least a certain tendency in philosophy one small step toward an explicit disavowal of the significance of racial difference.

Of course the medieval doctrine of species mentioned in the passage was in fact meant to deal with the problem of distinguishing between angels, not between races, and it says much about the preoccupations of Daniel's era that a doctrine such as this could be appropriated and racialized in this way. Daniel writes as if the distinction between human races had been a concern of philosophers all along, and as if Descartes is breaking with long tradition in implying that an African's soul is a thinking thing as much as any other, whereas in fact it would be more accurate to say that Daniel, writing at the very end of the seventeenth century, belongs to the first generation, or very near to it, that would consider it necessary to distinguish between races at all. Part of the perception of an urgency in doing so appears to have come from the perceived antiracist implication of dualist philosophies such as that of Descartes.

In later chapters we will explore in greater detail the argument that metaphysical dualism served as a bulwark against racism, while naturalism about the human self served to bring this bulwark down. The story is of course more complex than this, but Daniel provides some preliminary evidence for it. For the moment, however, we are principally concerned with the case of the African's lynching, and his soul's subsequent migration into the Cartesian camp, as an illustration of the complex ways in which gratuitous racism and philosophical argument are often interwoven in the early modern period, and of the importance of dwelling on an author's choice of examples. Daniel's story of the "Negro" is not like Aristotle's illustration of the theory of practical virtue by appeal to a diet of chicken. This fictional persona, rather, occupies a place at once in the history of racism and in the history of philosophy, and neither of these roles can be fully understood without adequate consideration of the other.

2.4. Conclusion

Writing about the history of racial thinking is, as we have been seeing, a difficult task, to the extent that (1) the thing itself, race, does not exist in a robust scientific or metaphysical way, and (2) it is often entirely unclear, when looking at texts from the past, whether different authors are writing about one and the same unrobust, imaginary thing. It is not like the case of, say, the history of the concept of "mammal," where we can easily translate antiquated classificatory terms (e.g., "quadruped"), and can be relatively assured that there is a real object of study that was of interest to both the historical authors and to us. The entire history of racial thinking, by contrast, has been based on a mixing of biological and political concepts, and in attempting to write about this history, one is also tempted to seek to unmix them and thereby to set things right.

And yet, as discussed in the previous chapter, race is also an extremely tenacious illusion, and to understand why this is so, it is worthwhile to try to follow out the thinking of the figures in history who have been drawn in by it, and have contributed to sustaining it. Again, even if it is an illusion, there are good reasons why this illusion took the form it did, in the historical period it did. These reasons have much to do with both social-historical factors, particularly the rise of the institution of slavery, and factors in the history of ideas, particularly the rise of the systematic scientific study of the order and diversity of nature. Another important factor, which we have not yet looked at in much detail, yet which is thoroughly intertwined with the others, is the encounter with the natural and cultural otherness of the Americas at the end of the fifteenth century.

Chapter 3

New Worlds

3.1. "I Had to Laugh Vehemently at Aristotle's Meteorological Philosophy"

In a vivid passage of *The Anatomy of Melancholy*, Robert Burton inquires as to the sources of natural diversity in general. He describes his interest as that of finding out "a true cause, if it be possible, of such accidents, meteors, alterations, as happen above ground."[1] He sees human diversity as fundamentally of the same sort as these natural productions, and goes on to ask "[w]hence proceed that variety of manners, and a distinct character (as it were) to several nations? Some are wise, subtile, witty; others dull, sad, and heavy; some big, some little . . . some soft, and some hardy, barbarous, civil, black, dun, white; is it from the air, from the soil, influence of stars, or some other secret cause?" Burton sees this sort of variety as just one instance of the general plenitude of nature, as is observed also in animals and plants: "Why doth Africa breed so many venomous beasts," he continues, "Ireland none? Athens owls, Crete none?" For Burton as for the Spanish more than a century before him, it is America that provides the richest examples of natural diversity: "Why so many thousand strange birds and beasts proper to America alone, as Acosta demands? . . . Were they created in the six days, or ever in Noah's ark?" Burton also wants to know how it comes to pass

> that in the same site, in one latitude, to such as are neighbors, there should be such difference of soil, complexion, colour, metal, air, etc. The Spaniards are white, and so are Italians, whenas the inhabitants about the Cape of Good Hope are blackamoors, and yet both alike distant from the Equator: nay, they that dwell in the same parallel line with these negroes, as about the Straits of Magellan, are white-coloured, and yet some in Presbyter John's country in Aethiopia are dun; they in Zeilan and Malabar parallel with them again black: Monomotapa in Africa and St. Thomas' Isle are extreme hot, both under the line, coal-black their inhabitants, whereas in

[1] Burton, *Anatomy of Melancholy*, book II, 43.

> Peru they are quite opposite in colour, very temperate, or rather cold, and yet both alike elevated.

And so on. Burton evidently sees the question of human diversity as including differences of temperament and differences of appearance alike. Moreover, he understands these as instances of a much more general question as to the causes of diversity in nature in general. He sees the sort of biogeographical surveying that was emerging in his era as potentially displacing premodern accounts, both biblical and Aristotelian, of the origins of natural variety. And he sees the biogeographical evidence flowing in from the Americas, in this case from Peru, as throwing into crisis many of the received views about human diversity that extended back to antiquity. Thus he cites with approval José de Acosta's observation, when considering the variety of climates in the New World and the kinds of animals and plants these climates supported: "I had to laugh vehemently at Aristotle's meteorological philosophy."[2]

When Acosta refers to "philosophy," he means first and foremost the natural-philosophical doctrines spelled out by Aristotle in the *Meteorologia*. For him, "philosophy" is largely coextensive with "meteorology" and related disciplines. Such an understanding lasted well into the seventeenth century. Thus we find John Evelyn in the *Fumifugium* describing "unwholsome vapours, that distempered the Aer, to the very raising of Storms and tempests; upon which a Philosopher might amply discourse."[3] This sense of "philosophy" is generally somewhat pejorative. It is not that Acosta and Evelyn fail to recognize the value of meteorological research; quite the contrary. Their objection is to reliance on received teachings about the natural world, particularly where these are confounded by new empirical evidence. As Acosta's work reveals particularly well, already in the Renaissance the greatest source of new challenges to "philosophy," understood in this special sense, was the encounter with America.[4]

[2]Burton, *Anatomy of Melancholy*, book II, 46.

[3]Evelyn, *Fumifugium*, 41–42.

[4]For a magisterial treatment of the impact of this encounter on the intellectual culture of early modern Europe, see Stephen Greenblatt, *Marvelous Possessions: The Wonder of the New World*, Chicago: University of Chicago Press, 1991. For other significant contributions to the Spanish Renaissance discussion of the nature and origins of the Native Americans, see Bartolomé de las Casas, *Apologética historia sumaria*, ed. Edmundo O'Gorman, Mexico City: Universidad Nacional Autónoma de México, Instituto de Investigaciones Históricas, 1967 [1551]; Gonzalo Fernández de Oviedo y Valdés, *La historia general y natural de las Indias*, ed. José Amador de Los Ríos, Madrid: Imprenta de la Real Academia de la Historia, 1851 [1535].

3.2. America and the Limits of Philosophy

The Spanish Jesuit naturalist and explorer José de Acosta traveled extensively in Central and South America in the late sixteenth century, and in 1589 published his influential *Natural and Moral History of the Indies*. This work was translated and published in French as early as 1598, and in English by 1604, and would prove to be an important source for authors interested in the problem of the origins and diversity of the human species. In the *History*, Acosta comes to the conclusion "that the first men of the Indies are come from Europe, or Asia," lest we contradict the holy scripture, which "teacheth us plainely, that all men came from Adam."[5] He goes on to implore that for animals and humans alike, "we must seeke out . . . the way whereby they might passe from the old world to this new."[6] In the case of animals, scripture requires us to limit the varieties of species to those that were included on Noah's Ark. Since the Bible does not give a complete list, however, it is possible to suppose that all the varieties of New World fauna left the Ark at Mount Ararat after the deluge.

But one still must account for their migration. Acosta conjectures (correctly) that the Americas are, or were, connected by land to the Old World at some unknown point. "Up until now," he notes, "we at least have no certainty of the contrary, since, toward the Arctic pole that we call the North, all the length of the earth has not yet been discovered and known, and there are many who affirm that beyond Florida the land extends very far to the North."[7] Yet if the accommodation of scripture is crucial for Acosta, he is at the same time, as we have seen, perfectly happy to reject Aristotle, that other great source of ancient authority, along with other ancient naturalists such as Pliny, when the evidence from the New World went against them. It was Acosta's own exploration of the equatorial zone of South America that led him to dispute some of the core claims of Aristotle's meteorology. In a chapter of the *History* titled "That the Torrid Zone Is Very Humid, and That on This Subject the Ancients Were Very Mistaken," Acosta describes his approach to natural history in general: "[W]e will first shew the trueth, as assured experience doth teach us, then will prove it, (although it be verie difficult) and will endeavour

[5] José de Acosta, *The Naturall and Morall Historie of the East and West Indies Intreating of the Remarkable Things of Heaven, of the Elements, Mettalls, Plants and Beasts Which Are Proper to That Country: Together with the Manners, Ceremonies, Lawes, Governments, and Warres of the Indians. Written in Spanish by the R.F. Ioseph Acosta, and Translated into English by E.G.*, London: Printed by Val: Sims for Edward Blount and William Aspley, 1604, book I, chap. 20, 64.

[6] Acosta, *Naturall and Morall Historie*, book I, chap. 20, 64.

[7] Acosta, *Naturall and Morall Historie*, book I, chap. 20, 67.

to give a reason, following the terms of Philosophie."[8] Thus experience comes first, and only subsequently can the fit between this experience and deductive philosophy be considered.

Here we may perceive what some historians of colonial Latin American science have identified as a sort of Iberian vanguard in the history of natural philosophy, where important developments we associate with the intellectual culture of northern Europe in the seventeenth century are in fact but echoes of developments in the Spanish and Portuguese worlds a century earlier.[9] In this particular case, Acosta's commitment to the facts, whether they respect the party line of philosophical tradition or not, may justly be described as a sort of proto-Baconianism. For better or worse, though, this spirit of inquiry would for the most part extend for Acosta only to the study of the natural world, which for him did not include the beliefs and manners of the non-European peoples inhabiting that world. This failure to deeply engage with native peoples was fairly characteristic of the early modern period, making ethnography in any proper sense of the term a science *manquée*.

Consider for example the characterization offered by Bartolomé de las Casas, the great Spanish defender of the dignity and right to freedom of Native Americans, of the views of his opponent, Juan Ginés de Sepúlveda: "[For Sepúlveda] the Indians are obliged by the natural law to obey those who are outstanding in virtue and character in the same way that matter yields to form, body to soul, sense to reason, animals to human beings, women to men, children to adults, and, finally, the imperfect to the more perfect, the worse to the better, the cheaper to the more precious and excellent, to the advantage of both."[10] In Renaissance polemics such as these, all of the ethnographic data that are trickling in from the Americas,

[8] Acosta, *Naturall and Morall Historie*, book II, chap. 3, 85.

[9] See in particular Jorge Cañizares-Esguerra, *Nature, Empire, and Nation: Explorations of the History of Science in the Iberian World*, Palo Alto: Stanford University Press, 2006. See also María M. Portuondo, *Secret Science: Spanish Cosmography and the New World*, Chicago: University of Chicago Press, 2009.

[10] Bartolomé de Las Casas, *In Defense of the Indians*, trans. and ed. Stafford Poole, DeKalb: Northern Illinois University Press, 1992 [1552], 12. For an excellent summary of the intellectual background to Spanish Renaissance debates on the origins of Native Americans, see Pagden, *Fall of Natural Man*. See also Lee Eldridge Huddleston, *Origins of the American Indians: European Concepts, 1492–1729*, Austin: University of Texas Press, 1967; Lewis Hanke, *Aristotle and the American Indians: A Study in Race Prejudice in the Modern World*, Indiana University Press, 1970; Hanke, *All Mankind Is One: A Study of the Disputation Between Bartolomé de las Casas and Juan Gines de Sepulveda in 1550 on the Intellectual and Religious Capacity of the American Indians*, Northern Illinois University Press, 1994. For a more global and comparative approach to philosophical anthropology in the Renaissance, see Frank Lestringant, Pierre-François Moreau, and Alexandre Tarrête (eds.), *L'unité du genre humain. Race et histoire à la Renaissance*, Paris: Presses Universitaires de Paris-Sorbonne, 2014.

however spotty and erroneous they may be, are consistently filtered through the mixed heritage of both Greek and Christian ideas about the boundaries of community. While Christianity demands a universalist or ecumenical conception of humanity, a presumption that in principle all human beings share in the same community, often the Greek distinction between civilization and barbarism is the more salient one in Renaissance European anthropology.

In the background to all such anthropological reflection in the era was the question of slavery, and whether, in particular, the inhabitants of the Americas could be deemed to be "natural slaves" in the sense spelled out by Aristotle in his *Politics*.[11] As the Greek philosopher explains in book I, there are some people for whom the condition of slavery "is expedient and right." He begins the argument for this conclusion by noting that there are many kinds of ruler and of subject, and he proceeds with a summary of his all-encompassing distinction between formal and material principles throughout the universe, or, as he puts it here, the duality "between the ruling and the subject element" that is present in all creatures. Those creatures in which the bodily aspect predominates are more suitably ruled by a despot, since the soul itself rules the body by a sort of despotic rule, and if soul is lacking or insufficient in a creature, then this rule must as it were come from outside. Aristotle concludes that those people are by nature slaves who differ from rational men in the same way as soul differs from body.[12]

Aristotle does not tell us, here, who the natural slaves are, but he does explicitly argue that there are some people, presumably non-Greeks, who are more like animals than others, which is to say closer to the material and nonrational principle of things, further away from the rational. This distinction maps fairly straightforwardly onto that made, for example by Herodotus, between the Greeks and the barbarians. As Pierre Vidal-Naquet explains it, this is also, at the same time, the distinction between "slaves of law and slaves of a despot."[13] Now, precisely what is at stake in the disagreement between Las Casas and Sepúlveda, which we have already discussed in passing, is whether the Native Americans are of such a nature that they must be ruled by the Europeans with despotic force. Sepúlveda would argue that everything about the natives falls on the inferior side of the series of oppositions outlined by Aristotle, in which that

[11] For a recent, compelling analysis of Aristotle's theory of natural slavery, see Joseph A. Karbowski, "Aristotle's Scientific Inquiry into Natural Slavery," *Journal of the History of Philosophy* 51, 3 (July 2013): 323–50.

[12] Aristotle, *Politics* 1254b16–21.

[13] Pierre Vidal-Naquet, *Le chasseur noir*; English translation: *The Black Hunter: Forms of Thought and Forms of Society in the Greek World*, trans. Andrew Szegedy-Maszak, Baltimore: Johns Hopkins University Press, 1986, 4.

between form and matter may be said to be primary and overarching. Las Casas argues by contrast that the Native Americans are exceedingly gentle—which is to say, inter alia, that they are not governed by passions, and thus also not subordinated to the despotism of the body:

> For the Creator of every being has not so despised these peoples of the New World that he willed them to lack reason and made them like brute animals, so that they should be called barbarians, savages, wild men, and brutes, as they [Sepúlveda et al.] think or imagine. On the contrary, they [the Indians] are of such gentleness and decency that they are, more than the other nations of the entire world, supremely fitted and prepared to abandon the worship of idols and to accept, province by province and people by people, the word of God and the preaching of the truth.[14]

The Spanish defender of the natives makes his case, however, at the expense of the Africans: he argues not that all of humanity is equal, but rather that the line is to be drawn elsewhere than his opponent would draw it.

Now it is ordinarily recognized that the Renaissance distinction between Christians and barbarians is not racial; it is sooner based on simple cultural chauvinism: the belief that anyone who speaks a foreign language is simply making sounds, repeating "bar-bar" incessantly.[15] But this is a more severe exclusion or dehumanization than might first be evident, to the extent that a widespread view holds that having access to language, to real language, is the same thing as sharing in the rational order of the cosmos, while lacking language is to fall outside of this order. A very close distinction, in turn, has it that the properly human is that which partakes of justice, which means, importantly, participation in a proper *oikos* or familial community, and also in an agricultural rather than hunting-based way of life.

The distinction between the different modes of production in different societies is more important than philosophers ordinarily take it to be, for the anthropological evidence suggest that the Greeks took "humanity" to be defined largely in terms of diet, which is of course in turn

[14]Las Casas, *In Defense of the Indians*, 28.

[15]Certainly, the distinction between civilized and barbarian, as reflected through language, continues well into the period of racial realism, and arguably continues to influence the very perception of racial difference. As Frantz Fanon describes the condition of the black inhabitants of the Francophone Caribbean, "[T]he more the black Antillean assimilates the French language, the whiter he gets—i.e., the closer he comes to becoming a true human being. We are fully aware that this is one of man's attitudes faced with Being. A man who possesses a language possesses an indirect consequence of the world expressed and implied by this language. You can see what we are driving at: there is an extraordinary power in the possession of a language" (*Peau noire, masques blancs*, 2).

traced back to the system of food production or procurement that characterizes a given society. As Pierre Vidal-Naquet explains, this conception of the boundaries of the human ecumene is evident already in Homer: "The travels of Odysseus are voyages outside the land of men, where he meets with gods, the dead, cannibals, or Lotus-Eaters. Man—that is, of course, a Greek—is one 'who eats bread.'"[16] Vidal-Naquet cites Hesiod as well, for whom humanity is distinguished from the natural realm largely in view of the limits it places on its own dietary practices: "Such is the law which Zeus son of Cronos established for men, that fish, beasts, and winged birds devour each other, since there is among them no justice." If there is any continuity between Greek philosophical anthropology and that of the European Renaissance, then one may easily imagine how this conception of humanity influenced the interpretation of groups of people who hunt and forage rather than consuming food that is the product of agricultural labor, and who practice ritualized anthropophagy rather than limiting themselves to the consumption of meat from ritually slaughtered members of other species. Native Americans are the people without justice, and this can be seen in their dietary habits, which are in turn the epitome of their mode of life.

In fact, there is tremendous continuity. One of the most common strategies for making sense of the complex ethnography of the Americas was to interpret it through the lens of ancient writing on the various barbarian nations. The Scythians of the frozen north (presumably Turkic peoples from around the Black Sea region) were most often invoked, since they had been held by the Greeks to be the very purest expression of untamed savagery. Often, however, defenses of the dignity and civility of the Native Americans took the form of an inversion of the expected analogy from the classical Greek world, placing the American peoples themselves in the role of the Greeks. The most remarkable and innovative example of this approach comes from the French Jesuit missionary Joseph-François Lafitau, who worked principally among the Iroquois and who in 1724 published an ethnographic study titled *Moeurs des sauvages amériquains, comparées aux moeurs des premiers temps* (*Customs of the Savage Americans, Compared to the Customs of the First Age*).[17] Lafitau's work is unprecedented, and would long be unrivalled, for its sensitivity to the explanations offered by the Native Americans themselves for their customs and beliefs. In his five years in Canada, Lafitau boasts, "I sought to learn all about the genius and the habits of these people."[18] But, he goes

[16]Vidal-Naquet, *Black Hunter*, 2.

[17]Joseph-François Lafitau, *Moeurs des sauvages amériquains, comparées aux moeurs des premiers temps*, Paris: Saugrain l'aîné, Charles Estienne Hochereau, 1724.

[18]Lafitau, *Moeurs des sauvages amériquains*, 2.

on, "I did not content myself with learning the character of the Savages, and with studying their customs and practices; in these practices and in these customs I sought after vestiges of the most distant Antiquity."[19] Lafitau notes that the study of foreign customs was appreciated and was given its rationale already in Homer's *Odyssey*,[20] thus implicitly identifying the Native Americans with the strange peoples encountered by the Greek epic's voyager. But, Lafitau continues, one of the strongest arguments against atheism is that all people in the world, when queried, will acknowledge the existence of a supreme being. And while the Iroquois in particular do not have revelation, the religion they do have, Lafitau thinks, resembles in important respects that of the pre-Christian Greeks: being outside of Christendom in space is interpreted on analogy to being prior to Christianity in time. "Not only do the people who are called Barbarians have a Religion," he writes, "but this religion has a very great conformity with that of the first times, with what they called in Antiquity the Orgies of Bacchus and of the Mother of the Gods, the mysteries of Isis and Osiris."[21]

But Lafitau is not principally interested in practices. He is rather concerned with interpreting, as charitably as possible, the meaning and function of native "mythology." Rather than dismissing it as "fables," Lafitau argues instead that it functions as a sort of "legislation,"[22] imposing "a certain uniformity" which "has a relation to the Truth in certain points of morality and in many legal observances, which involve Principles that are similar to those of the true Religion."[23] The standard line of European thinking about Native Americans had been to understand them as belonging to nature, and therefore also to the realm of injustice, to the extent that they live without law. And this absence in turn was inferred from the apparent fact that they had nothing comparable to the written legal codes, let alone the older holy writs with their various commandments, that could constrain the conduct of subsequent generations. Lafitau's argument is that oral culture is just as "legislative" as written culture, and therefore that Iroquois society, for example, is just as governed by law as, and no more a part of nature than, any given European society.

Without exaggeration, one could say that the preference of the written over the oral lies at the very heart of the European philosophical tradition. This tradition divides cultures up into the philosophical and the nonphilosophical on the basis of degrees, or perceived degrees, of sociocultural

[19] Lafitau, *Moeurs des sauvages amériquains*, 3.
[20] Lafitau, *Moeurs des sauvages amériquains*, 4.
[21] Lafitau, *Moeurs des sauvages amériquains*, 7.
[22] Lafitau, *Moeurs des sauvages amériquains*, 15.
[23] Lafitau, *Moeurs des sauvages amériquains*, 15.

complexity, and in turn sees many practical consequences for intercultural exchange as flowing from this basic division. Philosophy also defines itself as an essentially urban activity, to the exclusion of different, supposedly more "primitive" forms of life that could otherwise reveal radically different ways of thinking about basic philosophical questions. Thus in the *Nuova Scienza* of 1710, Giambattista Vico states the prejudice as clearly as it could possibly be stated: "First the woods, then cultivated fields and huts, next little houses and villages, thence cities, finally academies and philosophers: this is the order of all progress from the first origins."[24] Or, to cite a very similar example from Greek antiquity, where urban pride was immense and philosophy was seen as a crowning expression of the excellence of the polis, here is a passage from Plato's *Phaedrus*, as spoken by Socrates: "I hope you will excuse me when you hear the reason, which is, that I am a lover of knowledge, and the men who dwell in the city are my teachers, and not the trees or the country."[25]

For Vico, "the order of ideas must follow the order of institutions,"[26] and it is only a certain kind of institutions, namely, the crowning institutions of sociocultural development, that can give rise to recognizably philosophical ideas. The rest is myth, superstition, and what Lévi-Strauss would later call "the logic of the concrete."[27] Philosophy was, and remains, an activity associated with metropolitan centers, and with the institutions of the urban form of life. For this reason, although early modern globalization profoundly impacted the way European philosophers approached the questions of interest to them, it was excluded on a priori grounds that they could directly engage with the knowledge systems of non-European peoples as sources of philosophical insight. Generally, instead, native peoples were seen as capable of *providing* knowledge, but not of having knowledge themselves.

3.3. Native Knowledge

But what was the cause of this difference between Europeans and the peoples they encountered in the broader world? Excluding the polygenesis theorists (to be considered in detail in the following chapter), who argued for separate creations for Europeans, Africans, and Native Americans, most early modern thinkers offered one and the same explanation for the

[24] Giambattista Vico, *The New Science of Giambattista Vico*, 5th ed., ed. and trans. Thomas Goddard Bergin and Max Harold Fisch, Ithaca, NY: Cornell University Press, 1994, 78.

[25] Phaedrus 230.

[26] Vico, *New Science of Giambattista Vico*, 78.

[27] See Claude Lévi-Strauss, *La pensée sauvage*, Paris: Plon, 1962.

perceived condition of non-European peoples: decline, drift, forgetfulness. This account extends also to the non-Christian peoples of East Asia, who were not typically perceived as being inferior to Europeans with respect to their technological and political development. Thus Leibniz suggests that the Chinese are wise automata, who have preserved technology and statecraft, but have over centuries lost any historical memory of the theoretical foundations of their practical activity. To cite another example, the pioneering Orientalist and scholar of Semitic studies, Hiob Ludolf, in his *Historia Aethiopica* of 1681, contrasts the Orthodox Christian Abyssinians with the neighboring "Caffres," who, he maintains, "live without Law, without God, and without a king, and without any habitation more certain than where the night obliges them to sleep."[28] The Christians, by contrast, "have plenty of spirit, but they are ignorant of how to cultivate it, as they have almost entirely lost their knowledge, not only of the belles lettres and of the sciences, but also of the liberal arts."[29] Some decades earlier, in the *New Atlantis* Bacon had offered a similar, though fictitious, account in terms of forgetfulness of the condition of the Native Americans. He described a flood, more recent than the one braved by Noah, that had nearly wiped out the human population of the New World: "So as marvel you not at the thin population of America," a character in his tale advises, "nor at the rudeness and ignorance of the people; for you must account your inhabitants of America as a young people; younger a thousand years, at the least, than the rest of the world: for that there was so much time between the universal flood and their particular inundation. For the poor remnant of human seed, which remained in their mountains, peopled the country again slowly, by little and little."[30] Thus, for Bacon's narrator, the Americans were once in constant contact with the rest of the world, and participated in ocean travel and commerce with nations that over time forgot how to build ships entirely.

This account fits, in turn, with the general strategy in many early modern texts for explaining varying levels of sociocultural development. This strategy in turn parallels the one often deployed for explaining zoological and botanical diversity: there was an original perfect type, at the creation, and variation is to be accounted for in terms of deviation from that perfect type. But human beings, unlike plants, have more to lose than just magnitude or beauty, since as human beings move throughout the world into new climates and environments their degeneration also results in the

[28]Hiob Ludolf, *Historia Aethiopica*, Frankfurt, 1681, chap. 12.

[29]Ludolf, *Historia Aethiopica*, chap. 9.

[30]Francis Bacon, *New Atlantis*, in *The Works of Francis Bacon, Baron of Verulam, Viscount St. Alban, and High Chancellor of England*, London: J. Walthoe et al., 1740, 3:244.

Figure 1. Frontispiece to Willem Piso and Georg Markgraf's *Historia naturalis Brasiliae* (1648).

degradation of that mental faculty that philosophy long identified with the human essence, namely, reason.

What was left over in non-European peoples was a capacity to act in a way that practically duplicates or mimics reason, without any intellectual grasp of the reasons for such action. Non-Europeans were thought to have *savoir-faire* without *savoir*, know-how without knowing why. In his 1648 *Natural History of Brazil*, Willem Piso presents a striking picture of the native Brazilian as skilled brutes: "Unlike Mediterranean peasants," Piso writes, "they are truculent, brutal, without law, without religion, they roam in the manner of wild beasts, without any fixed or stable quarters, and here and there will lie in wait, with admirable knowledge and swiftness, for the fish or wild animals that will be their victuals, whether these be abundant or scarce. They know how to hurl javelins, without a bow, with the most admirable strength and with stupendous skill."[31] The perception of the simultaneous possession of skill and absence of reason gives rise for Piso and his contemporaries to the question of whether, and how, these peoples could themselves provide new knowledge to European naturalists, particularly in the domains of botany, pharmaceutics, and medicine, in which reliance on local knowledge often made the difference between life and death. One common strategy for explaining the possibility of learning from native know-how held that natural knowledge can as it were move up the chain of being, from animals, with their pure instinct, through savages and up to Europeans.

Yet, unavoidably, this attention to local knowledge carried with it a number of epistemological difficulties. Steven Shapin, to cite just one influential scholar, has emphasized the importance of local expertise in the acquisition of scientific knowledge in the early modern period, as for example the reliance of men of science upon divers for information concerning the effects of water pressure on the human body.[32] In many cases, particularly in the identification and classification of plant, animal, and mineral kinds not native to Europe, natural philosophers had no alternative but to rely on the local knowledge of members of cultures generally held to be by nature lacking sufficient rationality to fully participate in the scientific endeavor themselves. This reliance created a sort of epistemological tension: scientific claims are only as trustworthy as the scientist who communicates them, yet often science was constrained to rely upon

[31] Willem Piso and Georg Markgraf (the latter wrote later sections of the same work, but both are generally credited with coauthorship of the entire work), *Historia naturalis Brasiliae auspicio et beneficio illustriss I. Mauritii Com. Nassau . . . , In qua non tantum plantae et animalia, sed et indigenarum morbi, ingenia et mores describuntur et iconibus supra quingentas illustrantur*, Amsterdam: Apud Lud. Elzevirium, 1648.

[32] Steven Shapin, *A Social History of Truth: Civility and Science in Seventeenth-Century England*, Chicago: University of Chicago Press, 1994.

native knowledge if there was to be any advance at all in knowledge of the natural world beyond Europe's borders.

Botany, zoology, and to a lesser extent also chronology and meteorology are the domains for which the richest record remains of a history of appropriation and absorption on the part of Western science of non-Western knowledge, and of an epistemological tension of the sort just described. Mechanical physics may have been the paradigm science of the Scientific Revolution, but, in contrast with the less foundational sciences, there was little cultural exchange concerning the most basic properties of objects in motion. Ethnomechanics seemed a less likely source of valuable information to colonial Europeans than did, say, ethnobotany. Moreover, as discussed above, there is a good deal of compelling evidence that botanical and zoological knowledge was—and is—shaped by universal features of human cognition,[33] and that the differences between different cultures' taxonomical systems are underlain by deeper affinities. These could explain in part the relatively smooth exchange of knowledge in these fields in early modern exchanges between cultures: European systems of living nature may have shared certain universal features with, and so have remained more rooted in, the sort of folk science one could expect to find, for example, among Native Americans at the beginning of the Columbian era than would have been the case, for example, in the increasingly mathematized mechanical physics of the same era.

Even where Europeans in fact profited from the taxonomical knowledge of the native knowledge systems they encountered, they seldom felt a need to give credit. As we have seen, while the presence of human beings in the New World, along with unknown varieties of flora and fauna, presented a reason to revisit received tenets of ancient philosophy, most importantly natural philosophy, for many authors there was little thought of turning to these human beings themselves as a source of knowledge. Acosta, for example, with seemingly limitless curiosity in many domains, takes no interest in ethnology. He does offer a chapter on "What the Indians tell of their origins," but insists that "it is not worthwhile to pause to learn what the Indians habitually relate as to their beginnings and origins, for this resembles dreams more than history."[34] On a charitable reading, Acosta is simply confronting a real obstacle of ethnological data: taken at face value, they really do not offer much to the historian. Whatever attainments one might recognize in societies that do not keep extensive written records (and here the Incan *qipus* may be one of the few New

[33]For a fine overview of the scholarship in this area, see G. E. R. Lloyd, *Cognitive Variations: Reflections on the Unity and Diversity of the Human Mind*, Oxford: Oxford University Press, 2007.

[34]Acosta, *Naturall and Morall Historie*, book I, chap. 25..

World exceptions),[35] rigorous standards of historiography, conceptualized as the recording and dating of historical events, are not among these. In any case, Acosta is fairly typical in his approach to explaining human origins and diversity in the New World. He is, as we have seen, very suspicious of ancient authority other than scripture, and therefore of philosophy. He thus takes a thoroughly empirical approach to the description of the New World, bringing together evidence from botany, zoology, geography, geodesy, meteorology, and so on, while also rejecting any information that might be obtained from native informants.

Typically, when non-European local knowledge or forms of explanation enter into the work of European authors at all, it is for the sake of ridicule, or in the aim of positioning one's own view against a European opponent's, while likening that opponent's view to some traditional, non-European belief. Perhaps one of the best-known examples of this tactic is Locke's criticism of the metaphysics of substance in chapter 23 of book II of the *Essay Concerning Human Understanding*. Here, the English philosopher suggests that the idea that qualities inhere in substances is as incoherent as the claim made by an "Indian" that the earth rests on the back of an elephant, which in turn rests on the back of a tortoise. The origins of this reference, and even the precise meaning of "Indian" here, are unclear. The mytheme of the tortoise that supports the earth is found in both Native American and South Asian lore; if it weren't for the interpolated elephant, one might easily suppose that Locke is recalling an account of a widespread Native American belief, which had been documented, for example, in Jasper Danckaerts's description of a voyage to New York in 1679–80.[36] Whatever the case may be, it is clear that Locke does not consider it urgent to identify the true source of the belief, and is happy to leave "Indian" here in its generic sense, as applying to both the Indies. The Indies are the places with bountiful resources and land, but with belief systems that are seldom worth reporting on, except, as for

[35] See in particular Frank Salomon, *The Cord Keepers: Khipus and Cultural Life in a Peruvian Village*, Durham, NC: Duke University Press, 2004.

[36] Jasper Danckaerts, *Journal of Jasper Danckaerts, 1679–80*, ed. Bartlett Burleigh James and J. Franklin Jameson, New York: Charles Scribner's Sons, 1913, 78. Danckaerts reports of a conversation with an unnamed "Indian": "We asked him, where he believed he came from? He answered from his father. 'And where did your father come from?' we said, 'and your grandfather and great-grandfather, and so on to the first of the race?' He was silent for a little while, either as if unable to climb up at once so high with his thoughts, or to express them without help, and then took a piece of coal out of the fire where he sat, and began to write upon the floor. He first drew a circle, a little oval, to which he made four paws or feet, a head and a tail. 'This,' said he, 'is a tortoise, lying in the water around it,' and he moved his hand round the figure, continuing, 'This was or is all water, and so at first was the world or the earth, when the tortoise gradually raised its round back up high, and the water ran off of it, and thus the earth became dry.' "

Locke here, to the extent that they can cast the objectionable beliefs of one's opponents in a negative light.

As we briefly saw on the example of Piso's *Natural History of Brazil*, one way of dealing with the obvious practical sophistication of native knowledge systems was to make a fundamental distinction between know-how on the one hand and true knowledge based in reason on the other, and to argue that native peoples had only the former. The perception of the simultaneous possession of skill and absence of reason gave rise to the question of whether, and how, these peoples could themselves provide new knowledge to European naturalists, particularly in the domains of botany, pharmaceutics, and medicine, in which reliance on local knowledge could amount to the difference between life and death. Sometimes, the possibility of learning from native know-how was conceptualized as a sort of motion upward on the chain of being, from animals, with their pure instinct, through savages, and finally to Europeans.[37]

For example, zoopharmacognosy, or the ability of animals to treat their own ailments by seeking out remedies in nature, was widely recognized, but this did not require any acknowledgment of a faculty of understanding or reason on the part of the animals. Natives could in turn observe animals, and learn from them, as for example in seeking out quinine from the bark of the cinchona tree for the alleviation of fever. But it was up to the Europeans, who learned about this treatment from the natives—in the event, the Quechua-speaking people of Peru—to write treatises on quinine and thereby, from their point of view, to absorb this native know-how into knowledge properly so-called. Londa Schiebinger suggests that it was precisely in the colonial experience of native knowledge of nature that the reliance on ancient authority was most starkly called into question in the early modern colonial experience. She identifies an epistemological shift, which took place over the course of the sixteenth and seventeenth centuries, and which involved a move "away from Europeans relying on the "summa of ancient wisdom" (Dioscorides, Pliny, Galen) toward their valuing (or at least appreciating) the authority of native peoples encountered through global expansion."[38] At the same time, however, she acknowledges that "[a]ttitudes among Europeans," at least across the Caribbean, "were not uniform."[39] One widespread approach among these Europeans, as we have seen, which neither neglected native knowledge nor gave natives any credit for scientific discovery, interpreted their practical wisdom as arising from an instinctual, animal-like response to the exigencies of environment.

[37] Schiebinger, *Plants and Empire*, 75–76.
[38] Schiebinger, *Plants and Empire*, 75–76.
[39] Schiebinger, *Plants and Empire*, 81.

Schiebinger detects just this sort of thinking in the 1740 report of Charles Marie de la Condamine in *History of the Royal Academy of Sciences*. Here, Condamine relates how the Europeans came to know about the medical use of quinine (we are of course skipping rather freely in time here, yet with confidence that the same attitudes about native knowledge could be found in the previous century, and indeed are found in Piso's own work, though not so starkly stated):

> The usage of quinine was known by the Americans before it was known by the Spanish. And according to the letter cited by Sebastien Badus, written by the Genoese merchant Antoine Bollus, who traded at that place, the natives of the country long hid this treatment from the Spanish, which is very believable, in view of the antipathy that they still have today for their conquerors. As for their manner of using it, it is said that they infused it in water for one day, with the bark crushed, and gave the liquid to a sick person to drink, without the dregs. According to an ancient tradition whose truth I cannot guarantee the Americans owed the discovery of this remedy to lions, which some naturalists maintain are subject to a sort of intermittent fever. It is said that the people of this country, having remarked that these animals ate the quinine bark, used it in bouts of fever, which are rather common in that country, and recognized its salutary virtue.[40]

Condamine believes, in effect, that the natives of Ecuador and Peru learned about quinine by watching sick lions chew on the bark of the *Cinchona* tree. The Spanish in turn learned of the remedy from the natives, and the French from the Spanish, completing this bit of wisdom's movement up the hierarchy of beings.[41]

The widespread idea that native languages lacked terms for abstractions seemed to entail that, if anything could be learned from them, this would be concerning the concrete and particular, and there is perhaps no domain of knowledge more dependent on knowledge of particulars than botanical pharmaceutics. For many European explorers, as well as Europeans who interpreted the discoveries of the explorers from the comfort of home, native botanical knowledge could easily be seen as a branch of natural history complementary to botany itself, in a way that, again, ethnomathematics or ethnomechanics would not have been. Thus Jacobus Bontius remarks of the Malayans that "those who in other things are illiterate have an exact knowledge of herbs and shrubs."[42] This particular combination, of botanical sophistication and illiteracy, seemed to mean

[40]M. de la Condamine, "Sur l'arbre du Quinquina," *Histoire de l'Académie Royale des Sciences, Année MDCCXXXVIII*, Paris: Imprimérie Royale, 1740, 226–43, 233.

[41]Schiebinger, *Plants and Empire*, 81–82.

[42]Cited in Cook, *Matters of Exchange*, 203.

that one could learn from the natives, without any need to acknowledge that the natives themselves are learned. Thus Piso, as we have seen, does not put much credence in the native Brazilians' knowledge, even as he composes a massive tome on Brazilian medicinal plants that is deeply indebted to his observation of Brazilian medicinal practice.

This division of cognitive abilities is particularly clear in the curious history of the ipecacuanha root, which, alongside quinine, was for Europeans one of the most important discoveries of tropical medicine in the early modern period. Piso brought the root back to Europe in 1641, and wrote about it extensively in his *Natural History* seven years later. The root was introduced in Paris in 1672, and made famous when the physician Helvétius used it successfully to treat Louis XIV's dysentery. Later, in 1695–96, Leibniz would make what he himself held to be a significant contribution to medicine with his *Relation to the Illustrious Leopoldine Society of Naturalists concerning the New American Anti-Dysentery Drug, Attested with Great Successes*, written after reading extensively about, and evidently after conducting his own experiments with, the ipecacuanha root.[43] In his report, Leibniz offers extensive commentary on Piso's pharmaceutical data, yet he too has very little to say about the native peoples of Brazil.

There is at least an implicit anthropological commitment in Leibniz's promotion of his new antidysenteric, however: Brazilian bodies and European bodies are, for him, fundamentally the same, even to the extent that the illnesses of Brazilian bodies are not peculiar to their native environment. Thus Leibniz complains in his treatise that "there are in fact a number of people who deny that medicines of so much proven virtue could be obtained for the illnesses of every temperament and constitutions. Some of them condemn everything that is exotic as unsuited to our bodies."[44] Here, we see an implicit commitment to the fundamental uniformity of the human species, even with respect to the conformation of the body from one region to the next. What is more, his complaints about the stagnancy and inefficacy of European medicine appear to echo Piso's own derisive account of native Brazilian responses to illness. He writes that "medicine is an uncertain art, which sustains the credulity of men, like the great dream of the philosophers' stone."[45] What works for a Brazilian body works just as well for a European, and credulity functions in the one case just the same as in the other.

[43]G. W. Leibniz, *G. G. Leibnitii Opera Omnia*, ed. Louis Dutens, Geneva: Fratres De Tournes, 1768, II, ii, 110–19.

[44]Leibniz, *Opera Omnia*, II, ii, 111.

[45]Leibniz, *Opera Omnia*, II, ii, 111.

Leibniz proceeds to point out that the root's virtue could not be peculiar to the region in which it is found, since, after all, colonial bioprospecting is in the end a worthwhile endeavor only to the extent that the plants discovered have some use at home. As Leibniz ironically puts it, "[Piso], in fact, is not writing this for the sake of practicing medicine in Brazil."[46] Here Leibniz is subtly reminding his readers that, for better or worse, all human bodies are substantially the same, and it is only in virtue of this sameness that Brazilian medicine might be used to treat European illnesses. For Leibniz, the belief in the fundamental sameness of all human bodies is connected, as we will be seeing in later chapters, to the corollary question of the sameness of all human souls; or, more precisely, the question whether all human-seeming creatures may be said to have human souls in the same sense.

If responses to native knowledge systems were, as Schiebinger writes, not uniform among the travelers who encountered them, there seems to have been a relatively more uniform judgment of them by most of the figures who will be of principal interest in the chapters to follow: that is, first, authors who did not travel, but stayed in Europe and absorbed and reflected upon the reports coming back to Europe from throughout the world; and, second, authors who tended to prize abstraction as the highest attainment of human thought, or, more precisely, those authors whose work has been passed down to us today as being the work of philosophers. Here, Leibniz and some other curious exceptions aside, it was nearly universally agreed that native knowledge systems had nothing to offer.

This was a missed opportunity, to say the least. It would take several more centuries for the systematic study of native knowledge systems to take shape, and even now the precise value of such knowledge, relative to that offered by science in the strict sense, remains a hotly contested matter. It is worth briefly considering, however, that Leibniz, the early modern philosopher who perhaps came closest to an interest in native knowledge systems (as will be discussed at length in chapter 7), is also the philosopher whose system would prove to be the most fruitful point of comparison within the European tradition for twentieth- and twenty-first-century researchers engaged in the project of what might be called "comparative ontology." Philippe Descola, in his monumental *Par-delà nature et culture*, has done much to reconstruct the non-anthropocentrist ontologies of societies outside the modern West, focusing in particular on indigenous Amazonian systems, indeed people with beliefs that could well be continuous with those encountered, and largely ignored, by Piso three centuries earlier. The Makuna, Descola relates,

[46]Leibniz, *Opera Omnia*, II, ii, 111.

> say that tapirs groom themselves with roucou before dancing, and that peccaries play the horn during their rituals, while the Wari' suppose that peccaries make maize beer and that the jaguar takes its prey back home for his wife to cook. For a long time, this sort of belief was taken as testimony of a sort of thought that is resistant to logic, incapable of distinguish the real from dreams and myths, or as simple figures of speech, metaphors, or word play. But the Makuna, the Wari', and many other Amerindian peoples who believe this sort of thing are not more myopic or credulous than we are. They know very well that the jaguar devours its prey raw, and that the peccary ruins maize crops rather than cultivating them. It is the jaguar and the peccary themselves, they say, that see themselves as carrying out acts that are identical to those of humans, who imagine themselves in good faith to be sharing with humans the same technologies, the same social existence, the same beliefs and aspirations. In short, Amerindians do not see what we call "culture" as an appurtenance of human beings, since there are many animals and plants that are held to believe themselves to be in possession of it, and to live according to its norms.[47]

Similarly, when a hunter sings a song to his prey in order to woo it into "giving" itself, it is not that he has been unable to make the empirical observation that animals do not ordinarily respond to human natural language in the same way human speakers of that natural language do. Rather, every stage of the hunt, including the tracking and the slinging of darts, and other acts that can be recognized by an outside observer as expressions of practical reason, is embedded within a cosmology of perpetual exchange between all domains of the natural world. Descola asks rhetorically:

> When an Achuar hunter finds himself within shooting reach, and he sings an *anent* to the game, a supplication intended to seduce the animal and to assuage his mistrust with captious promises, does he suddenly lurch from the rational to the irrational, from instrumentalized knowledge to chimera? Does he completely change his register following the long period of approach in which he knew full well how to mobilize his ethological expertise, his deep knowledge of the environment, his experience as a tracker, all those qualities that enabled him to bring together almost by instinct a multitude of indices into a single thread that led him to his prey?[48]

It is not that the hunter suddenly shifts from practical-rational action to the merely "ceremonial" at the moment he begins singing, but rather the singing flows seamlessly from the same rationality that gives rise to the practices, and that is based on a belief in the constant cycling of im-

[47] Descola, *Par-delà nature et culture*, 187–88.
[48] Descola, *Par-delà nature et culture*, 125.

material life principles between the human and nonhuman domains. We can recognize and measure this cycling from the outside within the very limited terms of calories, but from within the cosmology that supposes that this exchange is itself constitutive of both individual human beings as well as of humanity itself, there is no reason why it should not also be manifested in verbal exchange, or communication in the usual sense, across domains.

The constant cyclical exchange rests, generally, on a metaphysics of the individual according to which every natural being, including every human being, is constituted out of the life principles of other natural beings. The predicament of the eater, and also what puts him most in danger of deep transgression through cannibalism, stems, as an Inuit informant would put it to the Danish-Inuit ethnologist Knud Rasmussen, "from the fact that the nourishment of men consists entirely in souls."[49] If this suggests to the student of Western philosophy a metaphysics of nested corporeal substances, she or he may not be entirely off track. Interestingly, Descola, following the precedent of Eduardo Viveiros de Castro and,[50] before him, Émile Durkheim,[51] sees Leibniz's metaphysics as providing a point of access to this sort of animist ontology. Leibniz, like the Makuna and the Wari', supposes that "that is a subject which finds itself 'activated' or 'agentized' by a point of view,"[52] and thus that the discontinuity of forms in nature is underlain by a deeper unity, and is explained by a difference of perspectives. Perspectivism, Descola explains, "is thus the expression of the idea that every being occupies a point of view of reference, and thus finds itself situated as a subject."[53] Descola, following Eduardo Viveiros de Castro, concludes that a Leibnizian perspectivism is "an ethno-epistemological corollary of animism."[54]

[49]Knud Rasmussen, *Intellectual Culture of the Iglulik Eskimos*, Copenhagen: Nordisk Forlag, 1922, 56.

[50]See in particular Eduardo Viveiros de Castro, *From the Enemy's Point of View: Humanity and Divinity in an Amazonian Society*, Chicago: University of Chicago Press, 1992.

[51]See Émile Durkheim, *Les formes élémentaires de la vie religieuse. Le système totémique en Australie*, Paris: Presses Universitaires de France, 1960 [1912], 386–87. Durkheim writes that for Leibniz, "the content of all monads is identical. All of them, in effect, are consciousnesses that express one and the same object, the world. . . . Except that each of them expresses it from its own point of view and in its own way. We know in what way this difference of perspective comes from the fact that the monads are differently situated the ones in relation to the others." For a similar account of Leibniz's philosophy in early twentieth-century social theory, see Maurice Halbwachs, *Leibniz*, Paris: Librairie Mellottée, 1907.

[52]Descola, *Par-delà nature et culture*, 197.

[53]Descola, *Par-delà nature et culture*, 197.

[54]Descola, *Par-delà nature et culture*, 202.

Every being, on this view, is an expression of exactly the same rational order. But heterogeneity or discontinuity of forms arises at the corporeal level. Different beings have different bodies, and so also different phenomenologies, since their perception of the world takes place through their bodily sense organs. This means also that they must conduct themselves in the world differently, that they will be nonidentical with respect to their agentive means, even if at a fundamental level all in the end have the same rational ends. Animals are pursuing fundamentally the same ends as humans, even if the different conformation of their bodies requires them to do this differently. There is no ontological gap between them and us, only circumstantial differences, or, to speak with Leibniz, different points of view. As we will see in chapter 7, it is precisely these points of view that ground not only Leibniz's deepest philosophical commitments, but also his account of human diversity. In both cases, the goal for Leibniz is to account for the unity of all people, and indeed of all things, while nonetheless interpreting the diversity of things and people as an expression of this underlying unity. Descola is right: Leibniz would have seen a mirroring of his own ideas, however differently expressed, in the perspectivism of the Makuna and the Wari'. The early modern European travelers who went to America and saw only nature, without an indigenous philosophy thereof, simply failed to detect what was all around them.

3.4. Conclusion

Most early modern authors, both those who stayed in Europe as well as those who traveled to the Americas, supposed that the native peoples of the Americas were rich with know-how, but lacked true knowledge or rationality in the deepest sense. At the foundation of this assumption is the idea that people living in "primitive" societies are not just living in greater proximity to nature, but in a direct sense are expressions or outcroppings of nature. They are natural beings like crystals or clouds, and so do what they do in accordance with the "reason" of nature, while still lacking reason in the sense of an ability to give accounts or to grasp the underlying principles of things. Living in this way, it is true, was occasionally seen as preferable, as the ideal state for humans, to the extent that it would cut out the deliberation and doubt that characterize full human existence. Such idealization of this state is familiar from its expression in Rousseau, and it has earlier iterations in the perception not only of "primitive" peoples by European authors, but also of animals. In all cases, as Descola sharply points out, the inversion of the usual value system, where the "natural" or "primitive" is now exalted rather than looked down upon, changes, literally, nothing. For in both cases a supposed difference between two different groups of human beings is exaggerated and essentialized, and neither "their" capacity for abstraction nor "our" embeddedness in nature is acknowledged.

The adjective "natural" was used as a synonym of "native," "indigenous," and "aboriginal" into the nineteenth century; we find it, as "les naturels," for example, in Condamine's 1740 description of the Peruvians. What are the theoretical commitments implicit in this lexical choice? One traditional way of understanding the "natural," as contrasted with the "spiritual," is that entities in the former category, though belonging to God's creation, are capable of being generated without God's direct involvement. A human being, even if her or his body is formed through a natural process of embryogenesis, is thus not naturally generated, insofar as God must spontaneously create (or traduce) the human soul at the moment of conception (or at some later stage of fetal development). To suggest that a human being is "natural" can thus mean that a person is generated entirely from nature, or is "born of the earth" (this was the original sense of "giants," *gigantes*, in Greek mythology).[55] Although the view was not held by a majority of thinkers, the implicit theory behind the understanding of Native Americans as "naturals" seems to be one according to which they are not specially created by God, but only share in creation in the same way other natural entities do. Thus, they do not descend from Adam and Eve. They are part of the creation, but they are not among God's children. They are held to be a reflection of divine reason in the way that the rest of the natural order is, but they are not themselves reasoning beings. This is in part what it means to say that in the era of European global domination, indigenous peoples have been deprived of their voice.

[55]Vico explains in the *New Science* that the first human inhabitants of the world, "[b]y long residence and burial of their dead . . . came to found and divide the first dominions of the earth, whose lords were called giants, a Greek word meaning 'sons of earth,' i.e., descendants of those who have been buried" (Vico, *New Science of Giambattista Vico*, 9).

Chapter 4

The Specter of Polygenesis

4.1. Libertinism and Naturalism from the Sixteenth to the Eighteenth Century

Are all people created in the image of God, or only some? Are some people generated by entirely natural means, and therefore lacking eternal, immaterial souls? Although the theological debates that gave rise to polygenetic accounts of humanity, according to which different human groups arise independently of one another, are very old, in the early modern period they were being played out in a new, globalized context. It was above all the discovery of the New World, as we began to see in the previous chapter, as well as the problem of accounting for its inhabitation, that provided a strong incentive for new articulations of polygenesis, according to which only some human beings are the children of Adam and Eve. On any non-polygenetic account of the dispersion of the human species, it had grown exceedingly difficult to think of human beings as essentially connected to any particular territory or other. It thus became easier for many to think of the New World inhabitants as having a separate creation. Francis Bacon writes for example in his essay "Of the Vicissitude of Things," first published in 1625, "If you consider well, of the people of the West Indies, it is very probable that they are a newer, or a younger people, than the people of the Old World."[1]

Many libertines who remained in the Old World, such as Lucilio Vanini and Giordano Bruno, and natural philosophers such as Paracelsus and Andrea Cesalpino, were ready to entertain the idea that barbarians are, like insects, "imperfect" and thus generated spontaneously out of the earth. The basic taxonomic division between perfect and imperfect animals is an ancient one: it supposes that there are some creatures, the products of sexual generation, that are generated in accordance with the particular formal principle of a species and thus that have their own internal perfection or teleology; in turn there are other creatures that are

[1] Francis Bacon, *The Essays, or Counsels Civil and Moral*, ed. Samuel Harvey Reynolds, Oxford: Clarendon, 1890, 383, cited in Richard H. Popkin, *Isaac La Peyrère*, 39.

generated haphazardly, and thus that are "imperfect" in the sense that there is no particular form at which the ordinary course of their development should in the best case arrive. Such an account was applied variously to both Native Americans and sub-Saharan Africans. Thus we find Vanini presenting the polygenesis theory in his *On the Marvellous Secrets of the Nature the Queen and Goddess of Mortals* (*De admirandis naturae reginae deaeque mortalium Arcanis*) of 1616, though cautiously framing his interest as a summary of the views of others: "Diodorus Siculus asserted that the first man was generated fortuitously from the mud of the earth. But Girolamo Cardano appears to have held the opinion, expressed in these words: Not only the tiny, but also the great animals are from putrefaction, no indeed it is to be believed that all have this origin, as besides it is known of mice, and fish are spontaneously generated in fresh waters."[2] In this text, written in the form of a dialog, an interlocutor named Alexander replies with evident sarcasm, "What an excellent argument Cardano offers, that if a mouse can be born from putrefaction, therefore a man can be too."[3] But Julius Caesar, speaking for Vanini himself, is unfazed, and continues with an argument that appeals to gross racist stereotypes while also, cautiously, broaching the possibility of natural continuities between the human and animal worlds: "Others imagine that the first man was generated from the putrefaction of apes, pigs, and frogs, since in their flesh and in their mores they are very similar. However certain more gentle atheists attest that only Ethiopians come from the genus and the seed of monkeys, since the color appears the same in the ones and in the others."[4] Neither the "gentle atheists" nor the less gentle ones with whom they are contrasted are named here, but the contrast itself is nonetheless very telling. A full-fledged atheist, on this account, is one who sees all of humanity as descended from animals; from their putrefaction to be precise: this is a version of the "Homo ex humo" thesis, "Man from dust," except that here this developmental process is not guided by divine concurrence or underlain by any species-specific

[2]Lucilio Vanini, *Iulii Caesaris Vanini Neapolitani Theologi, Philosophi, & Iuris utriusque Doctoris, de admirandis naturae reginae deaeque mortalium arcanis*, Paris: A. Perier, 1616, dialogue XXXVII, "De prima hominis generatione," 232–33. "Diodorus Siculus prodidit primum hominem fortuito e limo terrae genitum. Hieronimus tamen Cardanus huic videtur adhaerere sententiae, nam in haec verba prorumpit. Nec solum tam minuta, sed et majora animalia e putredinie, imo omnia credendum est originem ducere, cum iam de muribus constet, et pisces in aquis recentibus sponte generentur."

[3]Vanini, *Iulii Caesaris*, 233. "Egregium sane Cardani argumentum, mus e putredine potest nasci, ergo et homo potest."

[4]Vanini, *Iulii Caesaris*, 233. "Alii somniarunt ex Simiarum, porcorum & ranarum putredine genitum primum hominem, iis enim est in carne, moribusque persimilis. Quidam vero mitiores Athaei, solos Aethiopes ex simiarum genere & semine prodiisse attestantur, quia & color idem in utrisque conspicitur."

immaterial principle of development.[5] It simply happens, haphazardly and imperfectly.

There remains the question of how this system of generation and corruption got started in the first place. We are a far distance from the theory of natural selection, but arguably the cosmic or theological implications are the same: "atheism" is the view that the variety of natural beings is neither created nor designed, but rather issues forth spontaneously. A "gentle" atheist is one who keeps at least Europeans sheltered from such an account, while projecting natural origins outward, at a safe distance, upon geographically and culturally distant peoples.

A contemporary of Vanini, the Swiss naturalist and mystic Philippus Aureolus Theophrastus Bombastus von Hohenheim, more commonly known as "Paracelsus," is also worth mentioning as a defender of polygenesis. He is exclusively concerned with the origins of Native Americans, and says nothing of Africans. What is of concern to him is to account for the origins of people in such distant corners of the globe as the Americas, since he, like many others, finds it implausible to suppose that human beings migrated from southwestern Asia into the western hemisphere in only the few thousand years since creation. Thus he writes that

> the children of Adam did not inhabit the whole world. That is why some hidden countries have not been populated by Adam's children, but through another creature, created like men outside of Adam's creation. . . . I cannot refrain from making a brief mention of those who have been found in hidden islands and are still so little known. To believe that they have descended from Adam is difficult to conceive: that Adam's children have gone to the hidden islands. But one should well consider that these people are from a different Adam.[6]

Richard Popkin maintains, unconvincingly, that Paracelsus should not be counted among the polygenesis theorists, since the Swiss naturalist included Native Americans alongside "cases of nymphs, sirens, sylphs, salamanders, and so on," and since "these wild spirits are not other kinds of human beings," but are instead subhuman. Therefore, Popkin concludes,

[5] See Johann Jakob Scheuchzer, *Physica Sacra*, Ulm, 1731. Significantly, Vico for his part discerned an etymological and therefore conceptual connection between "homo" and "humus" ("soil"), but he saw the most relevant notion within this semantic cloud to be "humo," "to bury": human beings are the beings who bury their dead. See *New Science of Giambattista Vico*, 8.

[6] Paracelsus [Philippus Aureolus Theophrastus Bombastus von Hohenheim], *Astronomia Magna: oder Die gantze Philosophia sagax der grossen und kleinen Welt des von Gott hocherleuchten erfahrnen und bewerten teutschen Philosophi und Medici Philippi Theophrasti Bombast, genannt Paracelsi magni*, Frankfurt: Sigismund Feyerabend, 1571 [1537–38], 8–9.

"their separate creation from that of human beings does not involve a polygenetic theory."[7] But Popkin does not give any reason why Renaissance and early modern polygenesis theorists should have believed that separately created humans must be equal to Adam's descendants, and indeed most of the original sources, which Popkin generally treats adeptly, indicate clearly that those created separately were indeed conceived as inferior.

As has already been suggested, one aspect of polygenesis theory, at least of the sort promoted by Vanini and Paracelsus, that scholars have tended to overlook is that it is, in a sense, half right: it holds that a portion of the human race has natural origins, namely, Africans and Native Americans, while holding back from making the same claim about Europeans. There is thus a complicated relationship between the rise of naturalism on the one hand, and the rise of distinctly modern racism on the other. One way of seeing the legacy of Renaissance polygenism is that it amounts to a sort of half-step toward a naturalistic account of human origins: Vanini claimed Africans are related to apes, a proposition Darwin would supplement with the claim that Europeans are as well, and in so doing affirming, of logical necessity, Vanini's claim too. One might suppose that to claim of Ethiopians that they are born of the earth, or descended from apes, is perhaps to project onto a foreign group of people a heretical claim that one wishes to insinuate about one's own group.

To attribute natural origins to Ethiopians may thus be something like Descartes's description in *Le Monde*, under the guise of a fable, of a world formed through natural laws alone. If this comparison is strained—of course, one great difference is that the entirety of Descartes's *Le Monde* is devoted to developing this fable, whereas Vanini, again, spends only one sentence on the origins of Ethiopians—it nonetheless remains the case that in the sixteenth and seventeenth centuries polygenetic accounts of racial difference were in no sense "conservative." They did not aim to defend or deepen any status quo beliefs about human diversity. Quite the contrary, in fact, they violated the deeply held conventional belief that all human beings are descended from Adam, and that the proper reading of scripture leaves no opening for any humanoid yet nonhuman first parents of different races. In the vaguely polygenist remarks of Voltaire two centuries later (to be discussed below), we see the lingering hope of conjuring some of the libertine *frisson* that his polygenist predecessors had enjoyed when he posits that Europeans and Africans are of "entirely different races." But willy-nilly what the French Enlightenment thinker ends up doing is simply joining the chorus of all the wishful thinkers with a vested interest in maintaining the status quo of the institution of slavery.

[7] Popkin, *Isaac La Peyrère*, 34.

Things were different two centuries earlier. The sixteenth-century Italian astronomer and philosopher Giordano Bruno—like Vanini, executed by the Inquisition—provides a very interesting example of the sort of innovative and unconventional cosmology in which Renaissance polygenesis made sense. Bruno's belief that there could be descendants of separate, parallel Adams was in fact just one small corollary to his broader view that there is an infinity of worlds, and so an infinity of counterpart Adams. Bruno believes that precisely half of these Adams will not have eaten the fruit of the tree of knowledge, but since half of infinity is still infinity, it follows that there are infinitely many worlds that know only the prelapsarian state of grace. The existence of descendants of parallel Adams in parallel worlds admittedly does not help to account for human diversity in *this* world. But once such a heterodox and scripturally incorrect cosmology becomes thinkable, it is a rather small thing to suppose that in addition to the infinitely many transworld creations of the human species that occurred, there was also more than one creation of humanity within any given world.

There were of course many factors influencing the development of Bruno's infinity-centered philosophy, which would in turn be an important influence on the philosophies of Henry More, Leibniz, and others in the seventeenth century. The most important factor was probably the suite of astronomical discoveries that displaced the earth from the center of a neat, finite collection of concentric spheres. But one must also notice that geographical discoveries played a role in the multiplication of "worlds" in the Renaissance, and that there was an easy slippage between "world" in the sense of "universe" and in the sense of "continent." A clear illustration of the polysemy of "world" may be found, for example, in the Spanish-Inca political philosopher Garcilaso de la Vega's *Royal Commentaries of the Incas*, first published in 1609, in which the author devotes the first chapter of the first book to denying the impious view that there are "many worlds," and affirming that the "New World" is so called only because it was discovered recently, not because it is in any sense a discrete or independent reality. "If there are any men who imagine that there are many worlds," de la Vega writes, "there is no other response to offer them, unless they persist in their heretical belief until they are disabused of it in hell."[8] De la Vega, had he lived to see it, would have supposed that Bruno was finally being set right after his execution.

In the one case in which Bruno explicitly commits himself to separate creations within this world, it is not principally on a separate creation for those in the Old "World" and those in the New "World" that he focuses,

[8]Inca Garcilaso de la Vega, *Commentaires royaux sur le Pérou des Incas*, trans. René L. F. Durand, Paris: François Maspéro, 1982, 74.

but also on the separate generations of Ethiopians, Americans, "Pygmies" (not the ethnic group identified by that name, but rather the mythical beings of Homeric legend), and other beings more fabulous still. Bruno writes,

> There are many different species of men, the black generation
> Of Ethiopians, and the tawny ones, such as America produces,
> And the wet beings hidden away, living in the caves of the sea,
> And the Pygmies passing the ages in ever-inaccessible places
> And the citizens of the veins of the Earth, who stand guard
> At the mines, and the giant monsters of the South:
> Nor indeed do they hark back to the same origin
> Nor are they sprung from the same progenitor.[9]

A few pages on, Bruno notes, referring to the first parent of the Jewish race, "no one of sound judgment traces the Ethiopian race back to this protoplast."[10]

Bruno's census of various fabulous beings is very similar to Paracelsus's, and, pace Popkin, there is as little reason to disqualify the one as there is the other as a variety of polygenism on the grounds that it mixes the anthropological and the mythical. For Bruno, all of the types of being on this list were equally real, and they were for him all literally "races of men." The fact that Ethiopians and Americans occur on the same list as Pygmies and giants shows the limited extent to which, in the Renaissance, non-European peoples had begun to be incorporated into the European imagination. It is important of course to bear in mind that Pygmies and giants had been familiar figures in European lore for millennia, and that it would have made natural sense to assimilate reports about unfamiliar ethnic groups to the sort of motley collection, to explain the new and unknown by reference to the old and known, alongside unreal yet very well-known beings.

Vanini, Bruno, and Paracelsus all seek to account for the origins of certain non-European peoples in a way that separates them entirely from the history of Europeans. For the most part, moreover, these thinkers were not centrally concerned with human origins and human diversity

[9]Giordano Bruno, "De Immenso et innumerabilibus," in *Opera Latine conscripta*, ed. F. Fiorentino et al., Naples, 1879–91, vol. 1, pt. 2, 282. "Quia multicolores sunt hominum species, nec enim generatio nigra / Aethiopium, et qualem producit America fulva. / Udaque Neptuni vivens occulta sub antris, / Pygmeique iugis ducentes saecula clausis, / Cives venarum Telluris, quique minaerae / Adstant custodes, atque Austri monstra / Gigantes, Progeniem referunt similem, primique parentis / Unius vires cunctorum progenitrices."

[10]"De Immenso et innumerabilibus," 622, cited in Hermann Brunnhofer, *Giordano Bruno's Weltanschauung und Verhängniss*, Leipzig: Fues's Verlag, 1882, 202. "Aethiopum genus ad illum protoplasten nemo sani judicii referat."

as a theoretical problem. Vanini's comment on the relationship between Ethiopians and apes, again, takes up no more than a single sentence in an enormous tome that also contains an account of celestial bodies, the elements, and many other subjects. In the chapter in which this relationship is posited, the principal claim regards not racial differences among human beings, but rather the extent of spontaneous generation in the order of nature: Vanini wishes to claim that there is in principle no upper limit, with respect to size or "perfection," to the variety of creatures that might be generated out of the earth. As we will see in detail in chapter 9, by contrast, theoretical engagement in the eighteenth century with the possibility of separate origins for Europeans and for Africans is both much more sustained and much more recognizably racist. Voltaire's preoccupation with the subject, for example, is based on a deeply entrenched a priori commitment to the inferiority of Africans, while his detailed writings on the human epidermis show at least some interest on his part in providing evidence for his views. But the greater attention to the topic seems to be largely a result of Voltaire's interest in keeping up with the spirit of the times, namely, justifying one's views about human difference in relation to the most sophisticated science of anatomy, medicine, and even optics, if only to counter the prevailing monogenist account of his era. In the eighteenth century, you could no longer stir up any controversy by disputing the Aristotelian orthodoxy on spontaneous generation. By contrast, in the sixteenth century, one could casually let the claim that Ethiopians are related to apes appear in the course of a larger argument that oxen can be generated from the slime of the Nile, because it was this larger argument that would have appeared more controversial.

The naturalism at the heart of libertine theories of spontaneous generation would perhaps find its culmination in the materialistic monism of the eighteenth century. As Ann Thomson has insightfully noted,[11] in the eighteenth century it was often those of an anticlerical and materialist bent who were most eager to deny the fixity of races, even if in replacing this conception by a fluid continuum they also tend to see the human species as rife with inequalities. Typically, for a materialist, there is no fundamental or irreversible division between, say, Europeans and Africans, even if climatic, dietary, and cultural legacies have brought it about that, for now, the former are more advanced than the latter, in just the same way that certain breeds of horse, dog, or pigeon might exhibit superior traits. The supposed superiority of Europeans, here, is conceptualized naturalistically, as the result of processes fundamentally no different from what we see throughout the animal world, and without any role for

[11] Ann Thomson, "Diderot, le matérialisme et la division de l'espèce humaine," *Recherches dur Diderot et l'Encyclopédie* 26 (April 1999): 197–211.

providence. This division runs parallel with a further division between the polygenism of many of the defenders of divine creation, who hold that inferior races were created by God separately, and that subsequently they exist across a permanent natural divide. Materialist thinkers by contrast tend to suppose that there was no divine creation, that therefore no single race can be said to be the image of God any more or less than any other, and that racial divisions are simply the contingent result of natural, primarily climatic, factors.

Racial differences are nonetheless very real for the naturalist materialist: they are real in the way that, say, degrees of heat are real, rather than in the way that discrete natural kinds are real. Thus Denis Diderot explicitly argues that the division between kingdoms of nature, for instance between plant and animal, will lie forever beyond the reach of empirical science. Borrowing a comparison from mathematics, he describes the search for boundaries as a sort of asymptotic approach toward a limit. He argues that the faculties associated with the animal kingdom—action, sensation, and so on—disappear gradually as we descend toward lower life forms, only to vanish entirely at the point where the animal kingdom meets the vegetable; yet this is a "point that we will approach more and more through our observations, but that will forever escape us."[12] Similarly, although there is internal diversity within the human species, it is not the sort that consists in sharp boundaries between discrete subgroups.

For Diderot, physical causes provide the entire explanation for internal differences between different creatures. The body is not so much an unfolding or an expression of the internal capacities of what was once called a soul; it is rather a straightforward replacement for it. Education and custom are therefore central in Diderot's philosophy: they literally shape a person's physiology. Across several generations, limited or absent education within a population or a lineage will give rise to significant differences in appearance between this population and another.

In the *Encyclopédie* article on the human species, Diderot makes a case against polygenesis on the basis of the comparative anthropology of the northern and southern hemispheres: "From north to south one perceives the same varieties in the two hemispheres. All goes therefore toward proving that mankind is not composed of essentially different species. The difference from whites to browns comes from food, customs, habits, climates; that from browns to blacks, from the same cause. . . . There was therefore only one original race of men, which being multiplied and spread over the surface of the earth, has produced over time

[12]Denis Diderot, *Oeuvres complètes*, ed. J. Fabre, J. Dieckmann, J. Proust, and J. Varloot, Paris, 1975–, 5:388–89.

all of the varieties which we have just mentioned."[13] If we can take the *Encyclopédie* as indicative of Diderot's views for the period, then he may be classified as a monogenist who believes in gradual degeneration from the original human type as a consequence of migrations. This was indeed by far the most prevalent view of human diversity in the eighteenth century, with only a few Enlightenment thinkers coming out in favor of a thoroughgoing polygenesis theory.

Some commentators have identified a change in Diderot's views on race between the time of the *Encyclopédie* and the time of the contributions he is believed to have made to Guillaume-Thomas Raynal's *Histoire des Deux-Indes* of 1780,[14] where he offers one of the century's most compelling arguments against slavery. Here Diderot blames any apparent marks of intellectual inferiority among Africans on the historical fact that Europeans have blocked them by brute force from the means of gaining enlightenment.

Among defenders of monogenesis, who believe that human diversity can be accounted for in terms of differences in climate, diet, and so on, there are many, such as Kant and Hume, who nonetheless take the resultant differences between human populations as significant enough to justify claims of real racial inequalities, which in turn justify a political order that reflects these inequalities. Others, such as Diderot, see the fact that humans all have the same origins as sufficient in itself for grounding political equality, to the extent, again, that all differences in appearance or ability are contingent and reversible. Diderot's particular variety of *monistic* materialism holds, as we read in the *Rêve de D'Alembert*, that "there is only one great individual, the all."[15] This might seem to offer a solid metaphysical basis for political egalitarianism: everything is the same, therefore everyone is equal.

But again, Diderot also asserts that different sorts of creature can be more or less exalted as a result of the physical causes responsible for their physical conformation. Moreover, if in the *Encyclopédie* he emphasizes the hierarchical differences of physical conformation within the human species, in other, later works, he is intent on defending the full equality of all human beings, notwithstanding differences in physical appearance. In the *Encyclopédie* article, Diderot recycles a number of conventional prejudices about, for example, the "Laplanders" or Sami of Northern Europe, describing them not just as ugly, but also correlating to this physical conformation to the psychological traits of superstition and stupidity. But

[13] "Humaine espèce," "Humaine espece," Denis Diderot, et al., *Encyclopédie ou Dictionnaire raisonné des sciences, des arts et des métiers*, , Paris, 1765, 8:344.

[14] Guillaume-Thomas Raynal, *Histoire des Deux-Indes*, Geneva, Pellet, 1780.

[15] Diderot, *Oeuvres complètes*, 2:139.

the physiognomy and the psychology are both in turn the simple consequence of environmental forces. There is thus a contingent inequality between human populations, and indeed it is not different in principle from the sort of inequality that exists between any two given individuals. Here we have what Jacques Proust identifies as a "paradox," one common to many thinkers besides Diderot, where moral equality between human beings can take no root in a philosophical anthropology that denies physical and intellectual equality.[16] Yet, again, even if the paradox is widespread, it is particularly problematic for a materialist monist such as Diderot, for it is not at all clear what it is for him that binds human beings together in moral equality. There is, in stark contrast to Leibniz, no prior criterion of humanness lying behind the physiological and intellectual diversity of different human populations, in virtue of which every human individual may be said to be united as equal with every other human individual, notwithstanding the superficial expressions of diversity. If physiology just is a replacement for soul or essence, as already suggested, then there can be no unitary humanity behind the physiological diversity, and it is not clear in virtue of what shared humanity all individuals grouped as humans may be said to be morally equal, notwithstanding their physiological and intellectual inequality. If it is true, as Diderot asserts, that "it is very difficult to do good metaphysics and good morals without being an anatomist, naturalist, physiologist, and physician,"[17] then we may ask whether by his own standards Diderot's commitment to human moral equality amounts to "good metaphysics."

As we have seen, the history of the theory of polygenesis from the sixteenth to the eighteenth centuries provides a very vivid illustration of an important lesson in the study of the history of racism: that one and the same theory of human origins and human diversity can serve entirely different rhetorical purposes and different ideological agendas depending on the context in which it is espoused. Between sixteenth-century libertine polygenesis and its eighteenth-century echoes, however, we also see a number of very creative attempts to account for humanity's physical diversity and geographic diffusion in ways that nonetheless steer clear of libertine or antireligious implications. Some of these attempts would seek to ground polygenetic accounts of human origins in idiosyncratic readings of scripture. Let us turn in the following section to the most well known of these, the doctrine of so-called pre-Adamism.

[16]Jacques Proust, *Diderot et l'Encyclopédie*, 3rd ed., Paris, A. Michel, 1995, 417.

[17]Denis Diderot, "Réfutation de l'ouvrage d'Helvétius, intitulé *L'Homme*," in *Oeuvres complètes*, 2:322.

4.2. Pre-Adamism

Pre-Adamism seeks to ground polygenesis in a particular interpretation of the Bible, namely, in a reading of the cryptic claim at Romans 5:13 that "until the law sin was in the world." Pre-Adamists took this verse to assert that there was not only sin but also, necessarily, sinners, which is to say sinful *people* in the world before God gave the law to Adam. The earlier Renaissance polygenists, by contrast, had sought to base their account of the separate origins of non-Europeans in nonbiblical traditions: Vanini draws on the Aristotelian theory of spontaneous generation, Bruno and Paracelsus on other Greek sources (e.g., the "Pygmies" first mentioned in Homer) and on folklore. Renaissance polygenesis, again, while mingling its explanations with ideas that we today place in the category of the supernatural, nonetheless shows a distinctively naturalistic inclination, to the extent that it seeks to account for human beings (again, even if it leaves Europeans out of this account) as in some way or other generated out of the earth, or from some primordium within nature, rather than being directly crafted by God. The interest, as already suggested, was in describing non-European peoples as literally "autochthonous," to use another revealing term, which is to say "from the depths of the earth": a notion that we find in both Herodotus's and Thucydides's accounts of certain tribes that have occupied the same regions for all of known history.[18]

In the seventeenth century, the French theologian and millenarian Isaac La Peyrère, as well as a number of lesser-known figures, would seek rather to give an account of the origins of non-European peoples that remains fixed to the scriptural tradition.[19] In these accounts, preoccupation with "racial" difference is more or less absent. What is of greatest concern here is not so much accounting for either physical difference or perceived cultural inferiority, but rather accounting for (1) the diffusion of human beings throughout the world, (2) the fact that many of these human beings themselves offered alternative chronologies that contradicted the one familiar from Genesis, and (3) the apparently extra-historical (because extra-scriptural) status of the various widely dispersed groups of people. Thus in the 1644 *Relation de l'Islande* (*Relation of Iceland*), written more than a decade before he articulates the pre-Adamist theory, La Peyrère criticizes the view that the Icelanders, along with their revered "prince"

[18]See Herodotus, *Histories*, 4.109.1; and Thucydides, *The Peloponnesian War*, trans. and ed. Richard Crawley, London: J. M. Dent, 1910, 6.2.1.

[19]For a thorough treatment of the role of scripture in ideas about race from the early modern period to the present day, an aspect of the history of race that is here being treated only in passing, see Colin Kidd, *The Forging of Races: Race and Scripture in the Protestant Atlantic World, 1600–2000*, Cambridge: Cambridge University Press, 2006.

Odin, are descended from an errant faction of the ancient Roman army as follows: "[W]hat is the hope of being able to accept all of the fables that they tell about this Asiatic Odin, and what connection could such weak fables have with the age of Pompeius, which is an age so well known, and so historical?"[20] Here, La Peyrère is arguing that it is much easier to suppose that far-flung peoples—and here Iceland is thought to be a sort of stepping stone toward the New World, rather than an island securely within the European sphere—have their own histories, whether written or unwritten, rather than to strain to derive them from the available textual traditions beginning in the Mediterranean region.

In his 1655 *Prae-Adamitae*,[21] which would appear in English translation the following year under the title, *Men before Adam*,[22] La Peyrère cites Paul's letter to the Romans as support for his view that not all human beings are descended from Adam. The apostle's assertion that there was already sin in the world when Adam arrived provides La Peyrère with the apparent means of including all newly discovered peoples within the core textual tradition of Europe, even as it excludes them from the group of people who are bound by ancestry to this tradition. It makes sense of them, insofar as it finds a mention of them in the Bible, but at the same time it implies that they are not strictly speaking "of the book." The fact that his work would seek to ground polygenesis in scripture would however be far from enough to keep it free from controversy. La Peyrère was pressured into retracting his own views, but not soon enough to prevent his argument from making a profound impact.

There would be at least a dozen important treatises in the latter half of the seventeenth century seeking to refute La Peyrère's thesis. Three of these were of particular significance, all published in 1656: Philippe Le Prieur's *Animadversions against the Book of the Pre-Adamites*,[23] Johann Hilpert's *Disquisition on the Pre-Adamites*,[24] and the richly titled work of Antonius Hulsius, which translates as *No pre-Adamite being, or, a confutation of a certain someone's vain and Socinianizing dream, by which an anonymous author, on the pretext of the Holy Scripture, endeavored*

[20] Isaac La Peyrère, *Relation de l'Islande*, Paris: Louis Billaine, 1663 [1644], 67–68.

[21] Isaac La Peyrère, *Prae-Adamitae, sive Exercitatio super versibus duodecimo, decimotertio, & decimoquarto, capitis quinti Epistolæ d. Pauli ad Romanos. Quibus inducuntur primi homines ante Adamum conditi*, [Amsterdam], 1655.

[22] Isaac La Peyrère, *Men before Adam, or, A discourse upon the Twelfth, Thirteenth, and Fourteenth Verses of the Fifth Chapter of the Epistle of the Apostle Paul to the Romans: by Which Are Prov'd That the First Men Were Created before Adam*, London, 1656.

[23] Philippe Le Prieur, *Animadversiones in librum Prae-Adamitarum: in quibus confutatur nuperus scriptor, et primum omnium hominum fuisse Adamum defenditur*, Paris: Apud Ioan. Billaine, 1656.

[24] Johann Hilpert, *Disquisitio de praeadamitis, anonymo Exercitationis & Systematis theologici auctori opposita*, Amsterdam: Apud Johannem Janssonium juniorem, 1656.

not long ago to establish to the incautious that there were men in the world before the first Adam.[25] There were, in fact, far more refutations than defenses, and every defense was shrouded in either caution or anonymity. Popkin explains that virtually nobody in the seventeenth century "was willing, publicly, to accept the pre-Adamite theory or any form of polygenesis."[26] La Peyrère's thesis would however be at least cautiously praised by a number of prominent figures in the Republic of Letters, including Guy Patin and Marin Mersenne, both of whom saw in it a promising way of making sense of certain problems of biblical exegesis, with respect not only to Paul's epistle to the Romans, but also to the more well-known problem of accounting for the origins of Cain's wife in the book of Genesis.[27] Yet Patin and Mersenne expressed their opinions in private correspondence, and only on the basis of their knowledge of the precirculated manuscript of La Peyrère's *Prae-Adamitae*. After its publication in 1655, these key figures in the Republic of Letters seem to have backed away from the doctrine.

La Peyrère's thesis was challenged on many fronts, but most of all on the grounds that it did not in fact adequately account for events described in scripture. From the other direction, it was challenged on the grounds that there are other, less radical ways of accounting for both human diffusion throughout the globe as well as for the alternative chronological traditions of non-European peoples. We will turn to some of these accounts in the following section of this chapter.

William Poole, in contrast with Popkin, identifies a number of different sources of seventeenth-century pre-Adamism, not all of which were anonymous.[28] He emphasizes in particular the importance of the study of calendry and of comparative chronologies as evidence in favor of the view that there must have been Adams before Adam. Thus for example Thomas Herbert writes in 1638 of the problematic antiquity of Chinese history, "They say the World is aboue a hundred thousand yeares old after their Chronologies, and accordingly deriue a Pedigree and tell of wonders done ninetie thousand yeares before Adams creation."[29] Some

[25]Antonius Hulsius, *Non ens prae-adamiticum, sive confutatio vani et socinizantis cujusdam somnii, quo S. Scripturae praetextu in cautioribus nuper imponere conatus est anonimus fingens ante Adamum primum homines fuisse in mundo*, Leiden: Apud Johannem Elsevirium, 1656.

[26]Popkin, *Isaac La Peyrère*, 115.

[27]Giuliano Gliozzi, *Adam et le nouveau monde. La naissance de l'anthropologie comme idéologie coloniale des généalogies bibliques aux théories raciales (1500–1700)*, trans. Arlette Estève and Pascal Gabellone, Paris: Théétète, 2000, 445–46.

[28]William Poole, "Seventeenth-Century Preadamism, and an Anonymous English Preadamist," *Seventeenth Century* 19 (2004): 1–35.

[29]Thomas Herbert, *Some Yeares Travels into Africa & Asia the Great. Especially Describing the Famous Empires of Persia and Industant. As Also Divers Other Kingdoms in the Orientall Indies and Iles Adjacent*, London: Jacob Blome and Richard Bishop, 1638.

authors, particularly liberal Jesuits who were deeply familiar with non-European scientific traditions, would, shortly after the initial round of attacks on La Peyrère, begin expressing cautious sympathy for polygenesis on the grounds that no plausible reading of Genesis could possibly accommodate the vastly older calendrical systems of, for example, the Chinese and the Mexica.

As Poole notes, much of the evidence for alternative datings of the earth's distant past was being accumulated and reflected upon by Jesuit missionaries who considered it part of their missionary task to learn as much as they could about the scientific accomplishments of the peoples they sought to convert. The Jesuit Martino Martini, for example, in his *Sinicae Historiae Decas Prima* (*The First Decade of Chinese History*) of 1658, called attention to Chinese chronology's incompatibility with that of the Old Testament, and on this basis explicitly questioned biblical universality. For Martini as for other cautious sympathizers, it was chronology rather than phenotypic variety that presented the strongest argument in favor of multiple creations and against monogenesis. The most common approach in Jesuit circles remained, however, the monogenetic account that saw all civilizations as having secret and ancient connections to the Judeo-Christian scriptural tradition.[30]

4.3. Diffusionist Models

The pre-Adamite thesis was challenged simultaneously on the grounds of biblical exegesis, the implausibility of extra-scriptural chronologies, and, finally, the growing plausibility of diffusionist accounts of the population of the earth. Accounting for the details of global migration was a task that required the assembly and interpretation of extremely fragmentary geographical and linguistic data. One thing that did not seem to be terribly important to authors in the period, however, was accounting for the physical diversity of the human species in different parts of the world.

[30]Thus for example the Jesuit author Athanasius Kircher in 1667, *China monumentis*, Liber VI, caput III: "Characterum antiquissimorum Chinensium explicatio: Primaevi Sinae, uti dixi, Aegyptios, a quibus descendebant, secuti, scripturam compositione, sed figuris, ex cariis rebus natruralibus compactis, peragebant, quibut quot conceptus rerum, tot signa diversa respondebant" (*China monumentis*, Amsterdam: Apud Jacobum à Meurs, 1667, book VI, chap. 3, [repr., Frankfurt: Minerva, 1966]), 228. Similarly, in an anonymous Jesuit text, the *Confucius sinarum philosophus*, written between 1660 and 1670 and published in 1687, we read that three characters, "*zhou*, ship, *shu* a weapon and *min*, a household utensil, when put together make up the character *pan*. This is the name of the man the Chinese call Pan Gu, 'Pan of ancient times.' They believe that he was the first man on earth. Hence a European has surmised that the man we call Noah is called Pan by the Chinese" (187–88).

Authors did not, for the most part, touch upon what we would recognize today as questions of racial difference. As a telling example of this absence of perceived racial difference, in his *Origins of the American Peoples* of 1642, Hugo Grotius conjectured that the Native Americans must be of Norwegian origin, since, he thought, it is only the Germanic languages that have any clear similarity with the languages of the New World. The Dutch humanist arrived at this position, as the least implausible one, by a process of elimination: the only alternative to the hypothesis of Scandinavian origins, he reasoned, would be to hold "either that they existed from eternity, according to the opinion of Aristotle; or that they were born of the earth, as a fable tells us concerning the Spartans; or from the ocean, as Homer maintains; or indeed that they were created before Adam, as someone in France imagined recently." All of these views, Grotius reasons, "seem very dangerous for piety, while believing what I have said is not so at all."[31] Only diffusion through migration offers a way of accounting for human diversity while avoiding impiety. It is better to populate the New World with Norwegians than to suppose that the Native Americans have a separate origin, and there seems to be little concern at all about the counterargument that jumps out to us as obvious today: that the majority of the original inhabitants of the New World do not look like Norwegians. If there was a perception of racial difference, it was not for authors such as Grotius the primary explanandum of accounts of human diversity and geographical dispersion. Of course in the absence of rapidly circulating images, particularly photographic or cinematic images, the phenotypic differences between different populations would not be nearly as prominent an issue in, say, the attempt to account for which groups are related to which. But this does not fully account for Grotius's lack of concern; he at least knew from the reports of travelers that the typical description of a Native American differs significantly from that of a Norwegian. Nonetheless, Grotius evidently did not consider differences of appearance a significant impediment to the establishment of shared ancestry.

The problems of diffusion and chronology, as opposed to racial difference, are dealt with at length and in exemplary fashion by the English jurist Matthew Hale in his 1677 work *The Primitive Origination of Mankind*. Hale's contribution to the early modern polygenesis debate is of particular interest in view of its treatment of the problem of human diversity as a problem of natural philosophy rather than of biblical interpretation. Interestingly, while La Peyrère accepts the heretical doctrine of polygenesis, and Hale rejects it, it is the heretic La Peyrère who spends

[31] Hugo Grotius, *De origine gentium americanarum dissertatio*, Paris, 1642, 15, cited in Gliozzi, *Adam et le nouveau monde*, 450.

almost all of his effort explicating scripture and looking for a justification of his view in the holy writ, while the traditionalist Hale for his part devotes much of his energy in the *Primitive Origination* to providing a plausible naturalistic account of diffusion into, and biodiversity in, the New World.

Hale, like the libertine polygenists, explicitly identifies the possibility of the spontaneous generation of insects as relevant to our understanding of human origins. He seeks to provide an account of how it is that human beings could have arisen *ex non genitis*, that is, from elements or principles that were not themselves generated. He maintains that this production could have happened in one of three ways: it could have been "fortuitous or casual," it could have been "natural," or it could have happened "by the immediate Power, Wisdom, and Providence of Almighty God and his meer *Beneplacitum*."[32] All of these three ways, in any case, would require that the first creatures of any species had initially come into being in a different way than all of their subsequent descendants; they would have to be produced, namely, *ex non genitis*, or from things that are not themselves generated, whether these be eternal ungenerated atoms or God. Hale believes that this "Method of production of Men and perfect Animals is ceased" in the present age, "and their production now delegated ordinarily to Propagation," yet he considers the possibility that "in some places, and at some times, especially between the *Tropicks*, such a Pullulation of Men and Beasts may be supposed to be."[33] Hale rejects this possibility, but his characterization of the view of his opponents is significant, since it shows the widespread association in the early modern period (an association extending back to antiquity) between those parts of the world inhabited by Africans, and later also by New World natives, on the one hand, and the possibility of being born from the earth on the other.

According to Hale, the first generation of creatures was by "divine power and ordination," and not from any preexisting primordium or seed. But once the divinely ordained kinds were in existence, all future generation was by way of "propagation," which is to say by way of sexual reproduction: "[T]hat Prolifick Power of propagating was never delegated or committed to the Earth, or any other Casual or Natural Cause; but only to the Seminal Nature, derived from their Individuals, and disposed according to that Law of propagation of their kind, alligated as before to their specifical and individual nature."[34] If we were to suppose any

[32]Matthew Hale, *The Primitive Origination of Mankind, Considered and Examined according to the Light of Nature*, London: William Godbid, 1677, 256.

[33]Hale, *Primitive Origination of Mankind*, 257.

[34]Hale, *Primitive Origination of Mankind*, 305.

other kind of origination of humanity than that which proceeds through God's divine creation, "we may with as fair a Supposition imagin that a Man should be produced by the natural conjunction of Sheep or of Lions, or a Star be produced *ex putri materia terrestri*, as to suppose a Man to be produced accidentally, casually, or naturally."[35] In short, generation by fortuitous causes is just as much a disruption of the cycle of ordinary species reproduction as would be the monstrous birth of a creature of one species from parents of another. Hale finds far more reasonable the "divine hypothesis" that the first humans "had their Original from a Great, Powerful, Wise, Intelligent Being."[36]

But why not hold open the possibility that there had been several independent "originals" of this sort for different human populations? Earlier, in Section II of the *Primitive Origination*, Hale had noted that "[t]he late Discovery of the vast Continent of *America* and Islands adjacent, which appears to be as populous with Men, and as well stored with Cattel almost as any part of *Europe*, *Asia*, or *Africa*, hath occasioned some difficulty and dispute touching the Traduction [note the term] of all Mankind from the two common Parents supposed of all Mankind, namely *Adam* and *Eve*."[37] The greatest problems arise, Hale believes, "concerning the storing of the World with Men and Cattel from those that the Sacred History tells us were preserved in the Ark."[38] Thinking as a biogeographer, Hale recognizes that the study of cattle can help us to learn about the diffusion of men, and his proposal for studying the human past in part by looking at current biodiversity is remarkably advanced. There are, he notes, "divers perfect Animals of divers kinds in *America* which have none of the same kind in *Europe*, *Asia* or *Africa*." From this, he claims without explicitly citing the authors he has in mind, many people conclude that the Americans could not be descended from Adam. He summarizes their position as follows:

> That since by all Circumstances it is apparent that *America* hath been very long inhabited, and possibly as long as any other Continent in the World, and since it is of all hands agreed that the supposed common Parents of the rest of Mankind, *Adam*, *Noah* and his three Sons, had their Habitations in some Parts of *Asia*, and since we have no probable evidence that any of their Descendents traduced the first Colonies of the *American* Plantations into *America*, being so divided from the rest of the World, the access thither so difficult, and Navigation the only means of such a Migration being of a far later perfection than what could answer such a Population of so great

[35]Hale, *Primitive Origination of Mankind*, 316.
[36]Hale, *Primitive Origination of Mankind*, 316.
[37]Hale, *Primitive Origination of Mankind*, 182.
[38]Hale, *Primitive Origination of Mankind*, 182.

> a Continent: That consequently the *Americans* derive not their Original either from *Adam*, or at least not from *Noah*; but either had an Eternal Succession, or if they had a Beginning, they were *Aborigines*, and multiplied from other common Stocks than what the *Mosaical* History imports.[39]

Hale sharply disagrees, though he does acknowledge some ethnographic evidence that supports the polygenetic conclusion. Interestingly, he also takes seriously, at least in passing, the cultural traditions of Native Americans themselves—the "mythology"—as a relevant bit of evidence in determining how the New World came to be populated. Though the traditions of America "be mingled with some things fabulous," nonetheless "they seem to favour [the] Conclusion" of the polygenesis theorists.[40] Yet here again we see an example of an early modern European refusing to lend much credence to the accounts that non-Europeans give of their own place in the world. Hale's preferred sort of evidence is not ethnographic or scriptural, let alone mythological, but, again, biogeographical. He argues that "the Origination of the common Parents of Mankind were in *Asia*, yet some of their Descendents did come into *America*." He suggests in turn that the "transmigration" from Asia to America could have happened by land or by sea, but that the latter is more likely, since, though it could be possible that there are "some junctures between the North Continent of *America* and some part of *Tartary*, *Russia*, or *Muscovy*, yet none are known, unless the Frozen Seas in those Parts might be a means to transport Men thither."[41]

Hale also reasons that the current distribution and shape of land masses in the world is not the same as it once was. The prospect of continental drift and erosion had been widely entertained at least since Abraham Ortelius's *Thesaurus geographicus* of 1587,[42] but Hale appears to be among the first to use it to support an argument in favor of a diffusionist account of the unity of the human species. He argues at great length that "we can by no means reasonably suppose the Face, Figure, Position and Disposition of the Sea and dry Land to be the same anciently as now," and that "those parts of *Asia* and *America* which are now dis-joyned by

[39]Hale, *Primitive Origination of Mankind*, 183.

[40]Hale, *Primitive Origination of Mankind*, 183.

[41]Hale, *Primitive Origination of Mankind*, 189. It is interesting to note in passing here that one of the principal motivations Leibniz would have in gaining influence with the tsar of Russia would be to convince him to organize an expedition to determine whether Asia and America do in fact connect, a plan that would eventually be realized in Vitus Bering's second Kamchatka expedition of 1733–43.

[42]Abraham Ortelius, *Thesaurus geographicus*, Antwerp: Ex officina Christophori Plantini, 1587.

the interfluency of the Sea, might have been formerly in some Age of the World contiguous to each other."[43]

But what about the animals? Hale notes that, even if we suppose the human beings got there *ex industria*, still "it is not easily conceptible how Beasts, especially of prey, should be transported into *America* through those large Seas."[44] But he is not easily dissuaded from his diffusionist convictions, and he goes on to argue that there is a plausible way of accounting for the "Manner of Traduction of Brutes into *America*."[45] The variety of flora and fauna in the New World that are different from those of the Old World can be explained, Hale thinks, by appeal to degeneration (which we will discuss at length in the following chapter), by "the promiscuous couplings of Males and Females of several *Species*, whereby there arise a sort of Brutes that were not in the first Creation."[46] Hale suggests, for example, that the llamas of Peru might "be primitively sheep."[47] As with animals, so too with the variety of humans:

> Nay let us look upon Men in several Climates, though in the same Continent, we shall see a strange variety among them in Colour, Figure, Stature, Complexion, Humor, and all arising from the difference of the Climate, though the Continent be but one, as to point of Access and mutual Intercourse and possibility of Intermigrations: The *Ethiopian* black, flat-nosed and crisp-haired, the *Moors* tawny; the *Spaniards* swarthy, little, haughty, deliberate, the *French* spritely, sudden; the Northern people large, fair-complexioned, strong, sinewy, couragious. . . . And there is no less difference in the Humors and Dispositions of People inhabiting several Climates, than there is in their Statures and Complexions. And it is an evidence that this ariseth from the Climate, because long continuance in these various Climates assimilate those that are of a Forein extraction to the Complexions and Constitutions of the Natives after the succession of a few Generations.[48]

In sum, Hale derives the entire human species from traduction, and in order to do this he makes highly speculative conjectures about the migrations of peoples out of the Near East, and also about the way in which environments can change appearances without changing fundamental relations of kinship. Hale is interested in differences in the "constitutions" of human groups, but does not believe that these differences should be explained in terms of any basic racial classificatory schema. He is con-

[43]Hale, *Primitive Origination of Mankind*, 193.
[44]Hale, *Primitive Origination of Mankind*, 184.
[45]Hale, *Primitive Origination of Mankind*, 189.
[46]Hale, *Primitive Origination of Mankind*, 199.
[47]Hale, *Primitive Origination of Mankind*, 201.
[48]Hale, *Primitive Origination of Mankind*, 200–201.

cerned with "national physiognomy," not the typology of races. According to many scholars, the first such typology would be presented only in the decade following Hale's *Primitive Origination*, by the seventeenth-century French libertine philosopher and voyager François Bernier, who will be discussed at length in chapter 6.

Yet another fairly typical diffusionist work of the mid-seventeenth century, also providing a very rich overview of the state of scholarship in the era, is the German author Georg Horn's 1669 work *De originibus Americanis* (*On American Origins*). Horn writes of the various opinions on the origins of the inhabitants of the Americas. It will be useful, just to gain a quick idea of the multiplicity of views in the period, to run through Horn's own summary of the secondary literature on the topic of human origins and diversity. A partial list of the views he cites includes those of Christopher Columbus, François Vatable, and Henri Estienne, who hold that "America was inhabited from East India"; Gilbert Génebrard, who holds that "the ten tribes of Israelites went to Tartary and America"; Jan Gerartsen van Gorp and Lucius Marinaeus Siculus, who hold that "the Romans came to America"; Paolo Giovio, who holds, conversely, that "Mexicans came to Gaul. Americans were transported to our sphere by storms"; Martinus Hamconius, who holds that "the Americans come from the Frisians"; François de Bivar and Jacob Charron, who hold that "the Americans come from the French"; Abraham Mylius, "from the Celts"; Athanasius Kircher, "from the Egyptians"; Emmanuel de Moraes, "from the Carthaginians"; and Edward Brerewood, "from the Tatars."[49] And so on. For Horn and the authors who interest him, diffusion is the model for the origins of Native Americans that warrants attention; the question is not whether the indigenous people arrived there from somewhere else, but only *how* they arrived there. The prospect of true autochthony for Native Americans has largely faded from the picture.

[49]Georg Horn, *De originibus Americanis, libri quatuor*, Halberstadt: Sumptibus Ioannis Mülleri, 1669, 9. "Varia de Originibus Americanis opiniones. Columbi, Vatabli, Stephani. *America pro India Orientali habita. Hispanos a mauris fugatos in Americam venisse.* Theoph. Paracelsi. Ariae Montani. *Ophir an Parvajim, Iucatan an a Ioktan, Nomen Peru unae.* Postelli. *Origines ad Isthmum divisas*, Lerii. *Chamanas in America.* Genebrardi, *Decem tribus Israëlitarum in Tataria & America.* Tornielli, Acostae, Goropii Becani, Marinaei Siculi, *Romanos in Americam venisse.* Pauli Jovii. *Mexicanos in Galliam venisse, Americani tempestatibus in nostrum orbem ejecti*, Striffridi Pesri &Hamconii. *Americanos a Frisis* Fr. Bivarii, Jacobi Charron, *Americanos a Gallis.* Abrahami Mylii, *Americanos a Celtis Prisces nota Borealis America.* Nicolai Fulleri. *Americanos a Cushi posteris.* Athanasii Kircheri. *Americanos ab Aegyptis.* Emmanuëlis de Moraes. *A Carthaginensibus.* Edoardi Brerevvoodi *A Tataris Aliis ab Islandia.* Marei Lescarboti. *A Chamanaeis. Noacho nota America.* Hugonis Grotii. *A Norvvagis, Abissinis, Indis Oriental. & sinensibus* Johannis Latii. *A Tatariis & australibus.* Roberti Comtaei. *A Phoenicibus.* Roderici a Castro. *A **oenis.* Cottoni. *Iesuitae impietas. Difficultas quaestionis unde.*"

A major influence on the diffusionist model was exercised by Grotius, already discussed above. His 1642 *On the Origins of the American Peoples* is a good example of a sophisticated attempt to account for the settlement of the New World largely without appeal to evidence from the scriptural or other Western literary traditions. Grotius instead draws on biogeographical and toponymic clues in order to arrive at the conclusion that at least North America was settled not by an Asian or Pacific route, but rather via the North Atlantic. He considers, for example, the absence of horses in pre-Columbian America: "It is indeed well known that there were no horses in America before the Spanish. Scythia on the other hand was always full of horses, and nearly all Scythians ride horses. . . . If America and Tartary were connected, escaped horses, or horses that are freely grazing, would formerly have penetrated from Tartary into America."[50] Instead, as we have already seen, Grotius believes that the original settlers of North America must have been Germanic seafarers, or more precisely Norsemen. Grotius is committed to the Norse origins only of the Native Americans north of the isthmus of Panama, while for South Americans the argument is both less certain and less clear.[51]

The details of his theory are not important. What is significant is to note that Grotius, like Horn and Hale, defends diffusion of the human species by appeal to a Whewellian consilience of inductions. Some of the evidence he adduces appears absurd, such as that from toponymy, as Uto-Aztecan place-names do not have any non-coincidental resemblance to Germanic place-names; some of it seems highly sophisticated, such as that from the biodiversity of North America, as South American camelids really do share ancestry with Eurasian ones. The overall argument becomes much more plausible when we recall that the North Atlantic was well traveled since the Middle Ages, while a migration based on a hypothetical connection between Asia and North America (which would not be confirmed until 1733) could easily have seemed far-fetched in the seventeenth century. Interestingly, on at least one heterodox yet scientifically respectable theory of New World migration currently being discussed, the first settlers did in fact arrive by a North Atlantic route.[52] These settlers

[50]References are to a later edition incorporating responses to criticism: Hugo Grotius, *Hugonis Grotii de Origine gentium armericanarum dissertatio altera, adversus obtrectatorem*, Paris: S. Cramoisy, 1643, 5. "Constat enim in omni America ante Hispanorum adventum equos fuisse nullos. Scythia autem plena semper equis, & Scythae fere omnes equis vehi. . . . [S]i cohaerent America & Tartaria, equi aut fugitivi, aut libere pascentes, ex Tartaria in Americam olim penetrassent."

[51]For a detailed analysis of this treatise, see Joan-Pau Rubiés, "Hugo Grotius's Dissertation on the Origin of the American Peoples and the Use of Comparative Methods," *Journal of the History of Ideas* 52, 2 (April–June 1991): 221–44.

[52]See in particular Dennis Stanford and Bruce Bradley, "The North Atlantic Ice-Edge Corridor: A Possible Palaeolithic Route to the New World," *World Archaeology* 36, 4 (2004): 459–78.

would of course have vastly predated the earliest appearance of proto-Germanic peoples in Europe, but still it is interesting to bear in mind that in the scientific study of the settlement of the Americas, although today we have vastly more detailed information and more accurate dating techniques, the same basic theories of the patterns of migration remain in circulation.

4.4. Conclusion

The aims of naturalistic explanations of the origins of non-European peoples, and particularly of Native Americans, change significantly from the sixteenth and into the seventeenth century. Earlier, as we have seen, the main concern is to at least broach, if not to fully argue for, the possibility that human or human-like beings might not have been directly created by God, might even be truly autochthonous or born directly of the earth, and might be related by shared descent with nonhuman animal species. With authors such as Hale and Grotius, by contrast, the concern is now to account for the unity of the human species by proposing plausible patterns of migration from the Old World to the New. Autochthony is out of the question here, and it is at least implied that all Native American peoples have a scriptural pedigree. But it is not the principal aim of the monogenetic theories of Grotius and many of his contemporaries to establish the biblical lineage of the New World inhabitants. Rather, their primary purpose is geographic and demographic: to account for the global diffusion of a species that itself, by all available evidence, constitutes a proper unity. The Renaissance libertine polygenists had sought to go against scripture, while pre-Adamists such as La Peyrère sought to account for human origins by appeal to controversial theories that approached those of the libertines yet still remained grounded in scripture. But with Grotius and Hale, increasingly, accounts of human origins simply turned away from the matter of scriptural correctness as a central concern. Whether the human species had a single origin, or multiple ones, was now a question that could, in principle, be resolved on the basis of inductive inferences about the distant past. In broad outline, as we will see, this new biogeographical approach to human diversity was one of the key factors in the appearance of a distinctively modern way of thinking about human racial difference.[53]

[53]For an insightful phenomenological account of the geographical function of the race concept, in particular the way it helps to reify borders, see Robert Bernasconi, "Crossed Lines in the Racialization Process: Race as a Border Concept," *Research in Phenomenology* 42 (2012): 206–28.

Chapter 5

Diversity as Degeneration

5.1. The "History of Abused Nature"

For several centuries prior to the modern period, European naturalists were able to look to the ancient works on botany and zoology, particularly those of Aristotle and his successor Theophrastus, as providing an adequate account of the entire range of plant and animal kinds. Even where their local species differed from those of the eastern Mediterranean on which the ancient works were based, European systematists were able to interpret these local kinds as variations on the more basic (because studied by greater authorities) Greek types. This sort of subsumption into already established classificatory schemes became much more difficult in the early modern period. Certainly, as scholars have long recognized,[1] much of the classificatory work of the early modern period involved the reference of new, East and West Indian kinds back to familiar European kinds (thus, e.g., maize becomes "corn," and cougars become "lions"). Yet in many cases the new kinds encountered were simply too unfamiliar to be related back in this way, and native names had to be borrowed. In general, however, systematists sought to prepare themselves in such a way that no new encounter would force them to contend with something entirely unclassifiable by reference to the already available system. In contrast to the study, for example, of longitude and magnetism—in which physics, once an a priori science with universal scope, had to pay attention to local variations—in the case of biological classification a need arose of determining in advance the higher-order taxa of new species. Some scholars have noted that the first steps toward the systematic global classification of biological kinds came in the sixteenth century, when botanists such as Andrea Cesalpino sought to fix species as eternally self-perpetuating entities. As a consequence of the effort to develop a comprehensive and universal system of classification in the early mod-

[1] See, most notably, Alfred W. Crosby, *The Columbian Exchange: Biological and Cultural Consequences of 1492*, Westport, CT: Greenwood, 1972.

ern period, it was, as Scott Atran has argued, "necessary to fix a criterion for the species even in advance of future discoveries."[2]

One thing that has perhaps not been sufficiently emphasized in the recent literature on the history of early modern systematics is the vastly greater interest in the early modern period in the study of plants rather than animals. Edward Tyson laments as late as 1699 that "we've ransacked both the Indies" in search of rare and exotic plants, while neglecting what is most noble in nature—the animals.[3] Botany and medicine—and decidedly not zoology—were two branches of a common project until the seventeenth century, while zoology was scarcely on the agenda. Botany was preoccupied with finding *useful* plants, while there was no analogous perception of the usefulness of animals. To the extent that zoology begins to emerge in the sixteenth century, it does so in large measure as a consequence of a concerted campaign of principled arguments—fueled by the recent translation and publication of Aristotle's biological treatises by Theodor Gaza in 1483 and then, along with the rest of the oeuvre, in the great Lyon edition of these treatises of 1529–39—for the intrinsic value of coming to know nature itself independently of human concerns. In the early modern period, the study of animals, as opposed to the anthropomorphic use of them for the drawing of moral lessons, continued to require explicit defense as a project worth undertaking. In this respect, it could not have been more different from botany, the utility of which was never doubted by anyone who ran the risk of someday falling ill and needing a cure—which is to say by anyone.

In important respects, the paradigm brute had always been the quadruped, and species that deviated from this model gave rise to conceptual problems of their own. Thus we cannot speak of *the* concept of animal in the early modern period, but must always make precise just what kind of animal we have in mind. Some borderline cases of animal, such as the famous "zoophyte," were of interest to early modern naturalists such as Julius Caesar Scaliger because they seemed to embody features of both the plant and the animal kingdoms and thus to belie the standard cosmological view of nature as consisting in neat, hierarchically ordered levels of beings. More threateningly, the great apes (or, as they were generically called throughout the seventeenth century, the "orang-outangs") seemed to straddle the animal-human boundary in the same way that the zoophytes did the plant-animal one. If the new systematics prescribed the

[2]Atran, *Cognitive Foundations of Natural History*, 142.

[3]Edward Tyson, *Orang-outang, sive, Homo sylvestris, or, The Anatomy of a Pygmie Compared with That of a Monkey, an Ape, and a Man to Which Is Added, A Philological Essay Concerning the Pygmies, the Cynocephali, the Satyrs and Sphinges of the Ancients: Wherein It Will Appear That They Are All Either Apes or Monkeys, and Not Men, as Formerly Pretended*, London: Thomas Bennet, 1699.

subsumption of new kinds into familiar higher taxa, then why should the newly encountered great apes not be considered a variety of human beings? But if this is to be permitted, then what are the implications for the traditional identification of "man" as essentially rational, and of reason as essentially connected to speech? Particularly in the period surrounding the publication of Edward Tyson's monumental 1699 work, *Orang-Outang, or the Anatomy of a Pygmie Compared with That of a Monkey, an Ape, and a Man*, a major philosophical debate raged concerning the lower boundary of the human species, and the possibility that human beings do not differ essentially from the brutes, but only by degree. While there had been tales of satyrs and other beast-like men in antiquity, this newly invigorated modern debate about "orang-outangs," which drew in Leibniz, Locke, John Ray, Linnaeus, and many others, can be traced directly to encounters and reports coming back from Sumatra, Angola, and elsewhere. Here, it seemed that local variations posed a threat to the possibility of any universal claims about the nature of a given kind, including even the human kind.

As we have seen, early debates about human diversity—about the origins and nature of far-flung people—were not centrally focused on what we today would recognize as racial difference. This is not to say that the visible physical differences between different human groups were not also of interest, but the problem of origins took precedence. Indeed, as long as there remained a possibility of separate origins for separate human groups, then the question in need of explanation could not have seemed to be why, say, Europeans and Native Americans appear so different, but rather why, in spite of the fact that they do not descend from the same first parents, they ended up looking so remarkably *similar*. Occasionally, as for example in Bruno, as we have seen, physical differences—generally exaggerated—are taken as evidence of separate origins. For those however who are committed to monogenesis, there is a problem, and one that appears to grow in significance over the course of the modern period, of accounting for the causes and significance of the physical diversity of the human species. Environmental influence on moral and physical character, as we have already seen, is by no means new in the modern period. As early as Hippocrates the role of the environment was often adduced in causal explanations of human physical and temperamental differences. In the Hippocratic corpus itself we find, for example, a lengthy account of differences in skin pigmentation and moral temperament in different parts of the world. The differences between Asians and Europeans are described, interestingly, in a manner favorable to the former. Asians are gentle, Europeans bellicose (the exact opposite of the early modern stereotype), and this because of the way each group is influenced by climate and wind: "Asia differs very much from Europe, in my opinion, as to the nature of all the things that grow there, and of the Inhabitants. For they are much fairer and larger in Asia; the Country itself is milder than ours; and the manners of the People more mild and unactive."[4] Nicolas Malebranche, writing in the 1670s, continues in a more or less Hippocratic vein, even if the particular characteristics of the particular continents are now changed, arguing that the different qualities of air in different places bring about differences in natural character. He notes that "it is certain that the most refined air particles we breathe enter our hearts," and believes that this process is corroborated empirically from our daily observation of the "various humors and mental characteristics of persons of different countries. The Gascons, for example, have a much more lively imagination than the Normans. The people of Rouen, Dieppe, and Picardy, are all different from each other: and they all differ even more from the Low Normans, although they are all quite similar to one another. But if we consider the people of more remote lands, we shall encounter even stranger differences, as between an Italian and a Fleming or a Dutchman."[5] But Malebranche is offering here a late echo of a long tradition of what we have already called "national physiognomy," and one that is very much out of step with the sort of treatments of human diversity that were coming to occupy people's minds as a result of the massive global circulations of the early modern period.

Ancient accounts of global racial divisions often had a mythological character, involving a curse or a cataclysm that brought about a permanent separation between two groups. Thus the original blackening of Africans by burning or scorching—a one-time event, generally associated with a curse or misfortune, as in the Old Testament myth of the curse of Ham, or the Greek myth of Phaeton, who rode his burning chariot too close to the surface of the earth—is communicated to future generations by a physically comprehensible channel. But in this sort of myth what is emphasized is the miasmic and dynastic character of the differentiating traits. Degeneration is conceptualized here as a quick rather than gradual affair, as a one-time event rather than the sum of multiple events. But both those accounts of racial difference that appeal to a single traumatic event, such as a primordial scorching, as well as the gradual degenerationist accounts more common in the seventeenth and eighteenth centuries, remain similar to one another, to the extent that both understand racial difference to be the result of a certain part of the species taking a

[4]Hippocrates, *Upon Air, Water, and Situation; upon Epidemical Diseases; and upon Prognosticks, in Acute Cases Especially . . .*, trans. and ed. Francis Clifton, London: J. Wattes, 1734, 20.

[5]Nicolas Malebranche, *The Search After Truth*, book II, pt. 1, chap. 3, trans. Thomas M. Lennon and Paul J. Olscamp, Cambridge: Cambridge University Press, 1997, 95.

wrong turn, or being knocked off of the established course set out for humanity at the beginning.

In modern degenerationist thought, it is nearly always taken for granted that the modern people closest to the original type specimens are none other than Europeans, and, correlatively, the more different a given group appears from Europeans, the more that group may be judged to have degenerated. "Degeneration" and "evolution" both, in their original senses, are evaluative terms, the latter signifying progress from a lower stage to a higher one, and in this sense contrasting with degeneration or change in the opposite direction. In the seventeenth century, when common lineage between different kinds is taken to be a real possibility, it is generally supposed that the "lower" kind must have descended from the higher one. Thus the Baconian naturalist John Bulwer writes in 1650, "But by this new History of abused Nature it will appeare a sad truth, that mans indeavours have runn the clean contrary course, and he that [man] hath been so farr from raising himselfe above the pitch of his Originall endowments, that he is muchfallen below himselfe; and in many parts of the world is practically degenerated into the similitude of a Beast. The danger of man since his fall is more in sinking downe then in climbing up, in dejecting then in raising himselfe to a better condition or improvement of naturall parts."[6] According to Bulwer (who attributes the view to Plato, presumably drawing on a somewhat idiosyncratic reading of the *Alcibiades*), "onely the first men which the world possessed, were made by God, but the rest were made and born answerable to the discourse of Mans invention."[7] The author goes on to report on an interaction with a philosopher, regrettably unnamed, who has proposed that human beings descend from apes:

> In discourse I have heard to fall, somewhat in earnest, from the mouth of a Philosopher (one in points of common beliefe (indeed) too scepticall) That man was a meer Artificiall creature, and was at first but a kind of Ape or Baboon, who through his industry (by degrees) in time had improved his Figure & his Reason up to the perfection of man. It is (indeed) an old Observation of Pliny, that all the Race and kind of Apes resemble the proportion of men perfectly in the Face, Nose, Eares and Eye-lids: which

[6] John Bulwer, *Anthropometamorphosis: Man Transform'd, or, The Artificial Changeling*, London, 1650, introduction, n.p. Part of the treatment of Bulwer's work here was previously presented in Justin E. H. Smith, "'A Corporall Philosophy': Language and 'Body-Making' in the Work of John Bulwer," in Charles T. Wolfe and Ofer Gal (eds.), *The Body as Object and Instrument of Knowledge: Embodied Empiricism in Early Modern Science*, Dordrecht: Springer, 2010, 169–84, though in the course of making a very different argument about the place of Bulwer in the Baconian philosophical tradition.

[7] Bulwer, *Anthropometamorphosis*, introduction, n.p.

eye-lids these Creatures along of all four footed have under their eyes as well as above.[8]

This is not a theory of evolution by a natural mechanism, but of degeneration by an artificial mechanism, in particular, by unhealthy and immoral cultural practices. More than one hundred years later, George-Louis Leclerc de Buffon would continue to account for morphological drift in a population over time as a consequence of degeneration, though for him this would be part of a naturalized scientific theory, rather than the sort of occasion for moralizing it had been for Bulwer.[9] Thus Buffon writes in the *Histoire naturelle*, the first edition of which was published in 1749, "Not only the ass and the horse, but also man, the apes, the quadrupeds, and all the animals might be regarded as constituting but a single family. . . . If it were admitted that the ass is of the family of the horse, and different from the horse only because it has varied from the original form, one could equally well say that the ape is of the family of man, that he is a degenerate man, that man and ape have a common origin."[10] Buffon believes, like evolutionists, that human beings and apes are related by descent, but unlike the evolutionists he presumes that the common ancestor of both humans and apes was a human, and that the apes have, as a result, presumably, of climatic and geographical factors, deviated morphologically from the original population. While Buffon does not write in overtly normative terms, there is nonetheless a presumption that deviation occurs only in unfortunate circumstances, that the original represents the ideal type, rather than that new forms emerge simply as expressions of fitness in the face of new environmental exigencies.

Buffon is perhaps the thinker who is most associated with degenerationist thinking, though in fact when he defends it he is only elaborating on a conventional and inherited way of thinking about natural diversity. What is perhaps most original in Buffon is his unified account of the causes of variation throughout living nature, with human racial variation presented simply as a local instance of this vastly more comprehensive scheme. One of Buffon's long-standing preoccupations is the general degeneracy of American flora and fauna. Unaware of the recent extinction of Pleistocene megafauna in the western hemisphere, the French naturalist presents all of "animated Nature" in America as "weaker, less active,

[8] Bulwer, *Anthropometamorphosis*, introduction, n.p.

[9] For a very useful study of Buffon's racial theory and in particular its genealogical character, see Claude-Olivier Doron, "Race and Genealogy: Buffon and the Formation of the Concept of 'Race,'" *Humana Mente—Journal of Philosophical Studies* 22 (2012): 75–109.

[10] George-Louis Leclerc, Comte de Buffon, *Histoire naturelle, générale et particulière, avec la description du cabinet du Roy*, Paris, 1749, cited in Ernst Mayr, *The Growth of Biological Thought*, Cambridge, MA: Harvard University Press, 1982, 332.

and more circumscribed in the variety of her productions; for we perceive, from the enumeration of the American animals, that the numbers of species is not only fewer, but that, in general, all the animals are much smaller than those of the Old Continent. No American animal can be compared with the elephant, the rhinoceros, the hippopotamus, the dromedary, the giraffe, the buffalo, the lion, the tiger, etc."[11] This faunal degeneracy extends directly, in turn, to the condition of the Native Americans: "The American savage is feeble and has small organs of generation; he has neither hair nor beard, and no ardour whatsoever for his female . . . he is also less sensitive, and yet more timid and cowardly; he has no vivacity, no activity of mind, the activity of his body is less an exercise, a voluntary motion, than a necessary action caused by want; relieve him of hunger and thirst and you deprive him of all the active principle of all his movements; he will rest stupidly upon his legs or lying down entire days."[12] Buffon's enfeebled American parallels the demise of France's American colonial holdings.

We have seen that, according to the degenerationist thesis, morphological variation between populations is accounted for in terms of deviation from an original ideal type. The common presumption, again, was that the original humans looked more or less as modern Europeans do, and that non-European racial diversity was a result of deviation from the norm, stemming from the expansion of the human species into more extreme climatic and geographical zones. Physiological variability on this account is accompanied by variations in manners from one climatic region to the next. Typifying this parallelism, Buffon writes of Africa, "The climate is extremely hot; and yet the temperature of the air differs widely in different nations. Their manners also are not less various."[13]

[11] Georges-Louis Leclerc, Comte de Buffon, *Oeuvres complètes de M. le Cte.de Buffon*, Paris: Impriméric Royale, 1775, 3:179. "La Nature vivante y est donc beaucoup moins agissante, beaucoup moins variée, & nous pouvons même dire beaucoup moins forte; car nous verrons, par l'énumération des animaux de l'Amérique, que non-seulement les espèces en sont en petit nombre, mais qu'en général tous les animaux y sont incomparablement plus petits que ceux de l'ancien continent, & qu'il n'y en a aucun en Amérique qu'on puisse comparer à l'éléphant, au rhinoceros, à l'hippopotame, au dromadaire, à la giraffe, au buffle, au lion, au tigre, &c."

[12] Buffon, *Oeuvres complètes*, 203. "Le sauvage est foible par les organes de génération; il n'a ni poil, ni barbe, et nulle ardeur pour sa femelle: quoique plus léger que l'Européen parce qu'il a plus d'habitude à courir, il est cependant beaucoup moins fort de corps; il est aussi bien moins sensible, et cependant plus craintif et plus lache; il n'a nulle vivacité, nulle activité de l'âme; celle du corps est moins un exercice, un movement volontaire, qu'une nécessité causée par le besoin: ôtez-lui la faim et la soif, vous détruirez en même temps le principe actif de tous ses mouvements."

[13] Georges-Louis Leclerc, Comte de Buffon, *A Natural History, General and Particular, containing the History and Theory of the Earth, &c.*, 8 vols., trans. William Smellie, London: T. Kelly, 1781, 3:194.

Buffon believes that skin color is transmitted by parents from one generation to the next, but supposes that if a group of sub-Saharan Africans were transferred to Northern Europe, the "descendants of the eighth, tenth, or twelfth generation would be much fairer."[14] The climatological theory of racial variation was however complicated by a wide range of counterexamples. For example, as Buffon noticed, while the North American natives appear on this theory, as they should, to resemble the Tartars, the people of Mexico and Peru, "though like the Negroes they live under the torrid zone, have no similarity to them."[15] For Buffon, as for Kant after him, it is the temperate zone, including most of Europe, that is ideal for the flourishing of humanity. "[T]hat portion of the earth," Kant writes in his treatise *On the Different Races of Man* of 1775, "between the thirty-first and fifty-second parallels in the Old World . . . is rightly held to be that in which the most happy mixture of influences of the colder and hotter regions and also the greatest wealth of earthly creatures is encountered."[16] On Kant's "out of Europe" hypothesis, as for Buffon and Bulwer before him, the human species originally began in Europe, looking more or less as modern Europeans do, and subsequently radiated out onto the other continents, gradually taking on new racial traits.

The possibility of degeneration from an original ideal type threatened, for obvious reasons, to deal a blow to natural theology, that is, to the very idea that any natural being may be said to have a particular ideal state at all, from which all deviation must be conceptualized as a worsening. With the growing evidence for long, slow change in species, came a growing awareness of the possibility that there is no one way any given species ought to be. Natural theology—that is, the effort to establish the wisdom and power of God through observation of the workings of nature—arguably was even more in conflict with the idea of morphological change than was revealed theology. Many thinkers were adamantly unwilling to consider that the conformation of the bodily organs results from its environmental circumstances. Thus Isaac Newton for example argues that such things as wings on birds or "swimming bladders" in fish must be seen as "artificial," in the now lost sense of the word, namely, that they are the product of the "Wisdom and Skill of a powerful ever-living Agent."[17] In some sense, natural-theological arguments from design

[14]Buffon, *Natural History*, 3:200.

[15]Buffon, *Natural History*, 3:172.

[16]Immanuel Kant, *Von den verschiedenen Racen der Menschen*, in *Kants Werke*, Akademie Textausgabe (hereafter "AA"), Berlin: De Gruyter, 1968, 440–41.

[17]Isaac Newton, *Opticks: or, A Treatise of the Reflections, Refractions, Inflections and Colours of Light*, 4th ed., London: William Innys, 1730 [1704], 379. In seventeenth-century natural philosophy, often observations of how well organs fulfill functions are made in connection with classic formulations of the teleological argument, or argument by design, for

in the seventeenth century are even further from explanation in terms of natural selection than Aristotle's teleological biology had been: the latter attributed the intelligent design of organisms only to the proper functioning of a nature, which, Aristotle forcefully insisted, "does not deliberate."[18] Early modern natural theologians, in contrast, imagined that God had consciously and meticulously seen to every smallest detail of an organism's design. But such a belief was difficult to sustain in the face of growing evidence that the conformation of the animal or human body was the result of long, slow adaptation to environmental exigencies. This was of course a dawning awareness that had broad implications across the life sciences. But its implications were particularly important for the understanding of human racial diversity.

Degenerationism would increasingly, in the late eighteenth and early nineteenth centuries, be flipped on its head in favor of a conception of the relationship between species as one of a gradual climb upward rather than a slide downward. In his *Zoonomia* of 1794, Erasmus Darwin pondered, "[W]ould it be too bold to imagine, that all warm-blooded animals have arisen from one living filament, which THE GREAT FIRST CAUSE endued with animality . . . [and] with the power of acquiring new parts?"[19] Fifteen years later in his *La philosophie zoologique* of 1809, Jean-Baptiste Lamarck would develop the elder Darwin's suggestion into what we now know as the hereditary theory of Lamarckism, according to which traits acquired in the course of an animal's life may be passed on to subsequent generations.[20] Significant in the work of Lamarck and Erasmus Darwin is the absence of any supposition that environmentally induced morphological change must be change for the worse.

An earlier indication of the possibility of "evolution," in the sense of change for the better, had been Buffon's own remark that, perhaps, the domestic ass is not a degenerated horse but rather a more perfect one, and that the sheep, as a result of human solicitude, is in fact but a more delicate goat. Domestic animals, he suggests, "have at their origin a less perfect species of wild animals. . . . Nature alone not being able to accomplish what

---

the existence of an infinitely wise creator. Thus John Ray argues for the design of the eye as follows: "Because for the guidance and direction of the Body in Walking and any Exercise, it is necessary the Eye should be uncovered, and exposed to the Air at all times and in all Weathers, therefore the most wise Author of Nature hath provided for it a hot bed of Fat which fills up the interstices of the Muscles; and besides made it more patient and less sensible of cold than our other parts" (John Ray, *The Wisdom of God Manifested in the Works of the Creation Being the Substance of Some Common Places Delivered in the Chappel of Trinity-College, in Cambridge*, London: Samuel Smith, 1691, 182).

[18]See in particular Aristotle, *Physics* 199b28.

[19]Erasmus Darwin, section 39, "Generation," in *Zoonomia: Or, the Laws of Organic Life*, vol. 1, 2nd ed., London: J. Johnson, 1796 [1794].

[20]Jean-Baptiste Lamarck, *La philosophie zoologique*, Paris: Dentu, 1809.

Nature and man can do together."[21] Later, Erasmus Darwin's grandson Charles would draw much of the inspiration for his own theory of natural selection by watching English breeders "selecting" pigeons in order to improve upon certain desirable traits. One detects the difficulty involved, even for Charles Darwin—who is credited with accounting for the origins of species in terms of perfectly blind and indifferent selective mechanisms—in purging the notion of evolution of its original, evaluative sense.

What Buffon admitted in the limited case of domestic breeding, Erasmus Darwin and Lamarck would seem to generalize into a universal account of the gradual emergence of biological order and variety, namely, the universal connection of all life by a single "filament," as well as a mechanism of transmission of morphological change from one generation to the next. The possibility of such a single filament made it supremely difficult to continue thinking of humanity as uniquely created in the image of God. At the same time, though, as shared lineage with other animals, including apes, became thinkable, so too did greater and lesser proximity to the apes within the human species come to seem a real possibility.

5.2. Diet and Custom

The very possibility of taking on new traits as a result of climate, diet, or custom was problematic, since it strongly suggested that there was no single behavioral repertoire for which the human species had initially been outfitted, that the species was potentially unendingly malleable. By the late seventeenth century, evidence had begun rapidly accumulating to suggest not just that there were once morphologically very different species of animals roaming the earth, but also that some of the morphology of living beings originally served for the fulfillment of functions that today's humans and animals no longer have. There was, then, a problem of vestiges not only in the skeletons found in geological strata, but also in our own, living anatomy. Writing in 1700, for example, on the question of whether carnivorism is natural for human beings, John Wallis summarily acknowledges what revealed theology has to say about humanity's place in nature, and then moves on to what in fact interests him. "[W]ithout disputing it as a point in Divinity," he writes, "I shall consider it . . . as a Question in Natural philosophy, whether [meat] be proper Food for man."[22] Wallis begins this letter to Edward Tyson, on whether humans are naturally carnivorous, by noting that it is the

[21] Buffon, "La chèvre et la chèvre d'Angora," in *Histoire naturelle*, 5:60.

[22] John Wallis, "A Letter to Edward Tyson," *Philosophical Transactions* 22 (1700): 772.

> [o]pinion of many Divines that before the Flood, Men did not use to feed on Flesh, because of what we have in Gen[esis 9:3], where God says to Noah, (after the Flood,) Every moving thing that liveth, shall be meat for you, even as the green Herb have I given you all things: Compared with Gen[esis 1:29] where God says to Adam, I have given you every Herb bearing Seed, and every Tree in the which is the fruit of a Tree yielding Seed, and every Tree in the which is the fruit of a Tree yielding seed, to you it shall be for Meat.[23]

So much for scriptural considerations. Wallis abruptly turns from here to notice the structure of the teeth and the length of the colon, arguing that the latter is longer in herbivorous creatures such as sheep than it is in carnivores such as foxes, wolves, and dogs. Where do humans stand in this comparison? Interestingly, they are placed with the other primates: "Now it is well known, that in Man, and, I presume, in the Ape, Monkey, Baboon, &c. such Colon is very remarkable." The distinguishing feature of the human colon is the relative smallness of it, in proportion to the whole body, in contrast with the relative size of the fetal colon: " 'Tis true, that . . . in Man [it] is very small, and seems to be of little or no use: But in a Foetus, it is in proportion much larger than in persons adult." Wallis conjectures that this contraction is brought about by dietary factors: "And It's possible, that our Customary change of Dyet, as we grow up, from what originally would be more natural, may occasion its shrinking into this contracted posture."[24] In other words, our colons are more naturally like those of sheep, but meat-eating transforms them into something more closely resembling the colons of wolves.

This transformation, Wallis speculates, "Seems to be a great Indication, that Nature, which may be reasonably presum'd to adapt the Intestines to the different sorts of aliments that are to pass through them, doth accordingly inform us, to what Animals Flesh is the proper aliment, and to what it is not."[25] If nature has the power to adapt the human body to new and unforeseen circumstances, it appears to follow that there is no one way the human body ought to be. Ought we, for example, to be eating meat? This depends on the available food sources in our environment: our intestines will expand or contract in response to these sources, and the human body will, across the generations, learn to make do.

The question of diet lies at the boundary between physiology and custom: once eaten, the digestion of food is a bodily process like respiration or circulation, but the selection of which foods are to be eaten, how they are prepared, and so on, is a matter of culture. For this reason, perhaps,

[23] Wallis, "Letter to Edward Tyson," 772.
[24] Wallis, "Letter to Edward Tyson," 772.
[25] Wallis, "Letter to Edward Tyson," 772.

diet provides one of the most useful windows into early modern reflections on what it means to live in accordance with nature, and how we can know when we are departing from the state to which we are best suited. There are other human practices that appeared to constitute more direct interventions on the part of humans in their own relationship to nature. Some such cultural practices, such as shaving or tattooing, offer very revealing insight into the way human diversity was often conceived in the early modern period as resulting directly from cultural practices. The human species, on this understanding, is naturally a unity, while its diversity by contrast is "artificial," though not of course in the positive sense Newton had in mind when describing the bladders of fish.

John Bulwer, to whom we have already been introduced, delivers a remarkable screed against tattooing and related practices in his work, *Anthropometamorphosis, or, The Artificial Changeling*, first published in 1650. As the title suggests, the book is a compendium of half-true ethnographic data about practices of bodily modification from the far reaches of the known world, coupled with warnings about the dire moral and physical consequences of adopting these practices at home. Bulwer's preferred examples of tattooing come from the ethnographic materials to which he had access concerning the customs and appearance of Native Americans. He writes that "the Virginian women" "pounce and rase their Faces and whole Bodies with a sharp iron, which makes a stampe in curious knots, and drawes the proportions of Fowles, Fishes, or Beasts; then with painting of sundry live colours they rub it into the stamp, which will never be taken away, because it is dried into the flesh."[26] Bulwer returns to this same practice a few pages later, this time openly paraphrasing John Smith's *Generall Historie of Virginia* of 1624: "The Virginian women adorne themselves with paintings; some have their Face, Breasts, Hands, and Legs, cunningly embroidered with divers workes, as Beasts, Serpents, artificially wrought into their flesh with black spots."[27] Following the general pattern of the whole work, Bulwer does not see this sort of practice as limited to faraway and exotic cultures. He quickly follows up the American ethnography with the observation that "Our Ladies here have lately entertained a vaine Custome of spotting their Faces, out of an affectation of a Mole to set off? their beauty, such as Venus had, and it is well if one black patch will serve to make their Faces remarkable; for some fill their Visages full of them, varied into all manner of shapes and figures."[28] In general, Bulwer does not seem to

[26]Bulwer, *Anthropometamorphosis*, 252. See also John Smith, *The Generall History of Virginia, New-England, and the Summer Isles*, London: Edward Blackmore, 1632 [1624].

[27]Bulwer, *Anthropometamorphosis*, 257.

[28]Bulwer, *Anthropometamorphosis*, 261.

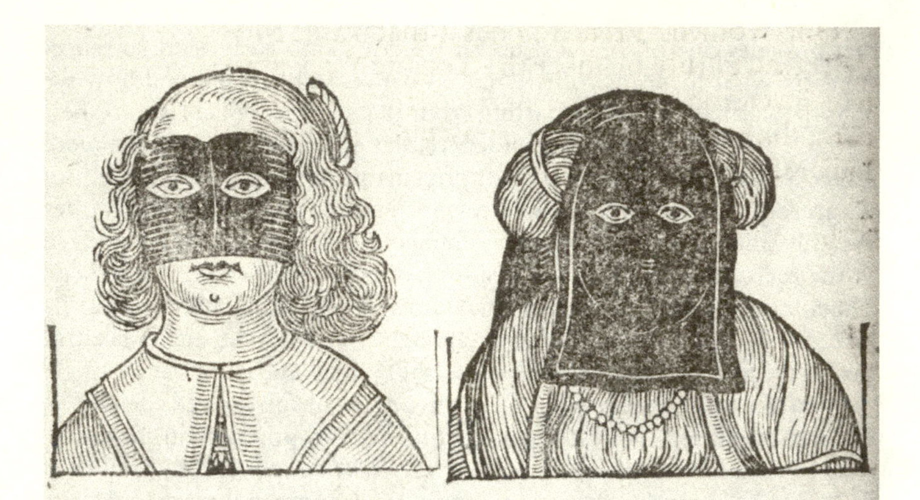

Figure 2. “A Virginian woman,” from John Bulwer’s *Anthropometamorphosis: Man Transform’d, or, the Artificial Changeling* (1650). Image courtesy History of Science Collections, University of Oklahoma Libraries; copyright the Board of Regents of the University of Oklahoma.

distinguish between different forms of body "painting," which include tattooed representations, or simply designs, or even the use of cosmetics to change the hue of the skin. Thus he denounces the Virginian women for the same "bravery" he sees in "the Ladies of Italy," who "to seeme fairer than the rest, take a pride to besmeare and paint themselves."[29] He complains of all of these "Nations," the Virginians, the Englishwomen, the Italians, and the Moors alike, "what needlesse paine they put themselves unto to maintaine their cruell bravery!"[30]

Interestingly, Bulwer must concede that not every form of bodily modification amounts to an unnatural intervention in the divinely instituted course of nature. Some things were set up in nature in order to guide human beings toward the right sort of conduct, including those excrescences of the body, such as hair and nails that require regular grooming. While Bulwer believes that for a man to shave his beard would be to unnaturally do away with the "naturall ensigne of Manhood, appearing about the mouth,"[31] he nonetheless will not go so far as to claim that clipping one's nails is an impermissible derailing of a natural process. Instead, he maintains that in all parts, "there is an appointed end, a certain commoderation of the quantity of parts to the actions of them, according to the faculties using the Organ in the Body." He maintains further that the "continuall increase [of the nails] in man is an Argument of a Divine Nature, a prerogative in which beasts cannot participate, and teacheth us charity to our Bodies."[32] Bulwer thus infers that, before the Fall, in the absence of iron tools, Adam must have kept his nails short by biting them.

Why though should the boundary be drawn between nails and facial hair? Why is it in keeping with God's appointed ends to curtail growth in the one case, but an abortion of the same ends to do the same in the other? Clearly, in the final analysis Bulwer's conception of what is natural comes out looking like nothing so much as a list of his preferred social conventions. What is interesting for our purposes, however, is his conviction that there is a proper way to maintain the body and that this way is dictated by nature. The full elaboration of which modification practices are unnatural and which, in contrast, are there in testimony to the divine nature, will for him amount to nothing less than "a corporall Philosophy."

But let us try to make clearer the relevance of this philosophy of the body to the history of racial thinking. For Bulwer the physiognomist, the soul makes the body in the most direct way possible: through active, willful

[29] Bulwer, *Anthropometamorphosis*, 260.
[30] Bulwer, *Anthropometamorphosis*, 253.
[31] Bulwer, *Anthropometamorphosis*, "The Introduction," scene XII.
[32] Bulwer, *Anthropometamorphosis*, 297.

intervention, whether of the sort practiced at a "nationall" level, or rather as the result of an individual caprice. The best body is the one had by a moral soul, which is to say the one left to develop as nature intends. Failure to respect this intention can lead even to the loss of a properly human nature. Above, we already saw Bulwer relating a certain unnamed "Philosopher's" opinion, that apes and baboons are in fact degenerated human beings. Bulwer is convinced that this opinion, in comparison with those of Plato and Galen, constitutes a symptom of the moral decline of the modern period. For, he thinks, if mutation of humanity over time can occur, it will not, as the Philosopher supposes, take the character of an ascent from beast to man, but rather, as for Buffon after him, the reverse, a descent into apelikeness. The core feature of degenerationism, for Bulwer, is that human variety is artificial, the result of human intervention in the ordinary working of nature. Judgments abound in Bulwer as to the ugliness or beastliness of different "nations," as "Nationall monstrosities," or the "corporall Errata's in every Region," which are in the end only the reflections or bodily expressions of cultural backwardness. Because the differences are cultural rather than essential, there is no reason in principle why "the most civiliz'd Nations" should not "vie deformities with the most Barbarous Nations."

The *Anthropometamorphosis* is an exhortation, not a taxonomy. It warns against conduct that will lead to change in physiognomy, and this most importantly within Bulwer's own culture; it does not anchor particular physiognomies to particular cultures for all time. In fact, Bulwer seems to wish to trace *all* differences in appearance between different groups of people to "artificial" defects. Thus, for example, he claims that "Low-Country-Men or Dutch of Belgia, have some what Long Heads," which "their Mothers cause, being carefull to bring them to it, laying them when they are Infants, and wrapt in swadling Cloaths in their Cradles, suffering them to sleep most upon their sides and Temples."[33] The racial difference of "barbarous" nations is likewise speculatively attributed to cultural practices: "The Men of Brasil have flat Heads, the hinder part not round but flat, which may very well be imagined to proceed from some Affectation or Fancie, that they have such a forme of the Head."[34]

Bulwer supposes that "every Nation, whether Civill or Barbarian, hath not only Peculiar Customes and Rites, but also Peculiar Affectations of Forme or Shape of their Bodies, which will be Abundantly discovered by a world of strange Artifices and Pragmaticall endeavours Practised in this History."[35] Here, the customs and rites are not contrasted with the

[33] Bulwer, *Anthropometamorphosis*, 7.
[34] Bulwer, *Anthropometamorphosis*, 8–9.
[35] Bulwer, *Anthropometamorphosis*, 6.

physiognomic variety; rather, the latter are identified as a sort of corporeal expression of the former. The most truly barbarous "nation" for Bulwer, finally, is not a place, but rather an epoch—the present day—which, he thinks, has witnessed a proliferation of new physiognomical deformities unknown in antiquity. Thus he complains that the "Geometricall pates of our Square-headed and Platter-faced Gallants, is a new Contrivance: For, these Fashions of the Head were not knowne and discovered in the time of Galen, nor the violation of this Artifice practised."[36] Even in the chapter on pigmentation, Bulwer surveys various authors reporting on the barbarous practice of altering skin color with cosmetics. He cites Leo the African, to whom we have already been introduced, on the practices of sub-Saharan Africans, alongside Native Americans and Asians, condemning them all for using artificial discoloration to signify different states (mourning, and so on). But there is no thought of, or interest in, the variety of natural pigmentations, and no suggestion that any particular natural skin color is preferable to any other. The natural is preferable, whatever it may be, while artifice is always ugly. This, we may conclude, is an extreme formulation, perhaps the most extreme possible, of the degenerationist theory. It traces all degeneration to harmful human practices, and condemns human physiological diversity, howsoever it may be manifested, simply on the grounds that it has wayward cultural practices as its cause.

It is difficult to judge whether Bulwer's views on human physiological difference are racist, by our standards, or not. On the one hand, he generally takes European physical appearance, or a certain version of it, to be the standard by which all human beings are to be judged. On the other hand, he sees any deviation from this standard as simply the corporeal expression of cultural difference, and moreover sees European bodies as just as susceptible to such different and undesirable expressions as are bodies elsewhere. Bulwer is a cultural chauvinist, to be sure, with a particular understanding of the way physical bodies literally embody or reflect the cultures to which they belong. But his hatred of difference is hard to construe as a hatred of those who are racially different in any meaningful sense.

5.3. Hybridism and the Threat of Ape-Human Kinship

On the standard degenerationist account of diversity within a species, as we have seen so far, the principal causes of this diversity are, first, climate and other environmental factors and, second, degenerate morals and practices. Another purported cause of diversity is hybridism, or mating with

[36] Bulwer, *Anthropometamorphosis*, 14.

Figure 3. The "Orang-Outang," from Edward Tyson's *Orang-Outang, sive Homo Sylvestris* (1699). Illustration engraved by Michael van der Gucht from drawings by William Cowper.

members of other species. In the Aristotelian zoological tradition and in folk science more broadly, it was generally not presumed that there are fundamental natural barriers to interfertility between members of different species. Rather, the only real barriers in the animal kingdom came from the unfeasibility of copulation: it was not that bears and mice were essentially incapable of having hybrid offspring together, only that they were unlikely to get the chance to try. When a human being was one of the two species involved, there was a boundary imposed not only by feasibility, but also by morality: it is not necessarily that a human being cannot generate offspring with a nonhuman species, but only that a human being *should* not.

The animal species with which human beings have historically been thought to be most likely to interbreed are those that have been held (sometimes, it turns out, on good evolutionary grounds) to be particularly akin to humans. This includes not just apes, but also, significantly, bears. When François Bernier, to whom we will return at length in later chapters, identifies the Lapps in 1686 as a distinct race of men, and tells us that they "partake much of the bear," he is, consciously or not, drawing on a common perception of Scandinavian peoples according to which they are literally descended from ursine ancestors. This is not simply a projection from the outside, but is indeed grounded in official royal lineages of Scandinavian rulers, as described for example in Saxo Grammaticus's twelfth-century *Gesta Danorum*,[37] and echoed as late as the sixteenth century in works such as Olaus Magnus's *Historia de gentibus septentrionalibus* (*History of Northern Peoples*) of 1555.[38] It is moreover worth noting that when Bernier is writing, and even as late as Kant, northern people—and indigenous Sami or "Lapps" were not clearly distinguished from the Germanic Danes and Swedes—were seen as extremely exotic, and generally in a negative sense. It was not hard for Europeans from more temperate regions to convince themselves of the truth of the ancient idea about descent from bears, though what had been a point of pride and nobility in the *Gesta Danorum* becomes a point of radical difference and strangeness from the outside perspective of a Bernier or a Kant.

It is interesting here to observe that, while the prejudice against Scandinavians would entirely die out (in part this would have to do with a clearer distinction between Germanic Scandinavians and Sami, while within Scandinavia prejudices of the former against the latter would

[37]Saxo Grammaticus, *Saxonis Grammatici Historia Danica, recensuit et commentariis illustravit Petrus Erasmus Müller; opus morte Mülleri interruptum absolvit Mag. Joannes Mattias Velschow*, Copenhagen: sumtibus Librariae Gyldendalianae, 1839–58.

[38]Olaus Magnus, *Historia de gentibus septentrionalibus, earumque diversis statibus, conditionibus, moribus, ritibus, superstitionibus, disciplinis, exercitiis, regimine, victu, bellis, structuris, instrumentis, ac mineris metaliicis, & rebus mirabilibus . . .*, Rome, 1555.

surely persist), it nonetheless stands as a remarkable point of comparison with much more familiar prejudices about Africans. In the one case as in the other, people inhabiting regions to the north and to the south of supposedly temperate continental Europe are held to mate with nonhuman animals (Africans with apes, namely, as we'll see). At the boundaries of the inhabitable world, the thinking appears to have been, the boundaries of the human species grow more permeable. The further one strays from the temperate center, the less the ordinary patterns of nature hold, and this breakdown in nature's predictability, of which there is perhaps no greater example than the law of reproduction holding that "like begets like," is at once understood, implicitly or explicitly, as a moral failure.

As Nietzsche reminds us, there was in antiquity a preoccupation with the figure of the "Hyperborean,"[39] who was often conceived as at most quasi-human, but generally not in a negative light. The sort of disdain for the Sami expressed by Bernier appears to be a preoccupation of the early modern or late medieval world. By contrast the idea that the boundaries of the human species grow more permeable in Africa is already ubiquitous in antiquity. It is not just human beings, but indeed all species, that become more susceptible to blending with other species as one moves into the hotter climes. Indeed, this belief appears to be the very origin of the motto, "Out of Africa there is always something new."[40] In the *History of Animals*, referring to "Libya" as a synecdoche for "Africa," Aristotle explains that because water is scarce, animals in Libya are more likely to meet at a single waterhole and, finding each other in such an intimate setting, are more likely to mate with one another. Aristotle's usual argumentative rigor is not on display here, but his echoing of this belief, and his attempt to make sense of it, at least suggests how widespread and deep-seated it already was in antiquity. The belief survives more or less unaltered into the Renaissance and early modern periods. Thus in his 1607 work, the *Historie of Foure-Footed Beastes*, Edward Topsell expresses consistent astonishment at "the manifold and divers sorts of Beasts which are bred in Affricke," and cites as an example the Gorgon,

[39]Thus in *The Antichrist*, Nietzsche writes, "Let us look each other in the face. We are Hyperboreans—we know well enough how remote our place is. 'Neither by land nor by water will you find the road to the Hyperboreans': even Pindar, in his day, knew that much about us. Beyond the North, beyond the ice, beyond death—*our* life, *our* happiness." See Friedrich Nietzsche, *Der Antichrist*, in *Digital Critical Edition of the Complete Works and Letters*, ed. Paolo D'Iorio, based on the critical text by G. Colli and M. Montinari, Berlin: de Gruyter 1967–, §1.

[40]See Caius Plinius Secundus, *Naturalis historia*, Berlin: Apud Weidmannos, 1867, vol. 2, libri vii–xv, liber viii, 55. "[U]nde etiam vulgare Graeciae dictum semper aliquid novi Africam adferre."

a "strange Libyan beast," with unmistakably hybrid features.[41] In contradistinction to the classical authors however, most early modern naturalists are principally focused upon reports of human interfertility with other animal species. And these as well they often project onto Africa.

One long-standing trope of reports of ape-human hybridity in Africa holds that it results from the particular libidinousness of male apes, and their particular desire for human females. Thus Edward Topsell tells us that "[m]en that have low and flat nostrils are libidinous as apes that attempt women."[42] Thomas Herbert, for his part, in his 1638 account of a journey around the southern tip of Africa, describes the inhabitants of the Cape of Good Hope as progeny of Ham, whose "language is apishly sounded (with whom tis thought they mixe unnaturally)."[43] Now all of this might just be a historical curiosity if it had not played directly into some of the most important debates in early modern philosophy. In the *Essay Concerning Human Understanding* of 1690, John Locke—no minor travel writer or forgotten naturalist—tells us, "If history lie not, women have conceived by drills."[44] Indeed, in the early modern period the increased exposure of Europeans to nonhuman higher primates seems to have intensified debates in Europe about the lower boundary of humanity, and about possible gradations within humanity. It is a historical fact that these debates worked their way into philosophical discussions of such seemingly elevated matters as the problems of nominalism and real essences.

A significant moment in the history of speculation about human-ape kinship was the anatomical study that Edward Tyson performed on an infant chimpanzee in London in 1698, and the publication a year later of his findings.[45] Tyson discovers so much solid empirical evidence for ape-human kinship in his dissection of the primate, that in the end he is able to ground his prior commitment to the absolute distinctness of the human species only by appeal to considerations that fall altogether outside of the empirical realm:

> The Organs in Animal Bodies are only a regular Compages of Pipes and Vessels, for the Fluids to pass through, and are passive. What actuates them, are the Humours and Fluids: and Animal Life consists in their due and regular motion in this Organical Body. But those Nobler Faculties in

[41] Edward Topsell, *The Historie of Foure-Footed Beastes, Describing the True and Lively Figure of Every Beast*, London, 1607.

[42] Topsell, *Historie of Foure-Footed Beastes*, 166.

[43] Herbert, *Some Yeares Travels into Africa & Asia the Great*, 18.

[44] John Locke, *Essay Concerning Human Understanding*, book III, chap. 6, 23.

[45] Parts of the treatment of Edward Tyson here, and of his influence on Leibniz, were previously developed in, and partially overlap with, Justin E. H. Smith, *Divine Machines: Leibniz and the Sciences of Life*, Princeton: Princeton University Press, 2011, although this treatment occurred in the course of making a very different argument.

> the Mind of Man, must certainly have a higher Principle; and Matter organized could never produce them; for why else, where the Organ is the same, should not the Actions be the same too? and if all depended on the Organ, not only our Pygmie, but other Brutes likewise, would be too near akin to us. . . . In truth Man is part a Brute, part an Angel; and is that Link in the Creation, that joyns them both together.[46]

Tyson maintains that in its physical resemblance to humans, and not just in its learned behavior, the "the Orang-Outang imitates a Man."[47] Now Tyson is not explicitly interested in the subvarieties of human being, other than to insist emphatically that his orang-outang is not one. But the appearance of the orang-outang on the European scene would have a profound impact on debates about the possibility of divisions within the human species, and also the possibility of gradations: relative proximity or distance to our nonhuman simian "imitators." Eight years before Tyson's study, as already mentioned, Locke had claimed in his *Essay Concerning Human Understanding* of 1680, in the course of buttressing his species nominalism, that

> "[t]here are creatures in the world that have shapes like ours, but are hairy, and want language and reason. There are naturals amongst us that have perfectly our shape, but want reason, and some of them language too. There are creatures, as it is said . . . that, with language and reason, have hairy tails; others where the males have no beards, and others where the females have. If it be asked whether these be all men, or no, all of human species? it is plain, the question refers only to the nominal essence."[48]

Daniel Carey has compellingly argued that the approach to human diversity in evidence here, as well as in the global scientific project of Locke's contemporaries in the Royal Society such as Robert Boyle and Henry Oldenburg, stands in sharp contrast to traditional Christian anthropology, which presupposed the unity of the human species on a priori grounds, and then set about accounting for any apparent differences in terms of the degeneration brought about by original sin. Locke and his contemporaries by contrast "advocated inductive investigation of human nature through description of custom, without *a priori* assumptions."[49] Variety within the human species comes to be understood along the same lines as the broader project of natural history, and this same variety, which once posed a problem for those committed to the anthropological

[46]Tyson, *Orang-outang*, 54–55.

[47]Tyson, *Orang-outang*, preface, n.p.

[48]Locke, *Essay Concerning Human Understanding*, book III, chap. 6, 22.

[49]Daniel Carey, *Locke, Shaftesbury, and Hutcheson: Contesting Diversity in the Enlightenment and Beyond*, Cambridge: Cambridge University Press, 2005, 18.

and philosophical model of humanity as the *imago Dei*, now served on the contrary as a manifestation of the endless creativity of God.

Yet, even if diversity within the human species is celebrated rather than abhorred, there still remains the question, at least as urgent to taxonomists as to theologians, as to just what range of diversity deserves to be recognized as diversity *within* the species, and where, by contrast, we must start classifying *beyond* the species. Locke had the freedom not to worry overmuch about this question, in view of his philosophical commitment to nominalism. But not all philosophers were as ready to give up on the idea of a sharply delineated class of all and only human beings.

Six years after Tyson's study, Leibniz would write his lengthy reply to Locke's essay, and would adamantly deny that there are gradations of humanity, that "man" is just a name. In support of his firm commitment to the view that there is no third term between ape and man (*inter hominem et non hominem tertium non datur*), he would rely heavily on the work of Tyson, though the English surgeon would not be mentioned by name in Leibniz's work. Thus Leibniz mentions "the case of the *Orang-Outang*, an ape that is outwardly so similar to a man . . . , and whose anatomy has been published by a learned Physician."[50] Leibniz interprets Tyson's anatomical findings along Augustinian lines. The Carthaginian philosopher had similarly argued that there can be no gradations between human and nonhuman, that every creature either is or is not human, and that this is a question that ultimately cannot be determined with certainty by a consideration of the conformation of the body. "Whoever is anywhere born a man," Augustine writes, "that is, a rational mortal animal, no matter what unusual appearance . . . or how peculiar in some part they are human, descended from Adam."[51] As for Augustine, for Leibniz too "monstrosity" or severe morphological irregularity has nothing to do with the possession of the special marker of humanity, the rational soul. This remains the case even when the morphology is so distorted as to conceal from outside observers whether the creature in question is a human or not. Leibniz for his part writes, taking his departure from Tyson's description of the orang-outang, "it would not be known if it is of the human race, and if a rational soul lodges within."[52]

Tyson, who had earlier hypothesized that porpoises were degenerated cows, nonetheless refused to consider the possibility of an ancestral relationship between human beings and apes, even after having conducted

[50]Leibniz, *Nouveaux essais sur l'entendement humain*, in *Die philosophischen Schriften von G. W. Leibniz*, ed. C. I. Gerhardt, Berlin 1875–90, 5:217.

[51]Aurelius Augustine, *The City of God*, trans. Marcus Dodds, Edinburgh: T&T Clark, 1871, vol. 2, book XVI, chap. 8, "Whether certain monstrous races of men are derived from the stock of Adam or Noah's sons," 117.

[52]Leibniz, *Nouveaux essais*, 1:402.

his anatomical study of a chimpanzee in 1698. He concluded that, in all relevant respects, the differences between human and ape anatomy appear to be trivial. Yet over the course of the following century, humans and apes would come to be placed together in taxonomic schemes, even when these schemes did not explicitly postulate shared ancestry or real kinship relations, but simply left the question of the objective grounding of taxonomic proximity unanswered. Thus Linnaeus writes to Johann Georg Gmelin in 1747, "It does not please you that I've placed Man among the Anthropomorpha, but man learns to know himself. Let's not quibble over words. It will be the same to me whatever name we apply. But I seek from you and from the whole world a generic difference between man and simian that follows from the principles of Natural History. I absolutely know of none. If only someone might tell me a single one! If I would have called man a simian or vice versa, I would have brought together all the theologians against me."[53] By 1758, in the tenth edition of his *System of Nature*, Linnaeus would indeed come out and explicitly call human beings "primates" (previously, somewhat confusingly, *Homo* had been listed among the "anthropomorpha," as if a human being could itself be "human-shaped"). The Swedish taxonomist further divides the human species into a number of subvarieties. We do not need to consider these here, but what is worth emphasizing, for now, is the growing awareness in the modern period that if there is to be any basis for placing an absolute boundary between human beings and apes, this would be, to use Linnaeus's own distinction, a matter of "words" rather than of natural history.

Much of the purported evidence for ape-human kinship in the early modern period was false: humans and apes are not, for example, interfertile. But even the presumption that they could be interfertile was likely based on the perception of features resulting from a real biological and evolutionary proximity. As long as degeneration remained the primary or even exclusive model by which such proximity between different populations of humans or primates could be conceived, apes could not be conceptualized except as degraded or worsened versions of human beings. But the possibility of this sort of kinship opened up another possibility, in turn, of thinking of different human groups, by reference to their degree of degeneration, as standing in greater or lesser proximity to the ape. Arguably, in much the same way the Renaissance libertines maintained that the Native Americans are pure products of nature, and not created by God, the late seventeenth- and early eighteenth-century observation of a possible kinship between humans and apes, yet a kinship that is greater

[53]Linnaeus to Johann Georg Gmelin, February 25, 1747, letter L0783, in *The Linnaean Correspondence*, http://linnaeus.c18.net/Letters/.

in the case of non-Europeans, may be seen as a filtering through racism of a scientific insight that is, at bottom, perfectly true. What is clear, in any case, is that the history of thinking about human-ape kinship is deeply interwoven with the history of the rise of modern scientific racism. Linnaeus, as we've already seen, would explicitly anchor the investigation of human proximity to apes within the systematic study of nature by placing *Homo* among the *anthropomorpha*, which also included monkeys, apes and, curiously, sloths; and in turn by subdividing *Homo* into *Homo europaeus*, *Homo asiaticus*, *Homo africanus*, and *Homo americanus*.

There is a double movement here: both an insertion of "man" into a broader zoological order, and a simultaneous division of "man" into constituent subgroups. In an important sense, these two movements must occur together in order for the idea of race to play a meaningful role in the scientific study of humanity's place in nature. So long as human beings are kept out—as they had for the most part been prior to Linnaeus—of the project of zoological taxonomy, on the grounds that they are not beasts, it remains difficult to see in what sense the racial distinctions could pick out any real class of entities. Humanity had been an all-or-nothing affair, based on the inherence or the absence of a soul, which was not a biological matter, and it is for this reason that subdivisions of humanity within any given system of nature remained unthinkable. However, for the most part, after Linnaeus's insertion of "man" into nature, and the consequent double thinkability both of taxonomic links to other animals as well as of taxonomic divisions within the species, the nature of the latter sort of distinction would remain very difficult to specify with any precision or clarity, even though a great number of eighteenth-century thinkers were actively devoted to the project of delineating the subtypes of the human species.

Stephen Jay Gould once wrote that "[c]lassifications are theories about the basis of natural order, not dull catalogues compiled only to avoid chaos."[54] Yet as we have already been seeing, whether they are the one or the other depends quite a lot on the theoretical ambitions of the classifier. Linnaeus does not give a single, consistent account of the reality he takes his system of nature to be describing: sometimes he stresses the straightforwardly economic usefulness of the binomial system, while at other times he describes the classificatory project as one of "count[ing] so many species as were given at the beginning."[55] But of course Linnaeus is not just counting. He is also ordering, and grouping kinds under higher

[54]Stephen Jay Gould, *Wonderful Life: The Burgess Shale and the Nature of History*, New York: Norton, 1989, 98.

[55]See Lisbet Koerner, *Linnaeus: Nature and Nation*, Cambridge, MA: Harvard University Press, 1999, 44.

order taxa on the basis of supposed affinities. For Linnaeus, these affinities cannot be accounted for in terms of shared kinship, since he remains committed to the view that there is a fixed number of kinds and that these were "given at the beginning." But the groupings themselves are significant, not least the placing of human beings within the natural system alongside the other anthropomorpha, and leaving it to other naturalists to take the further step and to assert that what justifies being placed in this higher taxon is indeed shared ancestry. In this respect, Linnaeus's contribution may be seen as of a pair with Tyson's: both refuse to accept the conclusion for which their own scientific projects produce, willy-nilly, such powerful arguments, namely, the conclusion that human beings are not timelessly separated off from the apes that appear to resemble them, but rather are connected by bonds of kinship.

Both thinkers were unprepared to reject the ancient and deep-seated rationalization of the evident similarities between humans and apes, according to which these similarities result from an "imitation," a spurious and diabolical counterfeiting of the human being: an aping of humans. Nonetheless, again, Linnaeus's own work in particular, to the extent that it places human beings within a system of nature, effectively forces the question of what it is to occupy a place within a system, what the relationship between the different elements of the system must be understood to be. Increasingly, the only plausible explanation appeared to be that this relationship was, precisely, kinship, no longer understood in terms of a degeneration away from the original type of the human being, but rather of a natural separation of equally natural beings, humans and apes, in a distant, yet still perhaps reconstructible, past. The human-ape relationship would no longer be just a source of pejorative accounts of the nature of the difference between European and non-European cultures (whether this difference is conceptualized in overtly racial terms, as in Locke, or in terms of degenerate cultural practices, as in Bulwer). It would be the source of increasingly plausible accounts of the origins, and therefore also the nature, of the entirety of the human species.

5.4. Conclusion

The theory of degenerationism, whether arising from climate, custom, diet, or hybridism, held that every species consisted at the outset only in ideal types, and that as a result of climatic, geographical, and perhaps even cultural factors, some of the members of these species came to deviate from the original stock, usually in unfortunate ways. Thus Buffon describes North American fauna as stunted and dwarf-like, and supposes that this has been a result of the severe climate: in 1787, an indignant Thomas Jefferson sent him a counterexample in the form of a giant moose carcass.[56] In America, the French naturalist believed, nature is tired. For Jefferson, by marked contrast, American nature is brimming with energy and excellence, and this extends even, or perhaps especially, to the Native Americans. The American statesman writes for example that the Native Americans "astonish you with strokes of the most sublime oratory; such as prove their reason and sentiment strong, their imagination glowing and elevated."[57] There are, plainly, questions of national identity, and indeed of nation construction, in play. The French empire is retreating from North America by the middle of the eighteenth century, while the burgeoning pre-revolutionary American identity is looking to stake its claim on the same territory. At the same time, Jefferson's personal prosperity is directly linked to the maintenance of the slave system, and not surprisingly we also find him comparing Africans to Native Americans unfavorably. These considerations go well beyond the scope of our present concerns, but it will be enough for our purposes to bear in mind that the perception of degeneration always remained anchored to preperceptual commitments, and this was no less the case for accounts of botanical or zoological diversity than it was for human diversity.

We have seen that around the beginning of the eighteenth century there was a growing debate about the possibility of greater or lesser degrees of humanity: relatively traditional thinkers such as Leibniz excluded the possibility on the grounds that being a human or not is a question that has to do entirely with the inherence of a human soul, and this in turn has nothing at all to do with physiological features. But somewhat more radical thinkers such as Locke (and many were of course far more radical than he) began to argue that "human" is in fact a vague category with no clearly defined boundaries. To this extent, the possibility of both categorizing "man" within a broader system of relations that included other higher primates, as well as categorizing the traditional members of the species "man" according to their perceived affinities to the higher primates, became thinkable at exactly the same time and for the same reasons. Physical anthropology and primatology share the same origins, but so do physical anthropology and racism in the distinctly modern sense.

[56]See Lee Alan Dugatkin, *Mr. Jefferson and the Giant Moose: Natural History in Early America*, Chicago: University of Chicago Press, 2009.

[57]Thomas Jefferson, *Notes on the State of Virginia* (1787), in Eze, *Race and the Enlightenment*, 99.

Chapter 6

From Lineage to Biogeography

6.1. Race, Species, Breed

In its earliest usage, "race" was first and foremost a term of animal breeders. Today as well, *race* remains the common, neutral term in French, to mention just one language, for "breed." This latter sense has historical and conceptual precedence: there were races of dogs and horses before there were races of human beings. The determination of race, here, was based on lineage. Physical features were relevant in identifying what the race of a particular dog or horse was, but these were external *signs* of the animal's racial lineage, and not themselves the features that established membership in a given race. "Race" first appears in early modern Italian (*razza*), perhaps a borrowing from an Arabic word meaning "origin" or "principle," and perhaps stemming from an Old French term, *haraz* or *haras*, associated with the breeding of horses (in particular, a "haras" or stud farm). The etymology of the term remains uncertain until today, but most candidates involve some notion of breeding, ancestry, or intergenerational transmission of distinctive traits.[1] The term arises directly in these modern European vulgates, without an equivalent ancestor in Latin. When the term is extended from animal breeding to the description of human populations, it joins *genus* and *natio* as close Latin synonyms. But these have clear and unambiguous descendants in the vulgates, and the appearance of a new term, as if out of nowhere, suggests that there was a new preoccupation in the early modern period with the practices and techniques that "race," as applied to animal breeding, initially described.

Of course, the selective breeding of animals by humans is at least as old as the earliest domestication of wild species, and there were many treatises on breeding, which included the idea of distinct breeds within a species, already in classical antiquity. Plato, for example, drew on the

[1] Since in its modern form the word appears first in Italian, etymological reference works in that language tend to be most complete. The authoritative *Vocabolario etimologico della lingua italiana* of Ottorino Pianigiani (Florence: Ariani, 1926), also mentions a possible etymological connection to the Latin *radix* ("root"), to the Old German *reiza* ("line" or "ribbon"), and to the Slavic *raz* ("mark" or "strike").

example of poultry breeding in the course of laying out his eugenic program in the *Republic*. If such a source is still not antique enough, we may also for the sake of completeness mention Jacob, who in the book of Genesis (30:25–31:16) conspires to breed speckled goats by having his mono-hued stock breed in an environment surrounded by speckled décor. The scripture gets the mechanisms of hereditary transmission wrong, but it also indicates a knowledge of the practical possibility of breeding for particular traits, and so of creating distinct subpopulations within a given species. This knowledge, we may justly suppose, arises in parallel with pastoralism, long before the beginning of textual history. Beyond such practical knowledge, however, the Renaissance and early modern periods in Europe witnessed a revolution in the theory of breeding as a result of a combination of factors, including developments in fields seemingly as separate as art theory and applied mathematics.[2] We do not need to investigate these new developments here, but need only bear in mind that it is at least potentially significant that the term that was newly applied to the study of human diversity in the late seventeenth and early eighteenth centuries had already been a new term not so long before in the sphere of domestic breeding.[3]

By "species," early modern authors could not have intended the meaning commonly attached to this term today—namely, that each race is an isolated reproductive group—any more than Pierre Gassendi intended this when he spoke of "Negroes" as a "kind" (see chapter 2 above). One significant point of apparent disanalogy between races and species lies in the biology of reproduction: there is no barrier to reproduction between different breeds or races of a given animal kind, whereas species are often thought of as being defined by the very fact that they are reproductively isolated. Thus, for Aristotle, the humble mule had seemed poised to bring about a rupture in the very order of nature. As the fruit of a coupling between parents of two similar but nonidentical kinds, it appears to violate that most basic rule of generation, that like must always beget like. For Aristotle, the "production of a mule by a horse" is thus something that "happens contrary to nature."[4] But nature solves its own monstrous

[2]See for example Carlo Ruini, *Anatomia del cavallo, infermità, et suoi rimedii*, Venice: Appresso Fioravante Prati, 1618.

[3]See Maaike van der Lugt and Charles de Miramon (eds.), *L'hérédité entre Moyen Age et époque moderne*, Florence: Sismel—Edizioni del Galluzzo, 2008. See also Staffan Müller-Wille and Hans-Jörg Rheinberger, *A Cultural History of Heredity*, Chicago: University of Chicago Press, 2012.

[4]Aristotle, *Metaphysics* VII.8.

problem by nipping this new generational series in the bud: she makes the donkey-horse hybrid infertile.[5]

The term "mulatto" appears to have been coined in Spanish in the fifteenth century to describe, on analogy to the equine hybrid, the child of mixed European and African ancestry. But surely, one supposes today, its coiners must have understood that the analogy is forced, since in the human case it has always been perfectly clear that nature has no interest in bringing the series to an end, and declines to make the human "half-breed" infertile. Thus we see here a fundamental difference between what we today call "race" and what we call "species," a difference that has always posed an obstacle to essentialist thinking about racial difference. For races, on our understanding, unlike species, nature does nothing to ensure that like continues to beget like; if "miscegenation" is frowned upon, the understanding is that it *should not* happen, not that it *cannot*.

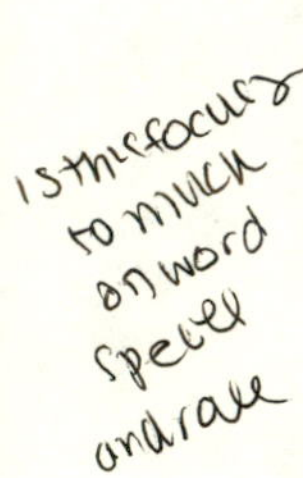

The textual evidence suggests that in the early modern period there was often not sufficient conceptual clarity in order to adequately reflect the points of disanalogy between race and species. For one thing, as we have seen, even if Aristotle has an explicit account of the unnaturalness of hybridity, interspecies fertility was generally presumed to be possible, *even if* it was unnatural, across a wide variety of species. In the Renaissance and early modern periods, the more interested a thinker was in questioning traditional claims about the order of nature, moreover, the more likely he was to emphasize the possibility of interspecies fertility. Insofar as reproductive isolation was often not part of the conception of a given animal kind, we tend to find a general unclarity in the way the concept of species is deployed, and this concept tends moreover to be used more or less interchangeably, beginning in the seventeenth century, with the concept of race. One could perfectly well overlook the obvious disanalogy between race and species—that there are no boundaries to interracial reproduction, while there are such boundaries to interspecies reproduction—since on the traditional understanding both sorts of reproduction were morally and thus "naturally" wrong, but not "biologically" (in our sense) impossible.

Some authors explicitly avowed that the difference between a racial boundary and a species boundary is difficult to determine, and may at best be provisional. Thus Leibniz writes in his *New Essays Concerning Human Understanding* of 1704, of the different "species" of great cat, "[M]any animals that have something of the cat in them, like the lion, the tiger, and the lynx, could have been of one single race and now could

[5] See Aristotle, *On the Generation of Animals*, 2.4 738b28. As Gareth Matthews points out, later in the GA (2.8 748a1) Aristotle intriguingly suggests however that the mule constitutes its own *eidos*. See Gareth B. Matthews, "Gender and Essence in Aristotle," in *Australasian Journal of Philosophy*, 64 (suppl., June 1986): 16–25, 23.

be new subdivisions of the ancient species of cats. Thus I come back continually to what I've said more than once, that our determinations of Physical species are provisional and proportional to our knowledge."[6] Leibniz supposes that variety of this sort among what we today would think of as different feline species is no different from the great diversity among *les races*, which is to say the breeds, of domestic dogs. A "race," in the sense in which Leibniz understands it here, is a *relatively* isolated reproductive community, but not one that is, as in Ernst Mayr's "biological species concept,"[7] isolated in view of its non-interfertility with other kinds. Thus a human group, such as "the Americans" or "the Ethiopians" (classifications Leibniz uses in a text to be discussed below), might be deemed a "race" akin to lynxes or poodles to the extent that its members *tend* to reproduce together, but this says nothing of the *possibility* of cross-fertility between lynxes and lions, poodles and bloodhounds, or Ethiopians and Europeans.

This, then, seems to have been the primary meaning of the term "race" when it was first deployed in the European vulgates in the mid-seventeenth century: a race is a lineage within a given kind of animals (or, later, humans) that tends, as a result of interbreeding across several generations, to share a number of the same traits. The primary application of this term, well into the eighteenth century, was to domestic animals that were selectively bred for certain traits (again, in today's French, the concept of race continues to have a currency in matters of animal husbandry equal to the one it has in matters of human diversity). "Species" occupied a distinct but nonetheless partially overlapping semantic field with "race," to the extent that the former term, as its etymology suggests, was concerned with the external aspect or appearance of a creature. Neither "species" nor "race" implied necessary reproductive isolation. The principal semantic difference between "species" and "race," where these in fact differed, had to do with the fact that the former focused on physical traits of creatures, while the latter also recalled to mind the lineage or generative series from which—to return to the deepest etymology of the term—these traits flow.

early use of race

6.2. François Bernier's Racial Geography

The French natural philosopher, diplomat, and disciple of Pierre Gassendi,[8] François Bernier, is often cited as the first author to use the term "race" to

[6] Leibniz, *Nouveaux essais*, 2:116.

[7] See Ernst Mayr, *Systematics and the Origin of Species, from the Viewpoint of a Zoologist*, Cambridge, MA: Harvard University Press, 1942.

[8] For Bernier's contribution to the development and propagation of Gassendian philosophy, see François Bernier, *Abregé de la philosophie de M. Gassendi*, Paris: Jacques Langlois,

designate different groups of humans with shared, distinguishing traits. Bernasconi and Lott, for example, write that "although many European travelers before Bernier noted the different physical characteristics of the various peoples they encountered, especially their skin color, he was the first to group these peoples specifically into 'races' on that basis. For this reason, Bernier's short article of 1684, entitled 'A New Division of the Earth,' is sometimes described as the first text in which the term 'race' is used in something like its modern sense to refer to discrete human groups."[9] This attribution becomes even more solidified in subsequent historical overviews, so that now it is often held to be not only a terminological innovation but indeed a conceptual one. Thus in the entry on "Race" in a standard philosophy reference work, we read that "[w]hile events in the Iberian peninsula may have provided the initial stirrings of proto-racial sentiments, the philosophical concept of race did not actually emerge in its present form until the 1684 publication of 'A New Division of the Earth' by Francois Bernier."[10] How we are supposed to get from the terminological innovation to the conceptual one is not at all clear, and certainly there is little evidence of any interest in developing "race" as a concept in Bernier's text. Of course, the boundary between terminological and conceptual innovations is not always perfectly perspicuous, but it is at least evident enough that in order for Bernier's work to be said to have involved any genuine conceptual innovation in the history of "race," it must be shown that the work is doing more than applying the term "race" in a novel way to describe a distinction that was previously described differently, but, more strongly, that a *novel distinction* is being made. In fact, in this case, neither the stronger nor the weaker sort of innovation may be said to occur.

Significantly, in a vivid illustration of the "Iberian curve" mentioned above, the Spanish-Inca philosopher Garcilaso de la Vega uses the term "race" several decades earlier than Bernier to describe the conceptualization of human diversity in South America. Even more significantly, he brings this up only to deny it, arguing that racial division is indeed an imposition of an artificial scheme by the Spanish colonizers, for the simple purpose of maintaining social order: "All these names [Creole, Negro, Mulatto, Cholo, Mestizo, Montagnard, Sacarunna, Quatraluo, Tresaluo] and many others I will leave aside were invented in my country in order

1671. Some authors, including Francisco Bethencourt (*Racisms*, chap. 15), have drawn attention to earlier works that argue for substantially the same biogeographical division of races as would later be found in Bernier. See in particular Alonso de Sandoval, *Un tratado sobre la esclavitud*, ed. Enriqueta Vila Vilar, Madrid: Alianza, 1987 [1627].

[9] Bernasconi and Lott, *Idea of Race*, 1.

[10] Michael James, "Race," in Edward N. Zalta (ed.), *The Stanford Encyclopedia of Philosophy*, Winter 2012 ed., http://plato.stanford.edu/archives/win2012/entries/race/.

to comprehend the mixture that had been made of races since the Spanish arrived there. By which we can well see that they introduced by their arrival a great number of things that were not there previously."[11] There are various ways of reading this passage. One is to suppose that for de la Vega there are preexisting races that enter into new combinations in colonial settings, necessitating the creation of new names. There are, he seems to imply, real races that lie behind the improvised categories that the new colonial setting generates. It is important to recall however that the Spanish-Inca author was born to a mother recognized as Inca royalty, and he was indeed bestowed with a royal title that remained valid during the bulk of his life in Spain. The preexisting "races" to which the author refers are thus likely not Europeans, Africans, and Native Americans, which get fractured in the colonial setting into finer-grained categories of Mestizo, Creole, and so on. Rather, the races are, one may suppose, lineages, including royal lineages, which the author conceives as relatively stable until the era of contact. Races are not biogeographical regularities, for de la Vega, but rather diachronic lines of descent. What is interesting in this passage, in any case, is, first of all, that the New World author, at the very beginning of the seventeenth century, already invokes the term "race" to describe divisions within the human species, and, moreover, that he displays a clear skepticism as concerns the project of rigorously taxonomizing subkinds of this species. He recognizes the artificiality and the merely local salience of supposedly naturalistic divisions. Thus long before Bernier gives us what is supposedly the first modern attempt to divide the human species into subtypes, we find an author, born in Peru, denying that such a project is worthwhile or even possible.

By his own lights, Bernier's work is novel in that it seeks to find a small number of basic human types throughout the world, rather than simply a long list of differences in facial and bodily traits from one canton or village to the next. "For although in the exterior form of the body," he explains, "and particularly of the face, men are almost all different from one another, depending upon the cantons of the earth in which they live, so that those who have traveled a great deal can often distinguish in this way, without mistake, each nation in particular, I have nevertheless remarked that there are above all four or five Species or Races of men, whose difference is so great that it could serve as a good foundation for a new division of the world."[12] Bernier describes his innovation in the article, previously mentioned, titled "A New Division of the Earth, by the

[11] Garcilaso de la Vega, *Commentarios reales de los Incas*, in P. Carmelo Saenz de Santa Maria (ed.), *Obras complétas*, Madrid, 1960 [1609, French translation by Jean Baudouin, 1633], book 9, chap. 31, 1277. I am grateful to Antoine Leveque for bringing this passage to my attention.

[12] Bernier, "Nouvelle division," 133.

Different Species or Races of Man," published in the *Journal des Sçavans* in 1684. "So far," he writes, "Geographers did not use any other criterion when mapping out the earth but that of the different countries or regions to be found on it. What I noticed in men in the course of my long and frequent travels gave me the idea to divide the Earth otherwise."[13] By his own account, Bernier's work ought to be seen as an innovation in that it does not simply divide populations up according to what might be called "national physiognomy," which is, again, a common practice going back at least to Hippocrates and still current in contemporaries of Bernier such as Malebranche. Instead, he tries to find a small number of more basic classifications.

The first "species" includes, roughly, those groups of people who today would be identified by linguistic criteria as speakers of either Indo-European or Semitic languages, which in turn, incidentally, map fairly closely onto those that are considered "white" for the purposes of census reporting in the United States. It includes, namely, "France and generally all of Europe, except a part of Russia. A small part of Africa, from the Kingdoms of Fez and Morocco, Algiers, Tunis, and Tripoli, to the Nile; as well as an important part of Asia, namely the Empire of the 'Grand Turk' with the three Arabias, all of Persia, the States of the Great Mogul . . . may be included in the first Species."[14] Bernier further identifies sub-Saharan Africa as inhabited by a different race or species: "I regard the whole African continent except the North African coast as previously described as the second Species."[15] Significantly, he does not see Native Americans, in contrast with Africans, as sufficiently different to warrant placing them in a distinct race: "As for Americans, in fact most of them have an olive complexion and their features differ from ours, but not enough to justify their belonging in a different species."[16]

The third "species" for Bernier are "Asians," which includes for him the inhabitants of "part of the kingdoms of Aracan and Siam, Sumatra and Borneo, the Philippines, Japan, China, Georgia and Muscovy, the

[13]Bernier, "Nouvelle division," 133. On Bernier and the history of race, see Pierre H. Boulle, "François Bernier and the Origins of the Modern Concept of Race," in Sue Peabody and Tyler Stovall (eds.), *The Color of Liberty: Histories of Race in France*, Durham, NC: Duke University Press, 2003, 11–27; Siep Stuurman, "François Bernier and the Invention of Racial Classification," *History Workshop Journal* 50 (Autumn 2000): 1–21. Parts of the summary of Bernier's racial theory given here appeared earlier in Justin E. H. Smith, "The Pre-Adamite Controversy and the Problem of Racial Difference in 17th-Century Natural Philosophy," in Marcelo Dascal and Victor Boantza (eds.), *Controversies within the Scientific Revolution*, Amsterdam: John Benjamins, 2011, 223–50.

[14]Bernier, "Nouvelle division," 134.

[15]Bernier, "Nouvelle division," 135.

[16]Bernier, "Nouvelle division," 136.

Usbek, Turkistan, Zaquetay, a small part of Muscovy."[17] Though Bernier acknowledges a wide variability of skin color among Asians, he distinguishes them from sub-Saharan Africans by noting that the latter are "essentially black," whereas darker skin color among non-Africans is a result of a contingent and reversible condition. Finally, the fourth species are the Sami, who, as mentioned above, are said to be "very ugly and partaking much of the bear."[18] He acknowledges only having "seen two of them at Dantzic; but, judging from the pictures I have seen, and the account which I have received of them from many persons who have been in the country, they are wretched animals."[19] The ranking of "Lapps" at the bottom of the scale of humanity would remain a commonplace throughout the eighteenth century, in Buffon, Maupertuis, Kant, and others.

Curiously, the last four pages of Bernier's very concise, seven-page article are devoted to extolling the beauties of the women of various nations of the world. He is convinced by his extensive travels that "beautiful and ugly [women] are everywhere to be found." He is particularly fond of women of the second species, and recalls seeing several of them in Ethiopia, "who were for sale, and I can say that one can see nothing more beautiful in the whole world; but they were extremely expensive, for they were being sold at three times the price of the others."[20] This is interesting material, but it is clearly more in the vein of boastful travel literature than racial science.

Is there anything new about Bernier's "New Division"? Bernier is, again, not the first person to use the French term "race" to describe a subdivision of the human species. Below, we will be discussing at length a text of Leibniz from the 1670s in which he uses the term in precisely this way, and with much more significant theoretical aims. In effect, Bernier seeks to tie race or species to physical traits, and to correlate the appearance of these traits with geographical factors. He is generally silent about the root causes behind the racial diversity of different regions, but at least opens up the possibility of a polygenetic account. Of course very many natural philosophers prior to Bernier had also argued for the role of climate and geography in the diversity of different human populations, but Bernier proposes that it is the different physical traits preponderant in different regions themselves that could determine the way in which different regions are bounded off from one another. From being seen as a superficial result of a group's migration into a given region, physical traits

[17] Bernier, "Nouvelle division," 136.
[18] Bernier, "Nouvelle division," 136.
[19] Bernier, "Nouvelle division," 136.
[20] Bernier, "Nouvelle division," 137.

of human populations are now, in Bernier's project, the very criterion by which regions are to be mapped out.

So neither does Bernier discover bioclimatic influence as the causal mechanism behind human diversity, nor is he the first to apply the term "race" to different groups that tend to display physical traits different from those of their neighbors. What Bernier is among the first to say is, first of all, that physical traits are rigidly correlated with races, *and* that races in turn are rigidly correlated with regions. This new division of the earth would indeed serve well as a template for modern racism: it does not ask about the causes of physical variation, or about their significance for understanding potentially deeper differences between human groups, differences, for example, of character or intelligence. Instead, it takes physical variation as a priori, as the biogeographical starting point for the determination of geopolitical order.

We have already observed that "race" denotes, in the first instance, a "flow," in the sense of the flow of traits or of kinship across generations. "Species" by contrast denotes the in*spec*table surface of a thing, its a*spec*t. It is the kind to which a thing belongs in virtue of its current, apparent conformation, rather than in virtue of its origins, of where it flows from. One way of understanding Bernier's innovation, somewhat paradoxically, is that he invented the modern concept of race by substituting the term "race" for what had previously been conceptualized under the notion of species. "Race" ceases to denote a potentially morphologically variable chain of descent, and comes to denote a fixed and bounded population of morphologically homogeneous individuals.

Just as no early modern pigeon breeder would have supposed that there were "essential" differences between different subpopulations of pigeon, indeed *had* to suppose that there could not be, in view of the descent of all pigeons from two original parents brought into existence at the creation, so too, anyone who chose to use the word "race" to describe human beings could not have been attempting to imply any more essential or biological, species-like differences between human subgroups than had been supposed prior to the term's extension. If anything, by reducing the difference between Africans and Europeans to one akin to that between breeds of domestic animals, an early modern author would have sooner been understood to be downplaying the differences, and to be committing himself to the changeability or reversibility of any given human population's physiological and temperamental traits through change of climate and diet or through interbreeding. But again, Bernier does not only adapt a term for animal breeding to human populations: if this were all he had done, his contribution to the history of racial science would have been no more significant than what we find in earlier authors, including for example Leibniz. What Bernier does, again, is to deploy a term more familiar from animal breeding, but to use it, now, in a way that distances it from diachronic considerations of lineage and "flow," considerations that had always been essential to breeders themselves, and instead focuses on the synchronic and spatial relationship between a group of people with a particular appearance, on the one hand, and the region that group inhabits on the other. This is sooner an impoverishment of a previously existing concept of race than it is the innovation of a new concept.

6.3. A Gassendian Natural Philosopher in the Court of the Grand Moghul

Few of the prominent early modern thinkers who wrote about human diversity traveled much at all, and far fewer ever set foot outside of Europe. For this reason, François Bernier's own life and work provide an interesting insight into the way significant exposure to non-Europeans cultures could influence a trained philosopher's reflections on the nature of human difference. Beginning in 1658 Bernier traveled extensively in Persia and in South Asia. He was fluent in Persian—at the time the lingua franca of the elites in northern India as well as of present-day Iran—and evidently translated the work of Descartes and Gassendi into this language. His most important task was as the personal physician to the Moghul emperor, Aurangzeb, and through his connection to the sovereign he was able to enter into contact with learned and luminary people, not just from among the Muslim elite, but also, on occasion, from the Brahmin class of people today called "Hindus." It is important to note here that the French traveler, upon arriving in India, had entered into a highly charged milieu of intense intellectual debate about the nature of religious faith and the possibility of religious syncretism. Bernier had begun his sojourn in India as part of the retinue of the great philosopher and heir apparent of the Moghul emperor, Dara Shikoh, who is best known as an advocate for the synthesis of Vedanta and Sufic Islam.[21] At around the same time Bernier was busy translating Descartes and Gassendi into Persian, Dara Shikoh was working on a translation of the *Upanishads* into the same language. Dara Shikoh's path to power was blocked by Aurangzeb, who had his more enlightened rival arrested and executed in 1659. Bernier crossed over into the service of Aurangzeb, and his prejudices about Vedanta and other schools of non-Muslim Indian thought seem to be inflected by those of his new host.

[21] See Annemarie Schimmel, *Im Reich der Grossmoguln: Geschichte, Kunst, Kultur*, Munich: C. H. Beck, 2000.

In an important treatise on the peoples, customs, politics, history, and resources of the region, translated into English and published as *The History of the Late Revolution of the Empire of the Great Mogol* in 1671,[22] the French traveler reflects on the work of an Indian poet who describes the "paradise" of "Indostan" at the confluence of four rivers: "the River Ganges on the one side, that of Indus on the other, the Chenau on a third, and the Gemma on the fourth." Bernier regrets that no scriptural or archaeological evidence can be found "that this was certainly the true Terrestrial Paradise, rather than that in Armenia."[23] In effect, Bernier here regrets that the authors of the Indian literary tradition do not explicitly lay claim to one of the most important places in scriptural geography, the region that supposedly hosted the Garden of Eden. If they had done so, it would have been plausible to grant it to them, given the actual topographical features of northern India, and given the relative closeness of this part of the world to the favored candidate in the South Caucasus. But without any explicit localization in ancient texts of Eden in South Asia, India lies beyond the pale of history. It does not identify itself with the texts that lie at the origin of everything that could in the seventeenth century be recognized as falling within the scope of history.

This extra-historicity of many of the earth's peoples, would in turn, in the high Enlightenment, serve as a very common rationale for what we have been calling, with Popkin, liberal racism. Many thinkers in the eighteenth century would argue that non-European peoples must be brought into the fold of European history in order to be able to ride the wave, so to speak, of historical progress. This is abundantly clear in Kant, who maintains that the lives of South Sea islanders, to the extent that they are spent outside of the fold of history, are literally not worth living.[24] And it continues to echo loudly in Marx, who maintains that the British installation of industrial looms in Bengal might have increased the misery of Bengali weavers for the time being, but at least it did them the service of moving them into a historical position from which social change

[22] See François Bernier, *The History of the Late Revolution of the Empire of the Great Mogol: Together with the Most Considerable Passages for 5 Years Following in That Empire: To Which Is Added a Letter to the Lord Colbert Touching the Extent of Indostan . . . and the Principal Cause of the Decay of the States of Asia*, London: Moses Pitt, 1671. See also Robert Bernasconi, "François Bernier and the Brahmans: Exposing an Obstacle to Cross-Cultural Conversation," *Journal for the Study of Religions and Ideologies* 7, 19 (2008): 107–17; Nicholas Dew, *Orientalism in Louis XIV's France*, Oxford: Oxford University Press, 2009.

[23] Bernier, *History of the Late Revolution*, 93.

[24] See in particular Immanuel Kant, "Recension des 2. Theils von J. G. Herders *Ideen zur Philosophie der Geschichte der Menschheit*," AA 8, 65.

could, for the first time, take place.[25] While Bernier may not have been the founder of scientific racism, as we saw above, it is nonetheless correct to identify him as an early exponent of liberal racism.

In his relations from the Moghul Empire, Bernier appears to perceive racial difference through the lens of political order, taking the ruling Moghuls as white, irrespective of their nationality, while identifying the "pagan" (i.e., non-Muslim) masses as uniformly brown: "[T]hose that are employ'd in publick Charges and Offices, and even those that are Listed in the Militia, be not all of the Race of the Mogols, but strangers, and Nations gather'd out of all Countreys, most of them Persians, some Arabians, and some Turks. For, to be esteem'd a Mogol, 'tis enough to be a stranger white of Face, and a Mahumetan; in distinction as well to the Indians, who are brown, and Pagans, as to the Christians of Europe, who are call'd Franguis."[26] India, unlike America and to some degree also sub-Saharan Africa, was of course part of the known world for Europeans since antiquity, and indeed was generally recognized as having been an important cradle and bearer of Eurasian history. Yet at the same time the fact that the majority of its population followed non-Abrahamic religious traditions put it in some respects on the same level as the New World. From an early modern European perspective, India was a place populated by strange heathens, yet ruled by close cousins. The Muslim elite was part of the same world as were the Christian Europeans; even if there was a long history of conflict between Christians and Muslims, it was still a shared history. The Hindus by contrast fell altogether outside of that history, and Bernier treated them, accordingly, as entirely foreign. In this, he is adopting a perspective that would be familiar and conventional for his Muslim hosts. Indeed, Bernier's contempt for Hindu beliefs and customs, mixed with occasional ethnological insights, echoes in surprising ways the tenth-century *Indica* of the Muslim geographer Al-Biruni.

In the course of describing Hindu customs of ablution, Bernier has occasion to make a revealing observation about the question of the universality and locality of religious traditions: "When I told them," he writes,

> that in cold Countries it would not be possible to observe that Law of theirs in Winter (which was a sign of its being a meer human invention) they gave this pleasant answer: That they pretended not their Law was uni-

[25] Karl Marx, "The British Rule in India," *New-York Daily Tribune*, June 25, 1853, in Karl Marx and Friedrich Engels, *Marx/Engels Collected Works*, Moscow: Progress, 1975–2005, 12:125. "English interference having placed the spinner in Lancashire and the weaver in Bengal, or sweeping away both Hindoo spinner and weaver, dissolved these small semi-barbarian, semi-civilized communities, by blowing up their economical basis, and thus produced the greatest, and to speak the truth, the only social revolution ever heard of in Asia."

[26] Bernier, *History of the Late Revolution*, 6.

> versal; that God had only made it for them, and it was therefore that they could not receive a Stranger into their Religion: that they thought not our Religion was therefore false, but that it might be it was good for us, and that God might have appointed several differing ways to go to Heaven; but they will not hear that our Religion should be the general Religion for the whole earth; and theirs a fable and pure device.[27]

Bernier is describing here what the nineteenth-century Orientalist Max Müller would later label "kathenotheism": the tendency he perceived in Hinduism to worship one god at a time, without promising singular devotion to the god that is the immediate object of attention. Such a practice of course opens up the possibility of what would appear from a European perspective as a variety of "relativism": worshipping one god, yet declining to deny that a person who worships another god is mistaken in doing so. Kathenotheism is of course a model of "toleration," which was contemporaneously being debated in Europe. It does not presume the mutual exclusivity of supreme beings; it does not suppose, as a philosopher such as Spinoza would rigorously argue, that there is room for only one God in this universe. The connection between the problem of toleration—that is, allowing that other groups of people are right to worship other gods—and the problem of human origins might not today be clear at first glance, but it would be hard for a thinker such as Bernier not to have seen it: from the seventeenth-century European point of view, it is but one small step from supposing that God had made possible several means of salvation, through several religions, to supposing that God separately created several groups of people, each with its own religion. After all, what it is to have the religion of the Judeo-Christian tradition just is, one might have supposed, to believe of oneself that one was descended from Adam and Eve. The only conceivable way for a person to fall outside of the scope of this tradition was to come from a separate line of descent. Interfaith respect—which Bernier disdains—thus practically requires a commitment to polygenesis.

At times, Bernier summarizes local beliefs with evident curiosity, while also compensating for this interest by an excess of dismissiveness. This is certainly the case with early European interest in Sanskrit learning in general, which, although this was a highly textual tradition with many accomplishments that could not but have been recognized by Europeans as philosophical and scientific, nonetheless was treated for the most part as a merely local, backward, and superstitious tradition. In Bernier's case, the knowledge of the Sanskrit-speaking Hindu pandits, to whom he was introduced by his Persian-speaking Muslim hosts, was a mere curiosity

[27] Bernier, *History of the Late Revolution*, 149–50.

in comparison to the respect-worthy philosophy of the Muslims, which ultimately traced its pedigree back to Aristotle, at the court of the Grand Moghul in Delhi. On this understanding, "philosophy" might be said to be a proper noun, the name of a series of comments on a set of questions extending back to a handful of figures in ancient Greece, and whatever does not belong to this series cannot be considered philosophy. For Bernier, this is a distinction that makes a difference, as philosophy in the narrow sense is for him the only tradition of thinking that upholds standards of reasoning and inference, and that expresses interest in the truth.

Like many of his contemporaries in the European Republic of Letters, Bernier's engagement with indigenous beliefs also serves a polemical purpose within the narrower European context. In his case, it is to defend a Gassendian, materialist theory of natural phenomena, by comparing European philosophers who disagree to Indian idolaters who fear eclipses with "childish credulity": "Those apprehending some malign and dangerous influence, and these believing that they were come to their last day, and that the Eclipse would shake the foundations of Nature, and overturn it, notwithstanding any thing that the Gassendi's, Robervals, and many other famous Philosophers could say or write against this perswasion."[28] Bernier recognizes that there are a "vast number and great variety of Fakires, Derviches, or Religious Heathens of the Indies," but has trouble describing the variety without lapsing into disdain. He describes the yogis, for example, who "make certain vows of Chastity, Poverty, and Obedience," yet who lead "so odd a life, that I doubt whether you can give credit to it." He goes on to give a remarkably vivid description of yogic practice, evidently one of the few such descriptions from a European traveler in the early modern period:

> You shall see many of them sit stark naked, or lie days and nights upon Ashes. . . . Their Arms were small and lean as of hectical persons, because they took not sufficient nourishment in that forced posture, and they could not let them down to take any thing without them, either meat or drink, because the Nerves were retired, and the Joints were filled and dried up: wherefore also they have young Novices, that serve them as Holy men with very great respect. There is no Megera in Hell so terrible to look on, as those Men are, all naked, with their black Skin, long Hair, dried Arms, and in the posture mention'd, and with crooked Nails.[29]

Bernier speculates that the yogis might be "a remainder . . . of that antient and infamous Sect of the Cynicks," yet in contrast with this school, Bernier concludes, their extreme practices are not supported by an interest

[28] Bernier, "A Letter Written to Mr. Chapelain," in *History of the Late Revolution*, 104.
[29] Bernier, "Letter Written to Mr. Samuel Chapelain," 131–32.

in, or even an awareness of, the ideal of reason. Instead, he finds "nothing in them, but brutality and ignorance." They seem to him "a kind of Trees, somewhat moving from one place to another, rather than rational Animals."[30]

One question that this passage, offensive and revealing as it is, leaves open to interpretation, much like similar passages from Burton, Evelyn, and Spinoza discussed in earlier chapters, is the precise significance of the adjective "black" in this context. Is it being used here as a "racial" term, or is it rather meant to describe the effects of a certain ill-advised or unfortunate way of life? Is it possible to distinguish fully between the two? We have seen, in the "New Division," that Bernier distinguishes the "essential blackness" of Africans from the contingent or circumstantial blackness of Asians, and the description of the practitioner of yoga seems to describe precisely such a case of circumstantial blackness. Yet it is not implausible that, as with the case of Spinoza's "black, scabby Brazilian," racial connotations might be intermingled with Bernier's own use of the term "black" to express disapprobation, and might also be read into the text by subsequent generations of readers. And indeed one thing is clear: for Spinoza and Bernier alike, blackness is bad. Whether this deployment of the term "black" has deeper roots in ideas about medicine, health, or environment rather than what we would think of as "race" does not change the fact that the negative valence of blackness in its pre-racial sense helped to facilitate and rationalize antiblack prejudice in the era of explicitly racial thinking.

Inquiring into Hindu belief is an enjoyable pastime for Bernier, and he is guided in this by his Muslim host:

> Do not wonder, if, though I know not the Hanscrit, the language of the learn'd, of which somewhat may be said hereafter, and which is perhaps the same with that of the old Brahmans) do notwithstanding tell you many things taken out of Books written in that Tongue. For you must know, that my Agah Danechmend-kan, partly out of my solicitation, partly out of his own curiosity, took into his service one of the famousest Pendets that was in all the Indies, and that formerly had had a Pension of Dara, the Eldest Son of King Chah-jehan, and that this Pendet, besides that he drew to our House all the most Learned Pendets, was for three years constantly of my conversation.[31]

Bernier reports that during his stay in Delhi he and his Muslim host would take breaks from discoursing on Harvey, Pecquet, Gassendi, and Descartes, which was his "chief employment for five or six years," in

[30] Bernier, "Letter Written to Mr. Samuel Chapelain," 134.
[31] Bernier, "Letter Written to Mr. Samuel Chapelain," 144.

order to take "refuge" in his pandit. The latter "was obliged to discourse and to relate unto us his stories, which he deliver'd seriously and without ever smiling. Tis true," Bernier concludes, "that at last we were so much disgusted with his tales and uncouth reasonings, that we scarce had patience to hear them."[32]

Remarkably, Bernier proceeds to give a fairly accurate summary of the contents of the Vedas, of the traditional caste system, and of the doctrines of the six principal schools of so-called orthodox (*āstika*) Indian philosophy.[33] But the rhetorical contrast between these "stories" on the one hand, and the learning with which he and his Muslim host are principally concerned, on the other, is unremitting. The removal of rationality from the explanation of what the yogis and the pandits are doing effectively unites Bernier's vision of "the Indian" with, for example, Locke's allusion to the "Indian" belief (again, it is not clear what sort of "Indian" the philosopher has in mind) that the world rests on the back of a tortoise: both see local or indigenous knowledge as useless in the pursuit of truth, since neither believes that this knowledge, having developed beyond the pale of philosophy, is underlain or driven by reason. It is a mere collection of stories; and the idea that stories might themselves point toward the truth, or be interesting adaptive responses to social or environmental exigencies, and in this sense might amount to "local rationalities," is still far from most European authors' minds. This idea, that there could be such a thing as local rationality, will of course be one of the core features of ethnography as a mature research program, but again, early modern philosophy, for the most part, though it is charged with questions of human nature and human variety, misses the opportunity to take a proper ethnographic and comparative turn.

6.4. Bernier and Leibniz

Bernier's role in the emergence of the modern conception of race has, as we have already seen, been somewhat overstated: his "new division of the earth" is principally motivated by questions of physical geography (as well as simple storytelling), and he is more or less silent as to the deep nature of the difference in appearance and character of the different groups of people he considers. The historical explanation of this is simple: he happens to have been read by Leibniz, who happens to have been read by

[32]Bernier, "Letter Written to Mr. Samuel Chapelain," 144.

[33]On the general outlines of early modern philosophy in India, indeed for a compelling argument that India experienced a distinctly "early modern" period in philosophy as in other domains of culture, see Jonardon Ganeri, *The Lost Age of Reason: Philosophy in Early Modern India*, Oxford: Oxford University Press, 2011.

Blumenbach, who in turn wrote *On the Natural Varieties of Mankind* in 1775, which served to codify the most basic outlines of racial classification, which would remain in place until the mid-twentieth century. Leibniz, for his part, though far from defending any theory of essential differences between different human races, did, as we will see presently, at least offer an explicit account of what a race is.

Even though Leibniz read the issue of the *Journal des Sçavans* in which Bernier's article appears, it does not seem to have made enough of an impression on him for him to retain Bernier's name. In a 1696 letter to the pioneering Swedish Slavist J. G. Sparwenfeld, republished in the 1718 *Otium hanoveranum*, a collection of various of Leibniz's letters and miscellanea, edited shortly after his death by Joachim Friedrich Feller, Leibniz writes, "If it is true that the Kalmuks as well as the Moguls and Tartars of China depend on the Grand Lama in matters of religion, it is possible that this says something about the relation among their languages and the origin of these peoples. It is simply that the size and constitution of their body is so different among them."[34] Here, Leibniz seems surprised that bodily morphology should be expected by anyone to correspond to linguistic kinship among different groups, since, true to form, Leibniz believes that language is far more important than "race" for determining ancestral relations. Nonetheless, he is aware of the interest of others in racial classification:

> I remember reading somewhere (but I cannot recall where) that a certain voyager divided human beings into certain tribes, races, or classes. He assigned a particular race to the Lapps and Samoyeds, a certain to the Chinese and neighboring peoples; another to the Negroes, still another to the Cafres or Hottentots. In America there is a marvelous difference between the Galibis or Caribbean, for example, who have a great deal of value and just as much spirit, and those of Paraguay, who seem to be children or youths all their lives. This does not prevent all human beings who inhabit the globe from being all of the same race, which has been altered by the different climates, as we see animals and plants changing their nature and becoming better or degenerating.[35]

The voyager in question is without a doubt Bernier. And this is not Leibniz's only mention of him; elsewhere in the *Otium hanoveranum*, we find another text, unfortunately undated, consisting in a nearly exact Latin

[34] G. W. Leibniz, *Sämtliche Schriften und Briefe*, Berlin: Berlin-Brandenburgische Akaademie der Wissenschaften, 1923–, I, xiii, 544–45.

[35] G. W. Leibniz, "Lettre de Mr. Leibniz à Mr. Sparvenfeld, (qui avoit soin de servir les Ministres étrangers qui étoient à Stockholm," in *Otium hanoveranum, sive Miscellanea, ex ore & schedis illustris viri, piae memoriae, Godofr. Guilielmi Leibnitii*, ed. Joachim Friedrich Feller, Leipzig: Impensis Joann. Christiani Martini, 1718, 37–38.

paraphrase of the contents of Bernier's "New Division."[36] In this text, he speaks of a "great voyager" (*magnus peregrinator*), who composed a work called the "Nova terrae divisio" that was published in the *Diario Eruditorum Parisino* (that is, the Paris *Journal des Sçavans*) of April 1684, at the instigation of the abbé Pierre Cureau de la Chambre. Leibniz goes on to present what might easily appear to be his own account of the proper method for dividing the different races of humans. But the text is interspersed with misspelled Italian expressions, motivating the conjecture that Leibniz composed it during his Italian voyage of 1689–90 for a colleague who could read no French, and amused himself in doing so by trying out his own elementary Italian skills. Evidently, Leibniz's reasons for writing it fall short of assertion of the truth of its claims.

Yet later, Blumenbach would incorporate elements from the *Otium hanoveranum* into his *On the Native Varieties of Mankind*, evidently failing to recognize that Leibniz is not stating his own views but rather summarizing those of Bernier. Blumenbach writes, "Leibniz established four orders of men of our continent [Leibnitius nostri continentis hominum quatuor ordines statuit]," citing, specifically, page 159 of the *Otium hanoveranum*, where we find Leibniz's summary of Bernier's "New Division."[37] Thus, the identification of Leibniz as a seminal thinker in the history of racial science appears to result from a simple mistaking of indirect statement for direct. For Leibniz, as we will soon see in detail, it is historical linguistics and not the study of morphological differences that will give us insight into the true lineages of the various human groups. As he concludes in the text just cited, adding his own view to the long summary of Bernier's system, "I should like for the regions [of the world] to be divided according to languages, and for this to be noted on maps."[38] This would be the consistent basis of Leibniz's mature thinking about

[36]G. W. Leibniz, "Pars altera, complectens meditationes, observationes et crises varias leibnitianas, gallico & latino sermone expressas," XLVIII, in *Otium hanoveranum*, 158–60. The text reads in part, "*Nova terrae divisio* per diversas hominum species vel generationes, quas magnus pergrinator misit Domino Abbati della Chambre, Parisino, extat in Diario Eruditorum Parisino A. 1684. d. 24. April. Res huc redit. Geographi terram per regiones (gubernationes potius) dividunt. Ego *quinque species vel generationes* observo. *Prima* continet homines Europae, parte Moscoviae excepta. . . . *Secunda* species est Africanorum: Grandia labra, naso scaffo o simo, paucissimis labia mediocria, nasusque aquilinus. . . . *Tertia* species implet partem regnorum Arakam, SIam, Sumatrae, Borneo, rum Philippinas vel Manillas, Japoniam, Pegu, Tunkinum. . . . His omnibus color albus, sed humeri largi, facies plana (viso piatto,) nasus exiguus (picciolo & schiacciato) oculi exigui & porcini, (lungi & incavati) & pauci in barba pili."

[37]Blumenbach, *De generis humani varietate nativa*, 297.

[38]Leibniz, *Otium hanoveranum*, 160. "Ego velim *regiones dividi per linguas & has notari in cartis*."

human diversity: it is rooted in language and culture, not, as Blumenbach would have wanted it, in physiology.

In the note on Bernier, however, Leibniz neglects to take up the project of a broad taxonomy of the most basic human kinds, and reverts to what we might call "national physiognomy": "In America there is a marvelous difference between the Galibis or Caribbean, for example, who have a great deal of value and just as much spirit, and those of Paraguay, who seem to be children or youths all their lives." This is an inauspicious introduction to Leibniz's theory of human diversity. He is guilty here of the sort of lazy, derivative transmission of hearsay that would later prove to be the most sophisticated reflection on diversity philosophers such as Hume and Kant would be able to muster. Yet Leibniz adds right away that national differences do not in any way reflect real divisions within the human species: "This does not prevent all human beings who inhabit the globe from being of the same race, which has been altered by the different climates, as we see animals and plants changing their nature and becoming better or degenerating."[39]

For Leibniz, it is neither blood nor soil, but speech, that reveals kinship. His new division of the earth would divide it up according to language families, not phenotypes. Beyond linguistic community, there is for Leibniz no meaningful classification of human populations short of their membership in the human race as a whole. As far as blood is concerned, everyone is related. This commitment was a fundamental principle of both Leibniz's political philosophy as well as his metaphysics. In the following chapter, we will explore in greater detail Leibniz's views on human diversity, as well as their place within his broader intellectual project.

6.5. Conclusion

Recall Bernier's account of his sessions with the Sanskrit pandit, of being "disgusted with his tales and uncouth reasonings." Bernier was not prepared to acknowledge the full working of reason in groups of people who follow intellectual and religious traditions that we might today call "non-Western," and this judgment moreover seems to have mapped for him, whether intentionally or no, onto a sort of racial distinction: the non-Muslim "natives," with their uncouth reasonings, are, as if by the same fact, also physically different. The idea by contrast of understanding or reason as the underlying, shared, though perhaps concealed essence of all human beings would be a commonplace of Enlightenment-era antiracism. Thus, in his 1789 memoirs, the freed slave Olaudah Equiano (also

[39]Leibniz, *Otium hanoveranum*, 37–38.

known as Gustavus Vassa), offers a straightforward echo of the Leibnizian view we have begun to adumbrate: "Are there not causes enough to which the apparent inferiority of an African may be ascribed, without limiting the goodness of God, and supposing he forbore to stamp understanding on certainly his own image, because 'carved in ebony'?"[40]

God does not forebear to stamp understanding on any human group, and to possess understanding is to be fully and equally human. The causes of phenotypic diversity, in turn, have simply to do with climatic and geographical forces. Equiano, for example, repeats the common view that the Portuguese in Africa become "perfect negroes" after a few generations. He believes, as some continue to believe today, that his own ethnic group, the "Eboans" or Igbo, are descended from Eastern Mediterranean Jews. He does not see the difference of skin color between West Africans and Jews as detrimental to this claim, commenting simply: "As to the difference of colour between the Eboan Africans and the modern Jews, I shall not presume to account for it."[41] He presumes however that whatever the cause of the difference, it is not for that reason the cause of any real divisions in nature, of anything that could serve as a basis for a new division of the earth. In this respect, in contrast with Bernier, Equiano offers not so much a biogeographical theory of race, as rather a biogeographical explaining away of race—at least where "race" is conceived, along Bernier's lines, as a natural way of dividing up different human populations rigidly fixed to different regions of the globe.

As we have seen, Bernier does not offer much by way of causal explanation for his new division. The key development in his work is rather that populations are taken out of relation to one another, and placed in a system of geographically based divisions. Bernier's proposal is thus a new division of the earth in a much deeper sense than he intended. It is indeed a new way of carving up the earth, a project that has occupied geographers since antiquity. But it is also a new way of carving up something, the human race, that until the modern period had been presumed to be possessed of a profound underlying unity.

[40] Olaudah Equiano, *The Interesting Narrative of the Life of Olaudah Equiano, or Gustavus Vassa, the African, Written by Himself*, ed. Werner Sollors, New York: Norton, 2001 [1789], 31.

[41] Equiano, *Interesting Narrative*, 30.

Chapter 7

Leibniz on Human Equality and Human Domination

7.1. Introduction

In the previous chapter the status of Bernier as the "inventor" of the modern race concept was called into question. Not only is the apparent terminological novelty not a conceptual novelty; it is not even a real terminological novelty. There are in fact plenty of texts prior to 1684 that use some local variant of "race" to describe human subgroups. As we have seen Garcilaso de la Vega provides a noteworthy example of such an early text, from 1609, yet there are many others still.

In a proposal to Louis XIV for the colonization of Madagascar, written in Latin in 1671, Leibniz mentions the "Ethiopians, Nigritians, Caribbeans, . . . and Hurons," describing all of these as different *genera*. The somewhat later appearance of the word "race" in Leibniz, far from signaling any conceptual innovation, stems simply from the fact that it was only toward the middle of his career (as a result of his sojourn in Paris from 1672 to 1676) that he began writing in French, and we can hardly expect him to be using any cognate of "race," a word that is not of Latin origin, when he is writing in Latin. "Race" had to wait to make its appearance not so much for the emergence of a distinctly modern conception of human difference, as simply for the vernacularization of European scholarly print culture. And once this started happening, Bernier was not the first on the scene.

In a note on a text by Wilkins most likely from around 1677, Leibniz identifies the terms *Race*, *genus*, and *Geschlecht* as synonyms, and defines them all as "generational series" (*series generationum*).[1] We will be returning to this text at length below. For now it is important only to establish the historical claim that Bernier is not the first author to use the term "race" to describe the diversity of the human species. These consider-

[1]Leibniz, *Sämtliche Schriften und Briefe*, VI, 4, 30–34. An earlier version of the argument of the present chapter was developed in Justin E. H. Smith, "'A Series of Generations': Leibniz on Race," *Annals of Science* 70, 3 (July 2013): 319–35.

ations might seem overly punctilious, but they are motivated by a concern to forestall any construction of a myth of origins for the modern race concept. No one invented race. The term is in part continuous with the long familiar usage of *genus* in Latin, as it refers to people, and it is already quite familiar in French as a term for describing breeds of domestic animals. When Bernier comes to use the term to describe human beings, what is most novel is not that he is dividing the human species into subtypes, but rather that he purports to be able to do so on an exhaustive, global scale.

Leibniz occupies a curious position in the history of racial thinking, and in general the views attributed to him, as we have seen, are based on a misguided understanding of his central concerns in matters of human diversity. Richard Popkin, to cite one influential scholar, understands Leibniz's interest in human diversity principally in relation to the problem of human origins. He maintains that Leibniz did not seem to be interested in the historical, chronological, and anthropological questions that provided the ammunition for the pre-Adamite theory (that is, as we have seen, one of the prevailing versions of polygenesis in the seventeenth century, according to which there were separately created human beings living before Adam's creation). Leibniz was, Popkin writes, "very much concerned to discuss other theologies outside of Christianity in terms of their ideological content, but not their differing claims about the facts of human history. . . . His efforts to unite the churches within Christendom, and then to unite them with Islam and Chinese religion, did not involve finding common historical ground, but rather common metaphysical and moral ground)."[2] However, pace Popkin, it is clear from a number of texts that Leibniz was in fact a committed monogenist, and that he believed that human diversity is a consequence of environmental influence over time. As we have already seen, in a text published in 1718, Leibniz notes that all human populations must belong to the same species even if they have undergone significant morphological transformations as a result of climate and habitat. These changes are always within a single species descended from the same first parents, and never result in speciation.[3] For Leibniz, there is only one origin for human beings, and subsequently the boundaries of the human species must remain rigidly fixed. In this respect, Leibniz might appear to be a moderate degenerationist, who believes that human diversity can be accounted for in terms of environmental pressures over time in different habitats. Yet for him, as we will see, the story of human diversity is somewhat more complicated.

[2] Richard H. Popkin, "Leibniz and Vico on the Pre-Adamite Theory," in Marcelo Dascal and Elhanan Yakira (eds.), *Leibniz and Adam*, Tel Aviv: University Publishing, 1993, 377–86, 381.

[3] Cited in Pagden, *Fall of Natural Man*, 138.

Leibniz's deepest metaphysical commitment—unity in diversity—might also be said to lie at the heart of his understanding of human diversity. Ordinarily, *unitas in diversitate* is taken as characterizing his theory of substance—the idea that a corporeal substance constituted out of sundry parts may nonetheless rightly be said to be *one* in virtue of the inherence in the bodily multiplicity of a single, unifying monad. My finger (or the infinitely many immaterial monads from which my finger results) is nonetheless truly and fully part of me insofar as its monads are under the domination of my soul. In another sense, we may say that Leibniz's universe is characterized by unity in diversity to the extent that every monad, though perfectly simple and unified, nonetheless represents within itself the entire order of coexistence of the infinitely many such simple substances that make up the world. In addition, there is a sort of unity among all of these perceiving substances to the extent that they are all simultaneously representing to themselves the same order, simply from different points of view and with varying degrees of clarity and confusion.

Can we, now, conclude from these considerations that Leibniz's account of human diversity is rooted in his metaphysical conception of *unitas in diversitate*? Peter Fenves has rightly noted that there is an apparent tension between Leibniz's "major philosophical speculation"[4] on the one hand, and, on the other, "the classificatory systems that depend on the idea of a generational series."[5] Fenves accordingly seeks to unravel the problem of why "Leibniz's name is often found in lists of those who were early proponents of a racial system of classification." One very interesting feature of Leibniz's notes on Wilkins—which we have already seen, and to which we will return in greater detail momentarily—as Fenves observes, is that it is here that we find one of the earliest occurrences of a variant of the term "monad." Leibniz writes, namely, that "the absolute lowest species" is a "species monadica," and identifies this as nothing

[4]See Peter Fenves, "Imagining an Inundation of Australians; or, Leibniz on the Principles of Grace and Race," in Andrew Valls (ed.), *Race and Racism in Modern Philosophy*, Ithaca, NY: Cornell University Press, 2005, 73–88. See also Fenves, "What 'Progresses' Has Race-Theory Made since the Time of Leibniz and Wolff?," in Sara Eigen and Mark Joseph Larrimore, *The German Invention of Race*, Albany: State University of New York Press, 2006, 11–22. In an earlier engagement with Fenves's work, I mischaracterized his account of Leibniz's views on a few important points. In particular, I mistakenly attributed to him the view that Leibniz supported Bernier's racial theory. In a subsequent article, I attempted to draw out some real points of difference between our interpretations, and to do so in due appreciation of Fenves's pathbreaking work on Leibniz on race (see Justin E. H. Smith, "'A Series of Generations': Leibniz on Race," *Annals of Science* 70, 3 [July 2013]: 319–35). The present chapter draws extensively on that article, and is similarly indebted to Fenves's important and careful work on Leibniz's theory of race.

[5]Fenves, "Imagining an Inundation of Australians," 73.

other than the individual being.[6] Fenves explains that the only "generative series in which a 'monadic species' can enter is that of its creator, who, for its part, transcends the series simply because it is, by definition, not an element of the class it creates."[7]

Yet we must be cautious about identifying such an intriguing early occurrence of the term "monadic" as having much in common with the theory of monads that would only gradually emerge over the next several decades.[8] It is not clear, in other words, that the mature metaphysical theory, according to which every monad is a world apart and has a causal link, by way of the creation, only to God,[9] is relevant to understanding Leibniz's account of race, species, or generational series.

Fenves rightly notes that, given the enormity of the corpus, with Leibniz one must always add "to my knowledge" to any claims that one makes. In this case, in particular, our account of Leibniz's views may be influenced by a deeper familiarity with "his immense series of historical, chronological, and genealogical studies."[10] Fenves is absolutely right here: what attention to these studies shows is that, first, Leibniz in fact is deeply committed to an understanding of "race" or "species" as the unfolding of a generational series; second, this account is not just different from, but indeed incompatible with, what we have been describing as "racial realism."

7.2. Chains: Leibniz on the *Series Generationum*

In an intriguing marginal note of the early 1670s on John Wilkins's work toward a universal characteristic, already mentioned above, Leibniz writes, "A *chain* is a series that is such that the beginning of the subsequent part is before the end of the prior part. *Race*, genus, Geschlecht.

[6] Leibniz, *Sämtliche Schriften und Briefe*, VI, 4, 31–32; Fenves, "Imagining an Inundation of Australians," 81.

[7] Fenves, "Imagining an Inundation of Australians," 81.

[8] While we cannot devote attention to the development of Leibniz's metaphysics, we can at least remark in passing that in broad outline Dan Garber's account, according to which the theory of monads begins to take shape only over the course of the 1690s, seems to us largely compelling. See Garber, *Leibniz: Body, Substance, Monad*, Oxford: Oxford University Press, 2010.

[9] For a paradigmatic statement of the worlds-apart thesis, see for example *Discourse on Metaphysics* § 14, in Leibniz, *Die philosophischen Schriften*, 4:440: "Each substance is a world apart, independent of everything outside of itself except God." This possibility of near-solipsism is in turn neutralized by other considerations in Leibniz's philosophy having to do with God's desire to create a world maximally rich in substances and in kinds of substance.

[10] Fenves, "Imagining an Inundation of Australians," 73.

A series of generations. *Genealogy* is the explication of this series."[11] The chain Leibniz envisions here must be distinguished straightaway from the more familiar "great chain of being." Leibniz's *catena* is not a *scala*, a scale or ladder of being, running from lower to higher: from worm to angel, as tradition had it. His chain is rather a chain in the sense that the elements of it are interlinked and are all, though numerically distinct, in essence the same. It is also a chain in the sense that one link begins before another ends, which is to say that offspring are born before their mother dies (and usually before either parent dies). This is a conception of shared kindhood that would continue to have tremendous purchase throughout the modern period. Thus in *Von den verschiedenen Racen der Menschen* (*On the Different Races of Men*), of 1775, Immanuel Kant notes, "In the animal kingdom, the natural classification into genera and species is based on the common law of reproduction, and the unity of the genera is nothing other than unity of the generative power, valid for a certain number of animals."[12]

In the eighteenth century, the idea of a hierarchical chain of being that flows from God's wisdom, and that places all natural kinds in a fixed and infinitely graded hierarchy in relation to one another, would become a poetic commonplace. Thus for example Alexander Pope writes, famously, of the "Vast chain of being! which from God began, / Natures aethereal, human, angel, man, / Beast, bird, fish, insect, what no eye can see, No glass can reach; from Infinite to thee." In his classic study of the idea of the chain of being throughout Western history, Arthur O. Lovejoy ascribes to Leibniz a central role, and sees Leibniz's version of the scale as, in effect, a prosaic expression of the very same image poetically expressed by Pope.[13]

In fact, pace Lovejoy, for Leibniz the *scala naturae* is not a chain at all in the strict sense, since its elements are not linked, let alone linked in such a way that the following link begins before the preceding link ends. A chain or generational series in turn, in contrast with the usual conception of the chain of being, is not a continuum. Its individual members, moreover, are all members of the same species, and thus neither is it characterized by the nominalism that thinkers such as Lovejoy have associated with the *scala naturae*. In the seriality of the chain, finally, in contrast

[11] G. W. Leibniz, "Definitiones notionum ex Wilkinsio" (1677–86[?]), in *Sämtliche Schriften und Briefe*, VI, 4, 30–34.

[12] Kant, *Von den verschiedenen Racen der Menschen*, in *Werke*, Bd. 9, Darmstadt, 1964 [1775]. For an extended analysis of the significance of this account of generation, see Justin E. H. Smith, "'The Unity of the Generative Power': Modern Taxonomy and the Problem of Animal Generation," *Perspectives on Science* 17, 1 (Spring 2009): 78–104.

[13] Arthur O. Lovejoy, *The Great Chain of Being: A Study of the History of an Idea*, New Brunswick, NJ: Transaction, 2009 [1936], chaps. 5–6.

to the plenitude of the scale, there is no conception of higher and lower, each link is essentially the same as each other link, even if there is a possibility of morphological variation (to which we will return below). There can be weaker links and stronger links, but the conception of linkage in operation here excludes the possibility that being weaker or stronger could imply being any less or more a member of the chain in question.

Lovejoy's sweeping intellectual history has been criticized on many fronts, but the difference between the technical notions of scale and chain has not been noted previously. It is true that we find both the phrase *catena rerum* as well as *scala rerum* or *scala naturae* deployed in various authors to describe what Lovejoy has in mind in the title of his study, but in Leibniz, at least, a scale and a chain are two very different things. The scale is hierarchical and nontemporal; the chain, temporal and nonhierarchical. It is in the temporal unfolding of the chain that morphological variation within a species occurs, even if this is always variation within the boundaries of a single, essence-sharing kind. At different points in a scale, by contrast, different individuals will be of essentially different kinds, but never through temporally unfolding changes from one generation to the next.

It should not be hard to see, already, why Leibniz's technical notion of chain, as used in the marginal note on Wilkins, precludes any possibility of racial inequality: it compels the view that, notwithstanding morphological variation in the course of the unfolding of the links, an identical essence is transmitted unaltered from link to link. Leibniz's infinite chains run horizontally, while vertically we find infinitely many gradations of higher and lower beings. As long as any two given beings are in the same horizontal chain, for Leibniz, they cannot be at different vertical positions in the scale. All human beings for Leibniz, finally, are links in the same horizontal chain, and therefore none occupy different positions higher or lower than the others in a vertical scale.

The conception of the members of a kind as bound together in a generational series or chain is by no means a *hapax legomenon* of the 1670s. Leibniz returns to it, in fact, throughout his career. In his debate with Georg Ernst Stahl of 1709–10, for example, Leibniz draws an explicit parallel between the successive, autonomous unfolding of the links of a chain, on the one hand, and, on the other, the preestablished harmony of soul and body:

> And just as all things among creatures, to the extent that they contain something of perfection, emanate from God and nevertheless derive in the natural course of things by the laws of nature from the preceding state to the succeeding one—God having most wisely constituted all things from the beginning so that they are born in turn from one another, bound in a

> sort of golden chain [*ut catena quadam aurea*]—so in the organic body of a living thing, in which the soul rules as a sort of singular governor, nothing happens outside the laws of the body.[14]

The golden chain Leibniz envisions here is, like the bodily machine of a corporeal substance,[15] an automaton, which is to say that all of its successive states (or, in this case, links) arise from its previous states; it contains within it the code or program, to speak somewhat anachronistically, of all of its future stages. Like the organic body of a corporeal substance, moreover, this chain is not a continuum, but rather consists in easily individuated, discrete entities. Indeed, there is nothing more paradigmatically discrete than the corporeal substance that constitutes a link in this chain.

This commitment to discreteness, again, contrasts sharply with the standard interpretation of Leibniz's iteration of the *scala naturae* doctrine, on which this *scala* is understood through the lens of Leibniz's commitment to the so-called principle of plenitude. According to this principle, between any two kinds there must exist a third kind, instantiated by at least one member. Naturally, again, such a view strongly suggests a variety of nominalism, on which every individual occupies its own kind, and between it and any other given individual there will be a third individual that possesses some features of each. But nor would it be right to call Leibniz a nominalist about biological species: in the *New Essays*, he argues clearly that the reality or irreality of species is something that can be determined only relative to different ontological domains. The analysis of the species of mathematical objects is one thing, for example, of inorganic physical objects another. "There is some ambiguity," Leibniz writes, "in the term *species* or *being of a different species*. . . . We can understand species either mathematically or physically. In mathematical rigor the slightest difference that brings it about that two things are not entirely similar, brings it about that they are *of different species*."[16]

In physical entities, in turn, there is rather more room for the definition of species by convention: "Men laying down the physical species," he notes, "do not adopt such a rigorous stance, and it depends on them to say whether a mass that they can themselves make to return to its original form remains in their view the same *species*." Here, there is clear justification for calling Leibniz a nominalist. However, at this point he proceeds to identify a third ontological domain, besides mathematical and physical entities, in which the definition of species involves entirely different con-

[14]G. W. Leibniz in G. E. Stahl, *Georgii Ernestii Stahlii Negotium otiosum: Seu Skiamachia adversus positiones aliquas fundamentales* Theoriae varae medicae, Halle, 1720, 4.

[15]For an extensive analysis of Leibniz's model of organic body, and how this is distinguished from corporeal substance, see Smith, *Divine Machines*, chap. 3.

[16]Leibniz, *Die philosophischen Schriften*, 5:288.

siderations, namely, the domain of plants and animals, where "we define the species through generation, so that this similar being, which comes or could come from the same origin or seed, will be of the same species."[17]

The fact that Leibniz is ordinarily taken to be a nominalist about kinds would seem to render at least problematic the claim that unity in diversity may be extended from the metaphysics to the anthropology—that, in other words, there can be any unity in a kind at all, even in a kind as obviously important as the human kind, in addition to the uncontroversial unity that characterizes individual substances. Yet if we can suspend our usual understanding of Leibniz as a nominalist, we will easily grasp why, for him, the metaphysical principle of unity in diversity might have seemed fitting for a characterization not only of the order of substances in general, but also of humanity.

Unlike other monads, human souls consciously act within the "kingdom of ends," that is, toward goals they recognize to be laid down in accordance with divine wisdom. The ability to do this—both the recognition and the action—is in the end what makes them human, and what differentiates them from other sorts of corporeal substance. Now the means of pursuing these ends will be very different in different cases, but the crucial thing, for Leibniz, is that no amount of apparent divergence, in behavior or in bodily conformation, can ever justify the judgment that the one or the other human individual is not a rational being operating within the kingdom of ends. For Leibniz, humanity is an all-or-nothing affair, and it is not dependent on any outward or visible signs. Humanity is thus unified through reason behind or underneath all the diverse ways it appears in different individuals and populations, just as a substance is unified through its entelechy behind or underneath all the articulations of its corporeal substance.

Leibniz supposes, in fact, that great morphological variation does not necessarily exclude shared kind membership. Thus for example, as already mentioned, in the *New Essays* he suggests that the lion, the tiger, and the lynx, could all be "of the same race," and could in fact be "new subdivisions of the ancient species of cats."[18] Leibniz concludes from this that we may end up having to place seemingly different kinds of animals within the same species, and for him even a great difference between such apparent species, morphologically and behaviorally, poses no tremendous problem, since a kind for Leibniz just is a generational series, and not a shared set of observable traits. Related to this, Leibniz supposes, in his response to Locke's view that there is no definitive answer to the question whether some ape-man hybrid shares in the human essence,

[17] Leibniz, *Die philosophischen Schriften*, 5:288.
[18] Leibniz, *Die philosophischen Schriften*, 5:296.

that being human is, again, a yes-or-no question, and that a creature in fact need show no outward signs of humanity at all in order to be human. What makes a creature human is its possession of reason or understanding, and these might very well lie concealed behind severe deformities for the entire life of a person. In this, to be sure, Leibniz differs from the defenders of the mid-twentieth-century consensus on race, who would in a certain sense be more Lockean in their wariness about the idea of a fixed human essence; but Leibniz anticipates them as well, in turn, to the extent that he, too, is wary of any attempt to read essence directly off of morphology or behavior.

Leibniz's view of the shared nature of each species, including the shared reason of humans, develops well before his confrontation with Locke's nominalism in the 1704 *New Essays*. In texts dating from the early 1680s, Leibniz develops a notion of the special office or *officium* of each species, which is the characteristic activity of it, the "for which" of the perpetual cycle of reproduction of the animal machine. Thus, as he intriguingly puts it, a spider is a web-weaving machine of quasi-perpetual motion,[19] while a squirrel is a jumping machine of quasi-perpetual motion.[20] This sort of reasoning may seem naïve, indeed Aesopian, but the point is that if we are looking for species essences, we will do far better to look in a given species' characteristic *activity* than in its characteristic appearance. In an intriguing letter of 1701, to A. C. Gackenholtz, Leibniz goes further than he had in the 1680s, and adds that contemplation pertains to human beings in the way that the special offices of other animals pertain to them: thus, a human being is, as Leibniz explicitly puts it, a machine of reason in just the same way a spider is a machine of web-weaving.[21] If the fulfillment of this function is not always evident in a given individual human being, or if one individual fulfills it less well than another, this in no way means for Leibniz that the one is less human than the other. Again, humanity is a threshold affair, determined exclusively by membership in the generational series that extends back to Adam and Eve.

Reason need not be clearly manifested in a given human being, and indeed may be undetectable over the entire course of that person's life as a result of disabilities or severe morphological differences that conceal it. But for Leibniz it is better to err on the side of caution, for priests to

[19]The animal is a *perpetual* machine to the extent that it requires, for the duration of its life, no outside machinist to see to its operation. It is only *quasi*-perpetual, however, to the extent that it does require outside fuel, in the form of nutriment it seeks out itself, something a *perpetuum mobile* in the strict sense would not need.

[20]See in particular the "Corpus hominis et uniuscujusque animalis est machina quaedam," Leibniz-Handschriften III, 1, 2, 1–2, in English translation in Smith, *Divine Machines*, appendix 3.

[21]Leibniz, *Opera Omnia*, II, 2, 171.

offer hypothetical baptisms for example, just in case a given creature happens to be human. Leibniz is committed to an all-or-nothing view of the divisions between species, and he seems more inclined to allow that the orang-outang falls on our side of the division than on the side of the animals. He appears to prefer to risk extending the boundaries of humanity too far, than to allow for the other possibility, where human beings go unrecognized as a result of an excessively conservative boundary. Better to accidentally call some apes human, than to dismiss some humans as apes.

Leibniz believes, against Locke, that species are set down once and for all by God, and that there can be no overlap between them. For Leibniz, it may ultimately be impossible to know whether an orang-outang possesses a rational soul or not. If it does, then it is wholly a human being, deserving of baptism and suitable for salvation, no matter how different it looks from a "normal" human being. There is a fact of the matter as to whether an ape is a man, or not. For Leibniz, this fact of the matter has to do with what Locke would call the "real essence" and what Leibniz himself sometimes calls the "interior nature." Again, there is for the author of the *New Essays* a clear criterion for distinguishing men from apes, even if it may be difficult to discern in practice on which side of the dividing line a given creature falls: that criterion is rationality, which, even if it is sometimes difficult to discern, ensures that all human beings are strictly speaking equal—equality does not flow from anatomical or behavioral features at all. In view of this deep commitment to the universal equality of different human groups, Leibniz will come to believe, as we will see further on, that human diversity is something to be studied as a cultural, rather than a natural phenomenon, even if cultural differences are best investigated on the model of natural history: languages, namely, as we'll soon see, should be collected and studied like so many flowers in need of classification.

Again, for Leibniz, anatomical, behavioral, or cultural differences cannot possibly be interpreted as signs of differing degrees of rationality, a faculty Leibniz in turn takes as the sole requisite for full membership in the human species. Thus, for Leibniz, it is literally impossible that there could be such a thing as inequality among human beings. To be human is to be in possession of a rational soul, and there is no sense in which a change in the conformation of the body could result in a diminished, reduced, or qualified possession. Since membership in the human species is, for Leibniz, grounded in the generational series, no amount of morphological variation can ever bring it about that one group of creatures descended from the first parents is any less a member of the human species than any other group.

As we have seen, Leibniz's is a strictly domain-specific nominalism. In the case of living beings, nominalism is out of the question, since he is committed to the unity of the links in a generational series, a unity ensured by the shared kindhood of all the members. This means that the plenitude of the *scala naturae* is a plenitude of chains, and not a plenitude of individuals. There is a chain between any two given chains on the scale, but the links in a single chain are not, relative to one another, characterized by the principle of plenitude. We might picture an infinity of chains laid out horizontally and stacked vertically one on top of another. The individual, horizontal chains consist in discrete and interlocked members, but the chains are not interlocked with the other chains higher or lower than they are in the scale. To be a link in a chain, of the sort described by Leibniz in his dispute with Stahl or in his notes on Wilkins, both cited above, is to be essentially identical with, and thus equal to, every other link in the chain. Individual human beings constitute such links in the chain that is the human species: they are the unfolding of this generational series.

7.3. Chains, Continued: Leibniz on Slavery

A commitment to shared essence flowing from shared membership in a chain, of the sort we see in Leibniz, does not however necessarily imply a commitment to political egalitarianism.[22] In the 1671 text with which we began this chapter, "A Method for Instituting a New, Invincible Militia That Can Subjugate the Entire Earth, Easily Seize Control over Egypt, or Establish American Colonies," written as an addendum to his better known *Consilium Aegyptiacum*, Leibniz sketches out a plan for training a new army of warrior slaves:

> A certain island of Africa, such as Madagascar, shall be selected, and all the inhabitants shall be ordered to leave. Visitors from elsewhere shall be turned away, or in any event it will be decreed that they only be permitted to stay in the harbor for the purpose of obtaining water. To this island slaves captured from all over the barbarian world will be brought, and from all of the wild coastal regions of Africa, Arabia, New Guinea, etc. To this end Ethiopians, Nigritians, Angolans, Caribbeans, Canadians, and Hurons fit the bill, without discrimination. What a lovely bunch of semi-

[22] For a compelling example of how one might in fact *derive* the right to seize land and keep slaves from a commitment to human equality, see Carlo Ginzburg, "Latitude, Slaves, and the Bible: An Experiment in Microhistory," in Angela N. H. Creager, Elizabeth Lunbeck, and M. Norton Wise (eds.), *Science without Laws*, Durham, NC: Duke University Press, 2007, 243–63.

> beasts! But so that this mass of men may be shaped in any way desired, it is useful only to take boys up to around the age of twelve.[23]

Leibniz proposes to segregate these prisoners according to language, which for him is the same as segregation by race or genus. In this way, unable to communicate with any warriors beyond their own small squadron, the warriors will be unable to plan an insurrection. "In every race [*genere*]," Leibniz writes,

> whoever is most trained in his squadron, which is to say among those who speak his language, shall challenge those who are the best trained in the other squadrons. The people [*gens*] that wins that year shall be the leaders. They will be able to strike terrible blows with their very powerful curved swords, to hit targets with their slings, and to rip things apart with their lances. They are to be trained to run races at such a speed as will be equal to that of horses. Which will come about first by pursuing them until they are able to touch the mane or the tail, and then freely [i.e., without horses]. . . . They shall learn to jump after the manner of the Tenerifeans, first jumping with the help of a lance . . . as far as human strength is able to reach, and afterward without these.[24]

This is an abhorrent proposal, to be sure (and there is no evidence that it is a "modest" proposal in a Swiftian sense, either), but it is crucial to emphasize that the basis for the slaves' mistreatment is sectarian and not racial: the fact that they are not baptized Christians, for the young Leibniz, is the key motivation for his withholding from them any moral concern. The basis for their segregation, in turn, is not that, say, Caribbeans form any natural or generational unity, but only that they form a linguistic community.

Leibniz goes on to describe the tremendous feats these warriors will perform with their lances:

> In the beginning they will alight from a higher place by the means of their lance touching the ground below; then they will leap horizontally on a level plane, and finally from below they will leap to the top. The will learn how to climb up smooth surfaces [*per lubrica klettern*]. . . . They shall become used to climbing however high their lance may be just by means of fixing their lances beneath them. They will learn moreover to carry the greatest and strongest lances, like Achilles, and like other ancients. Indeed, they shall learn to project them with great impetus toward a designated target, as well as of bringing one lance together with another if the one does not suffice for climbing. By means of this art they will easily conquer the

[23]Leibniz, *Sämtliche Schriften und Briefe*, IV, 1, 408.
[24]Leibniz, *Sämtliche Schriften und Briefe*, IV, 1, 408–9.

> mightiest European fortifications. They will be able to walk on their lances, as on stilts [*wie auff stelzen*].[25]

Where, now, did Leibniz learn so much about Canarian martial arts? In an anonymous travel report in Thomas Sprat's 1667 *History of the Royal Society*, we find the following description of a native Tenerifean "Guanchio" (today called Guanches), a description that Leibniz would reproduce four years later, sometimes nearly word-for-word:

> [An English traveler] himself hath seen [the Guanches] leap from Rock to Rock, from a very prodigious height, till they came to the bottom, sometimes making ten fathom deep at one leap. The manner is thus: First they *Tertiate* their Lance (which is about the bigness of a half Pike) that is, they poise it in their hand, then they aim the point of it at any piece of a Rock, upon which they intend to light (sometimes not half a foot broad.) At their going off they clap their feet close to the Lance and so carry their bodies in the Air. The point of the Lance first comes to the place, which breaks the force of their fall; then they slide gently down by the Stagge, and pitch with their feet upon the very place they first designed, and from Rock to Rock till they come to the bottome. Their Novices sometimes break their necks in learning.[26]

The account given in Sprat's volume, in addition to following the general cosmographical structure of countless other natural-historical texts of the era, moving from a description of the geographical features of the place, through the fauna and flora, and finally to the human culture of the region, also makes explicit the connection between the study of geography on the one hand, down to the deepest level which treats of the formation of terrestrial land masses, and on the other hand practical, political considerations about the worthwhileness of colonization, and the proper way to settle and inhabit a given region:

> [T]he whole Island being a ground mightily impregnated with Brimstone, did in former times take fire, and blow up all or near upon all at the same time, and . . . many Mountains of huge Stones calcin'd and burnt, which appear every where about the Island, especially in the Southwest parts of it, were raid'd and heav'd up out of the Bowels of the Earth, at the time of that general conflagration; and that the greatest quantity of this Sulphur lying about the Center of the Island, raised up the Pico to that height at which it is now seen. . . . And there remain to this time the very Tracts of the Rivers of Brimstone as they ran over all this quarter of the Island, which hath

[25] Leibniz, *Sämtliche Schriften und Briefe*, IV, 1, 409.

[26] "A Relation of the Pico Teneriffe," in Thomas Sprat, *History of the Royal Society*, London, 1667, 212–13.

> so wasted the ground beyond recovery, that nothing can be made to grow there but Broom. . . . These calcin'd Rocks resembling some of them Iron-Ore, some silver, and other Copper. Particularly at a certain place in these South-west parts called the *Azuleios*, being very high Mountains, where never any English man but himself (that ever he heard of) was.[27]

The theoretical considerations end here, and give way to a more mundane passion for treasure-hunting. Our traveler goes on to report that "a *Portuguez* told him, who had been in the *West-Indies*, that his opinion was, there were as good Mines of Gold and Silver there as the best in the *Indies*."[28] As David Abulafia notes, the Canary Islands, explored already by 1330 and fully conquered by the end of the fifteenth century, amounted to a sort of prelude to the big event: the domination of the New World, which would begin a century later and which would still not be entirely finished by the time Leibniz proposes his militia for, among other things, setting up an American colony.[29] When Leibniz proposes to France "an island off of Africa," such as Madagascar, this is evidently because the Canaries have already been taken by Spain.[30] But it is the Canaries, and the expansion of the Spanish Empire out from there, that serves as Leibniz's model and inspiration. And of course, eventually Madagascar would be colonized by the French: with long delay and after centuries of sovereign Malagasy rule, France would, in the 1885 "scramble for Africa," finally realize the advice that Leibniz had hoped to give to Louis XIV.

In the margin of the "Method," Leibniz jots down a few very revealing words: "Ianizari, Circassi, Pici Tenariffae descriptio in societatis Regiae Historia [Janissaries, Circassians, Description of the Peak of Tenerife in the *History of the Royal Society*]."[31] The janissaries are of course the prestigious infantry units charged with protecting the Ottoman sultan. The traditional practice was to kidnap them as children from their families in provincial regions of the Ottoman Empire, thereby minimizing the possibility of palace intrigues. Having no connections and no past in the palace, the janissaries were also without rivalries and factional loyalties. It is significant that they were on Leibniz's mind as a model for his own warrior slaves. One important thing to note is that, though technically slaves, and forced into their career against their will, the janissaries were

[27] "A Relation of the Pico Teneriffe," 206.

[28] "A Relation of the Pico Teneriffe," 206.

[29] See David Abulafia, *The Discovery of Mankind: Atlantic Encounters in the Age of Columbus*, New Haven: Yale University Press, 2008.

[30] One possible source of Leibniz's interest in Madagascar is Hieronymus Megiser, *Wahrhafftige gründliche une aussführliche so wol Historische alss Chorographische Beschreibung der überauss reichen mechtigen und weitberhumbten Insul Madagascar*, Altenburg, 1609.

[31] Leibniz, *Sämtliche Schriften und Briefe*, IV, 1, 408.

also an elite, and in terms of social position had nothing in common with the slaves exploited in the transatlantic trade. The Circassians, for their part, were the people of the Caucasus region traditionally prized for their physical beauty as slaves in the Ottoman Empire. From a comment in his *Directiones ad rem medicam pertinentes* (*Directions Pertaining to the Institution of Medicine*), also written in 1671, we know for certain that at that time Leibniz was reading the Venetian traveler Lazaro Soranzo's book, *L'Ottomanno*, and moreover the contemporaneous *Consilium Aegyptiacum* involves an extensive analysis of the traditional practice of kidnapping future janissaries.[32] From all these considerations, it is certain that Leibniz's interest in the institutions of Ottoman government is the most important model for his conception of slavery in the 1671 "Method."

Leibniz, as we have seen, defends a conception of shared kindhood as rooted in shared membership in a generational series. Now a wide variety of interbreeding populations might qualify, for Leibniz, as series of this sort, including what we would call "biological species," but also "nations" or ethnic groups, breeds of dogs or horses, and aristocratic lineages. The fact that reproducing across national boundaries is a distinct possibility, while reproduction across species boundaries is generally thought to be prohibited by nature, does not seem to have prevented Leibniz from thinking of all of these different generational series as series in the same sense. Again, a tendency to reproductive isolation in a given group does not imply any real biological boundaries between this group and others. A subspecies of birds might be reproductively isolated simply because other birds of the same species do not like the distinctive markings on its wings. In the "Method," Leibniz appears to take Ethiopians, Canadians, and Europeans to be in much the same situation: relatively reproductively isolated because mutually uninterested, even if biologically the same, and descended from the same first ancestors.

And yet, even if, as we have seen, Leibniz's metaphysics does not allow for essential inequality between human groups, it is nonetheless indisputable that he is far from being a full-fledged political egalitarian, and in fact accepts the justice and inevitability of the domination of some people by others. It is thus worth considering here Leibniz's account of domination in further detail, as well as considering how it may relate to some of his deeper philosophical concerns, since here we may gain some insight into how a philosopher, even one who does not believe in essential differences between human groups, may nonetheless provide theoretical sup-

[32]Lazaro Soranzo, *L'Ottomanno*, Ferrara, 1599. See G. W. Leibniz, "Directions Pertaining to the Institution of Medicine," in Smith, *Divine Machines*, appendix 1.

port for a political and economic system that functions as if there were such essential differences.

The notion of domination, as it is deployed in Leibniz's metaphysics, is well known to scholars concerned with the problem of the unity of corporeal substances. Leibniz introduces the concept of the dominant monad for the first time in a letter to De Volder of June 1703, where he distinguishes between "(1) Entelechy or soul, (2) primary matter or primitive passive power, (3) the monad that is completed by these two, (4) the mass or secondary matter, that is, the organic machine in which infinite monads concur, (5) the animal or the corporeal substance, which the monad dominating in the machine makes one."[33] Robert Adams defines the dominant monad as the one that "perceives more distinctly than any other monad in its body . . . an appetite or tendency for perception of the normal organic functioning of the body." In Leibniz's metaphysics, then, domination is grounded in the notion of perception, which is to say the representation of the order of coexistence of an infinite number of monads within the simple individual monad. It is precisely the clarity of perception that underlies the relation of domination: to dominate is to have a perception that is relatively more clear than that of the substances or individuals dominated. A comparison naturally suggests itself to what happens when a federal state disintegrates, such as Yugoslavia, whose provincial capitals become sovereign capitals at the moment of separation: a change that can ultimately be explained in virtue of the fact that the "perception" of what is going on in the former province by the politicians and bureaucrats in the former federal capital becomes reduced in clarity relative to the perception of the politicians in charge of the newly independent entity.

In the years that follow the letter to De Volder on monadic domination, Leibniz frequently utilizes the concept of domination to account for the role of the soul in relation to the infinitely many other monads from which the human or animal body results. On occasion he draws examples of groups of individual creatures that are dominated by a governor or sovereign in order to clarify his notion of domination. The most memorable of these are not drawn from the domain of geopolitics narrowly conceived, but rather from fantastic literature, which in the era often had a distinctly libertine tint. Thus in the *New Essays*, for example, Leibniz cites Cyrano de Bergerac's *Histoire comique des états et empires du Soleil*.[34] He writes, "For the rest I am also of the opinion that the *genii* perceive things in a manner that has some relation to our own, though they

[33]Leibniz, *Die philosophischen Schriften*, 2:252.

[34]Cyrano de Bergerac, Hercule Savinien. *Histoire comique des états et empires de la Lune et du Soleil*, ed. P. L. Jacob, Paris: Adolphe Delahays, 1858 [1655].

have the pleasant advantage, which the imaginative Cyrano attributes to certain animated natures on the Sun, composed of an infinity of small birds, which in transporting themselves according to the command of a dominant soul form all sorts of bodies."[35] In the story Leibniz mentions, Cyrano's protagonist describes several encounters, made during a voyage to the solar surface, with various composite beings. One such encounter is with a mass of little men who arrange themselves into a single giant, a golem that is given life only at the moment that their "king" jumps into its mouth and transforms it from a mere mass into a proper unity: "This whole mass of little men," Cyrano writes, "did not yet give any sign of life; but as soon as it had swallowed its little king, it no longer felt itself to be many beings, but one."[36]

We may appear to be rather far here from a concrete geopolitical example, such as the fall of Yugoslavia. The fantastic tale of the little king and its composite golem may, however, aid us in understanding the connection between political domination and metaphysical domination, and thereby to understand Leibniz's overarching understanding of the nature of composite unities, of both their origins and their nature. For Leibniz, the structure that characterizes all bodies in nature is of such a sort that all corporeal individuals are capable of participating in the constitution of other individuals of the same nature. Individual corporeal beings are at once "worlds apart," and at the same time they are not for that any less implicated in the constitution of greater corporeal beings. A corporeal being can indeed be fragmented ("deconstructed component-wise," in Robert Sleigh's terms),[37] but no being can pass from a corporeal to a noncorporeal state, any more than it could pass from existence to nonexistence tout court, nor yet in the other direction, from nonexistence to existence. The monads from which a corporeal being results are eternally implicated in an infinity of relations of subordination and domination, and when a new corporeal unity appears to come into existence, what in fact happens is that the relations of subordination and domination change in such a way that what was previously a "part" or a dominated region of another corporeal substance separates off and becomes a corporeal substance in its own right, which is simply to say that it becomes an infinite conspiracy of monads that acts in accordance with the principle of activity of a dominant monad, relatively unconstrained by the perceptual activity of monads that are extrinsic to it.

[35]Leibniz, *Nouveaux essais*, 2:chap. 23, §43; G V, 204.

[36]Cyrano de Bergerac, *Histoire comique*, 199.

[37]R. C. Sleigh, *Leibniz and Arnauld: A Commentary on Their Correspondence*, New Haven: Yale University Press, 1990, 126.

Key to Leibniz's understanding of domination is that it is always contingent and temporary: no individual corporeal substance will remain subordinated to another eternally. There is no absolute domination, but only relative domination. Interestingly, this aspect of Leibniz's theory of monadic domination corresponds precisely to his conception of domination in the political sense. In the important text, "Meditations on the Common Concept of Justice" of 1702–3, Leibniz explains that "the body of a man is the property of his soul and can not be taken away from him for as long as he is alive."[38] Taken out of context, this statement might be thought to have something to do with the metaphysical relation between body and soul. In fact it occurs in connection with an argument about the injustice of slavery.[39] For Leibniz, from the fact that a soul cannot be possessed absolutely by another (which, in metaphysical rigor, would mean that the possessing soul eternally perceives the possessed soul with a higher degree of clarity than the possessed soul perceives its own states), it follows that the body of one person cannot be possessed absolutely by someone else (which, strictly speaking, would be nothing other than for that other person's body to be a full and proper extension of the possessor's body).

Central to his argument against slavery is the belief that *ius strictum*, strict right or the right of the more powerful, cannot serve as an adequate foundation of jurisprudence. Two further elements are required: equity, which softens the rigor of *ius strictum*, and piety, which ensures that civic law is applied in conformity with divine law. Leibniz severely criticizes Thomas Hobbes for having stayed at the first level. For Leibniz the right of acquisition is founded in *ius strictum*, and if we were to stop there, with Hobbes, it would follow that the power that some people have over others, such as parents over children or masters over slaves, would become absolute and limitless. Leibniz deploys an example here from New World ethnography, albeit one that is crudely inaccurate, in order to argue that both of these sorts of relation, in the absence of equity and piety, become indistinguishable from one another, and can result—interestingly, given our principal concern here in the constitution of corporeal substances—in the *total* appropriation of subordinated others through cannibalism: "And if we wished to stop at *ius strictum* alone, the American cannibals would be justified in eating their prisoners. There are those among them who go further. They make use of their female prisoners to have children, and then they fatten and eat these children,

[38] G. W. Leibniz, "Méditation sur la notion commune de la justice," in *Mittheilungen aus Leibnizens ungedruckten Schriften*, ed. Georg Mollat, Leipzig, 1893, 41–70, 68.

[39] On Leibniz's philosophy of right and freedom, see in particular Luca Basso, *Individuo e comunità nella filosofia politica di G. W. Leibniz*, Rubbettino Editore, 2005; Patrick Riley, *Leibniz's Universal Jurisprudence*, Cambridge, MA Harvard University Press, 1996.

and finally eat the mother when she cannot produce anymore. Such are the consequences of the supposed absolute right of masters over slaves and of fathers over children."[40] Such a scenario, were it ever to happen in fact, would amount, in effect, to the combination of political and monadic domination: it would be an extreme instance of the master-slave relation, in which the limit case of *ius strictum* results in total bodily appropriation. Arriving at this limit case, it in turn would come about that the dominated subject is now dominated as a constituent of the corporeal substance united by the dominant monad that is the master's soul.

Leibniz accepts that in a certain sense children do belong to their parents, and at least according to the law of nations slaves belong to their masters in an analogous way. At the same time, he denies that the legal power that a master can have over a slave is in conformity with natural reason, and in fact he considers that the master-slave relation differs from the parent-child one in certain fundamental ways. Leibniz believes that even if the master-slave relation were in conformity with natural reason, there would be limits, in view of equity and piety, to what a master could do to his slave:

> If I would admit that there is a right of slavery among men that is in conformity with natural reason and that according to *ius strictum* the bodies of slaves and their children are under the power of masters, it will still remain the case that another stronger right is opposed to the abuse of this right. This is the right of rational souls that are naturally and inalienably free, this is the right of God who is the sovereign master of bodies and souls, and under whom masters are fellow citizens with their slaves, since the latter have the right of citizenship in the kingdom of God just as much as the former do.[41]

At most, Leibniz believes that what a master can have in a slave is not property, but rather only a sort of usufruct, and an extremely limited usufruct at that, to the extent that it must be exercised "*salva re*—so that this right could not go so far as to make the slave cruel or unhappy."[42] It would be a peculiar sort of slavery indeed that required the master to see to the slave's happiness and moral well-being, and it is highly improbable

[40]Leibniz, "Méditation," 69. For a rigorous and sustained investigation of the metaphysical background to early modern discussions of cannibalism, see Cătălin Avramescu, *An Intellectual History of Cannibalism*, trans. Alistair Ian Blyth, Princeton: Princeton University Press, 2009. See also Maggie Kilgour, *From Communion to Cannibalism: An Anatomy of Metaphors of Incorporation*, Princeton: Princeton University Press, 1990; Frank Lestringant, *Le cannibale: grandeur et décadence*, Paris: Perrin, 1994.

[41]Leibniz, "Méditation," 68.

[42]Leibniz, "Méditation," 68.

that the global slave economy could have been sustained for long if Leibniz's requirements had been taken seriously.

We seem to have strayed rather far from the metaphysical concerns with which we began, so let us now attempt to trace our way back. Again, for Leibniz, monadic domination can be understood in terms of relatively greater degrees of clarity of perception: the dominant monad perceives the states of the subordinate monad more clearly than the subordinate monad perceives its own states. At the phenomenal level, this greater clarity of perception translates into the exertion of active force; thus, for example, my soul exercises such force over the infinitely many monads from which my feet result, which causes me to walk, say, while none of the monads from which my feet result has much power over the states of my soul (except in moments of foot pain, when the appendage briefly has greater power over my soul than my soul has over it).

Political domination, in turn, can be understood as the power of an individual to constrain the bodies of others to do one's own will, again, to make them in a certain sense extensions of one's own body, to the extent that they are the executors of one's will. Leibniz does not say so explicitly, but it may be presumed that something like a greater degree of clarity of perception underlies domination here too: power is grounded in a more perspicacious grasp of the range of options for the motions of another person's body, a grasp that easily translates into the ability to control these motions. In the end, however, political subordination, including the extreme case of slavery, is not grounded in natural reason, and wherever there is subordination there is a parallel order in which any two given individuals are perfectly equal and independent of one another, while being dominated in exactly the same way by God. This feature of political domination has its correlate in the metaphysics as well, where Leibniz maintains that in the final analysis nothing would change for a given individual monad if in fact only it and God existed in the world, while all other intermonadic relations of domination and subordination were mere illusion.

There is of course one significant difference between monadic and political domination: the monads from which my foot results are not human spirits, whereas the monads that are the soul of the slave and the soul of the master both are. The monads from which my foot results are, like the vast majority of other monads in the world, "mere monads," endowed with only *petites perceptions*. This is important because there is no conceivable basis in natural reason for supposing that there is no absolute right to dominate a substance that in any event can only ever have an exceedingly dim perception of the world, well below the threshold of reason. Slavery is a case in which two individual corporeal substances enter into a relation of domination and subordination, even though both parties share equally in rationality and potentially enjoy equally clear perceptions. The parent-child relationship is somewhat different, however, in that it is a case in which a former proper part of a rational corporeal substance, originally a bare monad, separates off and becomes elevated to the status of rational being, though with the necessary cultivation and assistance of the rational being or beings from which it separates. Generation, birth, and child rearing are in effect an illustration of Leibniz's understanding of domination as a contingent and temporary predicament of substances: for a preformed animalcule to begin to develop as an embryo, eventually to be born, and finally to come into its own as an adult animal or human, is precisely to move out of total monadic domination—that is, domination as a proper part of the corporeal substance of one's parent—through a form of political domination—that is, the domination of a child by its parents—and finally, hopefully, into complete independence (save for one's subordination to God).

Having taken this fairly lengthy tour of Leibniz's theoretical engagement with slavery and domination, and of the place of the latter notion within his philosophy, let us now return briefly to the 1671 text on seizing an island off of Africa for the training of warrior slaves. The different groups of warrior slaves are to be segregated by *genus* or nation, which is for Leibniz exactly the same thing as separation by language. He sees such segregation as necessary simply because it will help, so he imagines, to prevent larger blocks of resistance from forming among the slaves. Although the slaves are taken against their own will, Leibniz's evident model for them, again, are the Ottoman janissaries, and he imagines that the ultimate outcome of this nonvoluntary undertaking will be the training of elite warriors, proud and happy with their plight. They are 'semi-beasts', but only in the sense that they remain, at the age of abduction, malleable and, so to speak, trainable, and not at all in the sense that they come from inferior or beastly nations. If there is any moral ground for taking only people from these nations, and not from Europe, it is that the Ethiopians, Canadians, and so on have not been baptized, and are thus barbarous not in virtue of any intrinsic inferiority, but only in virtue of their contingent life histories.

Leibniz will not return to any such overt apologetics for slavery of any sort in later texts, but already in 1671 the argument for slavery seems to borrow much more from Ottoman adventure stories than from contemporary apologetics for the transatlantic African slave trade. When Leibniz returns later to the topic of domination, and offers a more sophisticated theoretical account of it, his understanding remains, so to speak, more "Ottoman" than "transatlantic," that is, for Leibniz domination of one human being by another is acceptable, even though all human beings are fundamentally equal, only if this domination is justified by the circumstances of the two people in question, and only if it serves to edify or cultivate the abilities and moral well-being of the dominated. This is, arguably, an impossible desideratum, but it does not appear to involve even an implicit or unconscious commitment to essential inequalities between human groups, and so Leibniz's defense of domination does not, in the end, appear to be an instance of the liberal racism identified by Popkin as a prominent current throughout modern European thought.

For Leibniz, human reason remains fundamentally everywhere the same, even as it comes to be inflected differently as a result of different local exigencies. These different inflections are, precisely, the different natural languages of the world. The fact that these are only inflections of one and the same underlying reason sets Leibniz sharply apart from a thinker such as Locke, who is not prepared to commit himself to some underlying distinguishing feature of humanity that is everywhere the same, and correlatively is prepared to take physiological differences resulting from environment or purported hybridism as bringing about a sort of gradual shading of the human species into the world of nonhuman animals.

In metaphysical rigor, of course, Leibniz himself is committed to the view that the body is an unfolding or "explication" of what is enfolded or "implicated" in the soul, that the body is the phenomenal manifestation of the activity of the soul (in conspiracy with infinitely many other soul-like entities, or monads, all of which are in turn associated with phenomenal bodies). Leibniz is a dualist, then, only in the sense that he believes that we do not need to invoke the activity of the soul in order to provide an adequate causal account of any circumscribed sequence of bodily states. But this only makes his rejection of a racial hierarchy, and of any physiognomically based judgments about a person's moral qualities, all the more remarkable: he rejects the idea that any hierarchy of nobility could be established from an inspection of bodies. The disabled body that is the phenomenal unfolding of a particular human soul is the body of a soul that is no less human than the soul from which an able body unfolds. Likewise, a body that is dark-skinned is the phenomenal unfolding of a soul that is either human or not human, and if the former, then it cannot be placed hierarchically either higher or lower than any other. In metaphysical rigor, *every* body results from the confused perception of nonbodily substances; confusion just is the fundamental basis of bodily reality. But the perceivers themselves, representing the world in various confused ways (including the particular way in which they themselves are embodied), are all, in an important sense, equal (in the sense that they are all representing the same order of coexistence, from different points of view). To the extent that they are human souls, moreover, they are particularly alike, in that they are representing the same order of coexistence, in part by means of the faculty of reason.

Again, for Leibniz all human beings are equal simply to the extent that they are human, which is to say that they are in possession of a rational soul. The conformation of the body can in no way compromise this possession, since the rational soul is not bodily. "Racial" difference, for Leibniz, is understood as membership in a generational series or chain, and in the end every chain, whether a national chain such as that of Ethiopians or Russians, or rather a family chain such as the lineage of a given European noble family, is in the end only a subchain of the entire human lineage extending back to Adam. Leibniz conceptualizes races as generational series consisting in real individuals bound up with one another through real relations of descent. But to the extent that substantial individuality is scuttled in favor of a materialistic monism in later authors such as Diderot (as we will see in the following chapter), for whom generation is only separation, and results in neither the coming-into-being of a new substantial individual, nor the coming-into-dominance of a previously subordinate individual, it can simply make no sense at all to conceptualize race in terms of generational series or the explication of such series in genealogy. The alternative conception of race that imposes itself willy-nilly is a biogeographical and synchronic one, rather than, so to speak, a catenary and diachronic one.

We have by now extensively treated the cluster of terms that Leibniz identifies as synonyms of "race": "genus," "Geschlecht," "series generationum." But there is a further definition in the note on Wilkins that we have not treated directly: in this note, "genealogy" is defined as the explication of a generational series. By substitution of synonyms approved by Leibniz himself, it follows that, for him, race is a historical kind. It is to be understood or elucidated through historical investigation. Such a conception of historically unfolding generational series as lying at the basis of race is difficult to fit together with the conception of monads existing as worlds apart. If however we may discern any of the core philosophical commitments in Leibniz's account of human diversity, it is not that he is able to extend the conception of worlds apart into his thinking on diversity at the expense of the conception of generational series. It is rather that he conceives human diversity as the result of sundry local expressions of an underlying uniformity, with individual human cultures amounting to so many different inflections of an always-identical human reason. In this respect, Leibniz's account of race does not rest on the discreteness and irreducible singularity of monads, but rather, again, on another familiar Leibnizian idea: that of unity in diversity, or of multiplicity compensated by identity.

In the remainder of this chapter, let us consider in greater detail Leibniz's extensive writings on history, geography, and culture, in order to gain a more thorough understanding of the philosopher's subtle account of the nature of human diversity.

7.4. The Science of Singular Things

One of the great insights of early modern cosmology, and also eventually of mathematical physics, was that always and everywhere, the same laws hold as here. This was an insight of mechanical philosophers such as Descartes, who believed that the arc of a projectile such as a cannonball can be studied by the same methods and with the same presuppositions as the motion of the heavenly bodies; and it was also an insight of the less well-known chemical philosophers, who believed that microcosmic machines could be made to reproduce in miniature the workings of the whole world. Yet this distinctly modern view, that things are everywhere the same, was at the same time challenged by many of the reports coming in from around the world by explorers and naturalists. It is not just that people are different in the far Antipodes, but that the behavior of nature seems different there as well. There is thus a problem of what we might call the "local nuances of universal physics." Light seems to be refracted differently at the poles, as Kepler worried, to cite one well-known example. With the publication of William Gilbert's *De magnete* in 1600, to cite another example, magnetic variation came to represent perhaps the greatest challenge to those who wished to account for the local fluctuations of global natural phenomena. Problems such as these directly impacted debates about the relevance of individual observations to the discovery of regularities in nature: individual observations are also always *local* observations, and as science becomes an endeavor of global reach, such observations come to be both problematic and indispensable.[43] Perhaps no other early modern philosopher was more successful than Leibniz at bringing together the universalism of the mechanical philosophy with the localism or particularism that turned out to be necessary for the advancement of science.

As many scholars have noted, early modern globalization had an impact on all areas of science. This includes not just those, such as botany and indeed ethnography, in which the objects of study are different in one region than in another, but also astronomy and physics, whose objects of study are, one might think, fundamentally the same no matter where in the world one chooses to study them. Yet many of the remarkable

[43] See in particular Alix Cooper, *Inventing the Indigenous: Local Knowledge and Natural History in Early Modern Europe*, Cambridge: Cambridge University Press, 2007.

advances in the physical sciences made by the Jesuits in the early modern period, for example, were made possible by the fact that they were organized as a global network and could coordinate telescopic observations of the moons of Jupiter between research stations in Siam and Paris. The differences between the observations made from different positions contained important lessons not just about the orbits of bodies in the solar system, but also, for example, about the determination of longitude on Earth.[44] One might add that in optics as well, investigation of different parts of the world was thought to yield up different insights into the nature and properties of light. Particularly at the ice-covered poles, light was believed to manifest properties that could not be studied in more verdant climes.[45]

One of the most important practical aims that brought voyagers to the icebound parts of the world was the search for ocean passages to both the northeast and the northwest, and one of the most important theoretical aims attached to this search was the establishment of longitude. The difficulty of doing this was an omnipresent problem for early modern natural philosophy, and it, perhaps more than any other, forced natural philosophers to reckon with the local nuances of universal physics. The world could not simply be divided up into neat slices in an a priori way, using stationary models. Leibniz, for example, recognized as much in a 1716 letter to Peter the Great. "[I]n most places," he writes, "the magnetic needle does not face the North, but rather generally tends somewhat to the East or the West, and does so to differing degrees in different places."[46]

This letter is very revealing as a testament to the way in which the problem of local nuances had come to impact the early modern approach to the study of the physical world. Here Leibniz proposes that a system of observation of "latitude or height of the pole, and then of the declination of the magnet," be set in place, which would bring it about that one "would only have to look for the line on the magnetic globe where the magnet has the relevant declination, and to follow the line to the place where it comes underneath the present elevation of the pole, in order to

[44] See Florence Hsia, *Sojourners in a Strange Land: Scientific Missions in Late Imperial China*, Chicago: University of Chicago Press, 2009; Mordechai Feingold (ed.), *Jesuit Science and the Republic of Letters*, Cambridge, MA: MIT Press, 2002.

[45] See Adina Ruiu, *Les récits de voyage aux pays froids au XVII siècle: de l'expérience du voyageur à l'expérimentation scientific*, Montreal: Presses de l'université du Québec, 2007.

[46] G. W. Leibniz, *Sbornik pisem i memorialov Leïbnitsa otnosyashchikhsya k Rossii i Petru Velikomu*, ed. V. I. Ger'e, Saint Petersburg, 1873, no. 239. An earlier version of the argument of this chapter, both as concerns Leibniz's interest in "the science of singular things" and as concerns his application of this science to the study of the Russian Empire, was developed in Justin E. H. Smith, "Leibniz on Natural History and National History," *History of Science* 50, 4 (December 2012): 377–401.

determine . . . longitude."[47] Because the declination regularly changes, "such observations could be made anew from time to time," and "new magnetic maps or globes would only need to be made every 5 or 6 years, which could then serve for as long a time. And thus it would be just as good as if the secret of longitude were revealed."[48]

Leibniz believes that with this system in place, "a certain order will become apparent in the variation itself, and posterity will finally arrive at a greater knowledge of this mystery, so that new observations will no longer have to be made so frequently, but instead the variation will be able to be seen in advance, in which case the long-studied problem of longitude would have its much sought-after solution."[49] In short, local fluctuations require not just local measurements, but also massive coordination, and this, it was hoped, would eventually lead to a system of scientific data collection that would facilitate a global comprehension of the phenomenon in question, and even an ability to predict its future course. Here, Leibniz is promoting a conception of science as a massively collective project that had begun already with the Baconian program a century earlier. But what Bacon could not entirely foresee, perhaps, was that even a branch of natural philosophy as global, and in some sense as abstract, as geodesy would require the collection of data from the field, as well as attention to local variations in the object of study.

The need to pay attention to the variety within the order would in fact extend for Leibniz to nearly all domains of inquiry. A historian by profession, Leibniz pursued his research in this field with great interest, and explicitly saw it as inseparable from his broader intellectual project. The perception however that Leibniz was better suited to a priori speculation than to the investigation of particulars is one that hounded Leibniz even during his own lifetime. Thus he complains to the Swedish Slavist Johan Gabriel Sparwenfeld in a letter of 1698 that "people criticize me when I attempt to take leave of the study of mathematics, and they tell me that I am wrong to abandon solid and eternal truths in order to study the changing and perishable things that are found in history and its laws."[50]

If we were to attempt to summarize Leibniz's understanding of what history is, we might do so by saying that for him history is *the science of singular things*. In a 1708 draft of a proposal to Peter the Great for the classificatory system to be used in the eventual library of the Saint Petersburg Academy of Sciences, Leibniz identifies history, alongside mathematics and physics, as one of the three "Realien" or distinct domains

[47]Leibniz, *Sbornik pisem i memorialov*, no. 239.
[48]Leibniz, *Sbornik pisem i memorialov*, no. 239.
[49]Leibniz, *Sbornik pisem i memorialov*, no. 239.
[50]Leibniz, *Sbornik pisem i memorialov*, 38.

of science, namely, the one that "involves the explanation of times and places, and thus of singular things [*res singulares*]," including "the descriptions and attainments of kingdoms, states, and peoples."[51] Besides what might be called civil or political history, another central branch of history—again, understood as the science of singular things—was the study of plants, animals, and minerals, as well as magnetic variation, infant mortality rates, and in general anything that requires the amassing of data concerning singular events. In all of these projects, it was necessary to contend with regional or local inflections of universal phenomena. In the study of magnetic variation, for example, one could not treat the earth, as one might in mathematical geodesy, as a perfect globe with more or less the same properties wherever one might look; instead, the variations in the distance between magnetic north and true north differed from one place of measurement to the next, and indeed from one measurement to another measurement taken at a later time in the same place.

The result was a picture of the earth that was fundamentally characterized by diversity, and that thus was an object of study that, in order to be fully understood, had to be approached through a preliminary amassing and recording of vast quantities of data. On Leibniz's division of the different domains of inquiry, magnetic variation, botany, comparative linguistics, and so on are all "historical" endeavors in the broad sense, to the extent that they are all included within the science of singular things. Many other of Leibniz's projects required a similar approach, and his own, explicit division of the sciences indicates that he understood these different projects as being in important respects part of one and the same overarching project. If we are to believe Leibniz's complaint to Sparwenfeld, this overarching project, the science of singular things, was of central importance to Leibniz's self-conception as a natural scientist and as a thinker.

In another text, a preface to a work on the history of the Germans, Leibniz identifies the pleasure of knowing singular things as one of the ultimate purposes of the study of history: "We aspire to three things in history: the first is the pleasure of knowing singular things; next are the especially useful precepts for life; and, finally, recalling the origins of present things from past ones, as all things are best known through their causes."[52] To the extent that we wish to speak of Leibniz as a historian in this broad sense, we are, then, speaking of Leibniz as an empiricist. If we

[51] Leibniz, *Sbornik pisem i memorialov*, 96.

[52] Leibniz, "Praefatio accessionibus historicis quibus potissimum continentur scriptores rerum germanicarum" (1700), in *Opera Omnia*, IV, 2, 53–57, 53. "Tria sunt expetimus in Historia: primum voluptatem noscendi res singulares, deinde utilia inprimis vitae praecepta; ac denique origines praesentium a praeteritis repetitas, cum Omnia optime ex causis noscantur."

think of the mechanical revolution as motivating the view, among other things, that nature can be understood according to a few simple laws that hold in the same way everywhere, and that, for example, the earth can be understood according to the laws that would govern the motion of large spherical bodies, then, as we have seen, Leibniz rejects at least this aspect of the revolution to which he was a sort of heir, moving, again, for example, from Cartesian geodesy to geography: the study of regional variations, and of all the singular things behind these variations that make it impossible to conceptualize at least certain domains of nature in terms of "solid and eternal truths."

But if we are prepared to see Leibniz as an empiricist, he is nonetheless a defender of what might be called "rationalist empiricism." This only sounds like an oxymoron: perhaps somewhat like "democratic republicanism," these are two perfectly compatible tendencies, which are only subsequently transformed into political parties of sorts whose members are not allowed to mix. Leibniz is a rationalist empiricist to the extent that he believes that the amassing of data about singular things in the end amounts to an enriching or buttressing of the a priori commitment to a picture of nature as a rationally ordered and harmonic whole, and of individual beings as unique expressions of that same rational order. Rationalist empiricism, or "history" in the broad sense in which Leibniz has described it to Peter the Great, then, is precisely the search after order within variation, which in turn is an inflection of his most basic commitment, already sketched out in previous sections, to unity in diversity.

7.5. Mapping the Diversity of the Russian Empire

Whether Leibniz's proposals for the accumulation of natural knowledge of singular things involved prospecting at the local or at the global level, and what sort of knowledge of singular things he deemed important, was often determined by the scope of the empire of the ruler he was hoping to impress. He writes, tellingly, to Peter the Great's councilor, Aleksandr Golovkin, in 1712, that he is interested in obtaining "some desiderata . . . from Russia itself. And here I have no particular interest of my own, but all of my requests serve toward the public good and toward the glory of His Majesty."[53] In this period, Leibniz's curiosity about singular things from Russia seems insatiable. In a note consisting in scattered observations on Russia, which may be dated based on content to between 1705 and 1712, Leibniz writes of sundry curious plants and animals he believes are to be found in Russia, such as "the root of the famous kidney vetch,"

[53] Leibniz, *Sbornik pisem i memorialov*, 275.

which "has been observed in Russia, growing in Siberia, as well as the *Voltschnoi Koren*, which is called [in Latin] the *lupina radix* or wolf root, which is said to have the greatest [medicinal] virtue in healing wounds, and some claim that it is beneficial to chew it for the healing of wounds." He notes that "among fish the sturgeon stands out, [as it is] considered one of the Russian delicacies"; and he describes "the *vichochol* [i.e., the Russian desman], which is a large aquatic mouse that gives off a pleasant odor. Its furs are placed in the case in which linen garments are kept, thanks to which they acquire their odor."[54] Peter's empire was rich in resources and, from Leibniz's point of view, full of scientific and economic potential. It was also full of singular things, which Leibniz believed to be of independent scientific and indeed philosophical interest: large aquatic mice, too, are reflections of the order of nature.

As discussed in chapter 3, a central motivation for the colonial expansion of Europe throughout the world in the early modern period, including Russia's push eastward and its absorption of a great portion of Eurasia, was the prospect of natural resource extraction, and this practical end required for its success a deepened theoretical grasp of the real variety of things in nature. Above we were introduced to Londa Schiebinger's account of the early modern global system of "bioprospecting": the communal effort to discover and classify plants and animals (though mostly plants) throughout the world (mostly those parts of the world left unstudied by the ancients), particularly with an eye to discovering the useful kinds, and bringing them back to Europe for commercial gain. While Schiebinger's work has been very useful in drawing attention to a widespread early modern phenomenon, in a number of respects what she describes fails to adequately capture the full significance of the prospecting activity of a figure such as Leibniz.

Leibniz always remains receptive to the distinct scientific opportunities that different regions open up, and in the case of Russia in particular he perceives an opportunity to study a vast empire whose features—from its surface area to the number of different ethnic groups it counted among its subjects—were particularly useful in gaining the sort of universal scope to which he was committed on philosophical grounds. His motion from the local context of earlier prospecting endeavors (e.g., silver prospecting in the Harz Mountains in the early 1690s) to the transcontinental context (comparative ethnolinguistics of the Russian Empire in the 1700s and 1710s) is at once both a continuation of the same generic activity, but also a shift to a level that he considers rather more adequate to the project of science as he sees it: one that must take into account a maximum of variety (the best of all possible worlds being for him in the end nothing

[54]Leibniz, *Sbornik pisem i memorialov*, 47.

other than the maximal set of compossibles) and the broadest possible range of diversity. Of course, Leibniz is also prospecting for a new job, and he is optimistic that the tsar might offer him the sort of benefits he desires as a lifelong courtier; but this career prospect also happens to fit nicely with his intellectual ambitions.

One of the most remarkable insights of Leibniz's activity as a prospector was his recognition not only that one must go out in search of *res singulares*; one must simultaneously, and in a parallel fashion, "collect" the local words for the singular things.[55] In this respect, Leibniz may be seen as an early proponent of what would much later come to be called "ethnotaxonomy." He recognizes the importance of the study of the names of plants in local languages, as well as the value of toponymy and hydronymy for tracing the history of the migrations of peoples. In part this is because he understands that local knowledge will be useful in resource extraction and in the attainment of utility, but also in part because he believes that ethnotaxonomy can help to uncover the structures of the human mind, as revealed in the naming practices by which different groups of people carve up and order the natural world. Ethnotaxonomy, finally, can help to reveal the features of things in the natural world themselves. As he notes in the *New Essays*, "In time all the languages of the world will be recorded and placed in the dictionaries and grammars, and compared together; this will be of very great use both for the knowledge of things, since names often correspond to their properties (as is seen by the names of plants among different peoples), and for the knowledge of our mind and the wonderful varieties of its operations."[56] Here, then, we are able to discern the unity of bioprospecting and ethnoprospecting in Leibniz's mind, and nowhere do these two projects come together so clearly as in his late-career interest, which began in earnest in 1697 and lasted until his death in 1716, in the Russian Empire, not so much in Russia itself, as in the diversity of plants, animals, geographical features, climate, and human cultures under Russian imperial rule.[57]

[55]On Leibniz's contributions to comparative linguistics, see in particular Sigrid von Schulenberg, *Leibniz als Sprachforscher*, Frankfurt, 1973.

[56]Leibniz, *Die philosophischen Schriften*, 5:318.

[57]There are several useful overviews of Leibniz's relations with Peter the Great and of his interest in Russia. See in particular Moritz C. Posselt, *Peter der Grosse und Leibnitz*, Tartu: Friedrich Severin Verlag, 1843; V. I. Ger'e, *Leïbnits i ego vek*, Saint Petersburg: V. Golovin, 1868; Ernst Benz, *Leibniz und Peter der Grosse*, Berlin: De Gruyter, 1947; V. I. Chuchmarev, "G. V. Leïbnits i russkaia kul'tura 18 stoletiia," *Vestnik istorii mirovoi kul'tury* 4 (1957): 120–32; Vladimir Katasonov, "The Utopias and the Realities: Leibniz' Plans for Russia," in *Leibniz und Europa, Vorträge VI. Internationaler Leibniz-Kongress* 2 (1994): 178–82; Vladimir S. Kirsanov, "Leibniz' Ideas in the Russia of the 18th Century," in *Leibniz und Europa, Vorträge VI. Internationaler Leibniz-Kongress* 2 (1994): 183–90; Gerda Utermöhlen, "Leibniz im brieflichen Gespräch über Rußland mit A. H. Francke und H. W. Ludolf," in

But Leibniz also sees his ethnoprospecting activity as bidirectional, and is interested not only in collecting the languages of nontextual cultures, but also in disseminating a text of particular importance to him, the sacred scripture of Christianity, to these same cultures. Transmitting Christian sacred texts into the world of nontextual peoples would serve the dual purpose of forcing Christian concepts into, for example, Samoyed, while at the same time providing a standard grid for comparing samples of different languages among themselves. The specific method Leibniz proposes, involving transcriptions in Cyrillic with interlinear Russian translations, would be comparable to the standardization of the pressing of botanical holotypes, where a single plant is pressed and dried and then used as the standard or representative of the entire species.[58]

In a letter to A. C. Gackenholtz, written in 1701 and later published as "On Botanical Method," Leibniz notes that particular features of plants, such as the serration of the leaves, provide a fixed criterion of comparison: in each case, we can be sure we are looking at the analogous parts of the different specimens to be compared. This system, in the study of either plants or languages, is not a "natural" system, in the sense that it does not guarantee that one will be picking out the features of the specimen that provide a window to its essence or interior nature. Classification will always retain a degree of arbitrariness, even if one system of classification might be better suited to certain ends than others. Leibniz writes in the same letter to Gackenholtz, "I do not so much condemn the recent botanists' effort to assist memory by any method of dividing into classes that they judge the more suitable." In fact, he quite admires this approach, even if "in advance of knowing the interior constitution of these machines of nature, no accurate method can be instituted, nonetheless a certain substitute method may be employed for the sake of our comprehension and progress in theory."[59] Leibniz goes on to explain that "although I support an ordering into classes according to one certain criterion that is widely variable, I would reckon that other criteria for ordering and comparing plants should not be neglected."[60] We cannot know, at the beginning of any investigation, the kinds of plants from their in-

Leibniz und Europa, Vorträge VI. Internationaler Leibniz-Kongress 2 (1994): 304–9. For a very useful account of the diplomatic and political stakes of Leibniz's service to, or effort to serve, various sovereigns, see André Robinet, *Le meilleur des mondes par la balance de l'Europe*, Paris, 1994. None of these studies, however, attempts to place Leibniz's interest in the flora and fauna, as well as in the non-Russian peoples of the Russian Empire, within the context of his more global and general interest in political geography.

[58]On the history of the standardization of type specimens in botany, see Lorraine Daston, "Type Specimens and Scientific Memory," *Critical Inquiry* 31, 1 (2004): 153–82.

[59]Smith, *Divine Machines*, appendix 5, § III.

[60]Smith, *Divine Machines*, appendix 5, § VI.

terior natures. But we can contrast plants featuring serrated leaves with plants having smooth-edged leaves. We can determine the medicinal or nutritive properties of both varieties, and later we can map the regions in which the different varieties can be found. Such a map would immediately reflect a reality of interest to any natural scientist, and would be of obvious use to any merchant or physician.

Leibniz's treatment of ethnoprospecting as of a pair with bioprospecting suggests, perhaps, an implicit presupposition that linguistic communities grow up in their distinct regions in a way more or less analogous to plants. Each kind of thing can be collected and compared to related neighbors. Languages can be placed into families in view of deeper affinities underlying differences in the lexicon, or in the number of cases employed; plants can similarly be classed together in spite of variation from region to region in the length of the roots or the prominence of the flowering part. Yet it would be a mistake to suppose that this parallelism between plants and peoples in Leibniz's thought commits him to anything like organic nationalism, which sees national groups as essentially rooted in definite regions, and sees political rights to self-determination within these regions as flowing from this rootedness. Obviously, Leibniz believes in the right of the Russian Empire to rule subjugated territories; he envisages no Samoyed nation-state in the future.

In this respect, Leibniz is certainly no communitarian in the manner of Johann Gottfried Herder (who nonetheless drew inspiration from Leibniz's ethnolinguistic work, and who will be discussed at some length in chapter 9). He takes for granted the existence of transnational empires, in which less mighty nations are by rights dominated by the stronger ones. In fact, as Leibniz explains to Golovkin in 1712, compilation of data on the linguistic diversity of the *Empire des Russes* would be useful for "marking the extent of the empire." Diversity is to be promoted not in the interest of promoting the political rights of the individual communities, but rather as a reflection of the empire's vastness and power. Arguably, this approach would remain fundamentally unchanged right up through the Soviet period as well, which celebrated regional costumes and dances, but always in a way that worked to the greater glory of the political formation that had brought them all together into a single union.

Leibniz's official charge as privy councilor to Peter the Great, a role to which he was appointed in 1712, after much vigorous campaigning, was to help advance the sciences in Russia. His work in this role ultimately led to the foundation of the Saint Petersburg Academy of Sciences shortly after his death. Scholars have often supposed that Leibniz's interest in Russia was strictly instrumental: first, he saw it as a bridge to China, which was truly a terra incognita worthy of study, unlike the relatively close and familiar Slavic world; and second, he hoped to be able to gain political influence as an advisor to the tsar. Certainly, Leibniz became interested in Russia not out of any preoccupation with this particular ethnolinguistic region, but rather as part of a much more wide-focused perspective: he wanted nothing less than an ethnolinguistic map of all of Eurasia, and he recognized that the way to gain the relevant knowledge of a broad portion of the relevant geographical area was through the mediation of the tsar.[61]

The dream of such a map was rooted, again, in Leibniz's belief that language is the principal criterion of distinction between different human groups. Thus in response to François Bernier's attempt to provide a "new division of the earth" in terms of the basic "races" of people, Leibniz, as we already saw in chapter 6, offers his own, alternative criterion of division: "I should like for the regions [of the world] to be divided according to languages," he notes, "and for this to be noted on maps."[62] In a 1703 letter to Peter the Great's war councilor, Heinrich von Huyssen, Leibniz describes the project of making ethnolinguistic maps as itself a branch of geography: "Among other curiosities," he writes, "geography is not the least, and I find fault in the descriptions of distant countries to the extent that they do not take note of the languages of peoples." It is because of this oversight, Leibniz thinks, that "we do not at all know the relations between them, nor yet their origins."[63] In an undated letter to another of the tsar's councilors (probably composed between 1705 and 1710), Leibniz writes, again, "I have long wished to have *specimina linguarum* that are in the Tsar's territory, and of those bordering it, in particular I would like to have the Our Father written with interlinear translations, as well as certain common words that are used in the languages."[64] Similarly, to Peter, in December 1712, he expresses his desire for a "translation of the Ten Commandments, the Our Father, the Apostles' Creed, in the particular languages of the peoples who live in Your Majesty's broad kingdom or who are neighbors to it, together with a small dictionary of each language";[65] to Golovkin, a month earlier, he requests "samples [*des Essais*] of all the languages other than Slavonic that are to be found in the great Empire of the Tsar and in neighboring states. These samples would

[61]Leibniz's project for an ethnolinguistic map of the Russian Empire dovetailed with an intense interest in Russia in that period for geographical surveying of a more traditional variety. See, e.g., D. M. Lebedev, *Geografiia v Rossii petrovskogo vremeni*, Moscow: Izdatel'stvo Akademii Nauk, 1950. For an unsurpassed documentary account of Peter's sundry scientific projects, including geography, and including many in which Leibniz had a hand, see F. V. Tumanskiĭ, *Sobranie raznykh zapisok o sochinenii, sluzhashchikh k dostavleniiu polnogo svedeniia o zhisni i deiatel'nosti Petra Velikogo*, Saint Petersburg, 1787.

[62]Leibniz, *Otium hanoveranum*, 158.

[63]Leibniz, *Sbornik pisem i memorialov*, 51.

[64]Leibniz, *Sbornik pisem i memorialov*, 50.

[65]Leibniz, *Sbornik pisem i memorialov*, 286.

consist in a translation of the Decalogue, the Lord's Prayer, and the Apostles' Creed in each of the languages, written in Russian [Russiques] characters, with the Russian version written word for word interlineally";[66] to the Ryazan Metropolitan Stefan Iavorskiĭ, also in November 1712, he expresses his wish for "the Catechisms, which at the same time will be able to push the peoples along toward the true faith and piety that are being taught, as well as the Decalogue, the Lord's Prayer, and the Apostles' Creed; these are to be written in the language of the people who are to be converted, with each language [also] written in Russian characters, and with a Russian version inserted between the lines."[67] From one Russian correspondent to the next, Leibniz expresses his desire in identical or very similar language: he wants interlinear samples of the Lord's Prayer in as many languages as possible.

This plan for an ethnolinguistic survey of the Russian Empire would amount to a rigorous application of Leibniz's general view that the study of language, as opposed to texts, amounts to the ultimate frontier of philology. Thus in the *New Essays*, as already mentioned, he looks forward to the day when all of the sacred texts of every literate culture in the world will have been exhaustively studied, and scholars can move on to the study of the languages themselves, which always precede the composition of texts. "When the Latins, Greeks, Hebrews and Arabs shall someday be exhausted," he writes, "the Chinese, supplied also with ancient books, will enter the lists and furnish matter for the curiosity of our critics. Not to speak of some old books of the Persians, Armenians, Copts and Brahmins, which will be unearthed in time so as not to neglect any light antiquity may give on doctrines by tradition and on facts by history."[68] With these textual traditions mastered, Leibniz thinks that the real work will have just begun: "And if there were no longer an ancient book to examine, languages would take the place of books, and they are the most ancient monuments of mankind."[69] Leibniz's concrete efforts at obtaining samples of the languages of the Russian Empire, then, are in fact but one small part of this eventual post-textual human science. Sacred texts have no particular priority in the order of knowledge; indeed,

[66]Leibniz, *Sbornik pisem i memorialov*, 275.

[67]Leibniz, *Sbornik pisem i memorialov*, 278. Iavorskiĭ was the Metropolitan of Ryazan and a graduate of the Kiev-Mogyla College, which had been the major center of philosophy in the greater Russian world prior to the Petrine Enlightenment. He is the author of at least one significant philosophical work, the *Filosofskoe sostiazanie* [*Philosophical Contention*] of 1693, originally a lecture course given at the college, of which Leibniz may have been aware. See Yu. F. Samarin, *Stefan Iavorskiĭ i Feofan Prokopovich*, Moscow: Tipografiia A. I. Mamantova, 1880.

[68]Leibniz, *Die philosophischen Schriften*, 5:318.

[69]Leibniz, *Die philosophischen Schriften*, 5:318.

texts themselves are only one part of the vast learning of human cultures that can be extracted by a diligent and systematic human science. The Samoyeds have their own ancient monuments, which constitute a part of world history whether they know it or not. Language is a reflection of human reason itself, which is, like the great texts of literate civilizations, a reflection of the rational order of nature.

In part, Leibniz's interest in "primitive" languages could be motivated by a belief that they are better at capturing the essences of things. He is a staunch opponent of the Adamic theory of language, according to which the first language spoken by the first man was one that perfectly captured and reflected the essences of the things themselves, but at the same time he does believe that at its most primitive language is to some extent able to reflect essences, insofar as words are coined onomatopoetically. A false but interesting etymology offered by Leibniz in the *New Essays* provides a nice example of this: Leibniz believes that the word "quek"—which in old German signifies that which is living and which has modern cognates such as "Quecksilber," "erquicken," and the English "quick" (as in "the quick and the dead")—comes directly from the sound made by frogs.[70] In this and other cases, a word may be said to be primitive or natural when it reflects something real about at least the audible aspect of a thing's existence.

In any case, while all languages might have their deepest roots in the sounds of nature, the long span of time since their first appearance has brought it about that there are greater affinities between some languages than between others, and for Leibniz a crucial part of ethnolinguistics is to map out not only languages, but also language families. While linguistic affinities between presumably nontextual cultures are harder to discern, they are not for that reason any less important. He confidently spells out, in a 1708 letter to Johann Christoph von Urbich, the tsar's privy councilor in Vienna, the linguistic affinities between Russian, Polish, and Czech, on the one hand, and between Finnish and Hungarian on the other. But he adds, "I do not know how . . . many other particular languages are related, for example those of the Samoyeds, the Siberians, the Mordvins, the Circassians, and the Cheremysh."[71] Leibniz is virtually alone among his contemporaries in taking the languages of nontextual peoples as themselves repositories of learning. He is convinced that each language is equally expressive: Samoyed, again, is just as suited as Latin to expression of the Ten Commandments. But at the same time each language is for Leibniz uniquely fitted to an individual people's local environment and its exigencies. The Ten Commandments can be translated

[70]Leibniz, *Die philosophischen Schriften*, 5:261.
[71]Leibniz, *Sbornik pisem i memorialov*, 88.

from Latin into Samoyed with no loss, since they have universal validity, but Samoyed might still be better suited to the naming and description of some local moss or shrub, and cultural distinctness consists for Leibniz in the sum total of the ways in which a group's linguistic practices bring about a fit between that group and its habitat.

Of course, in his dealings with members of Peter the Great's inner circle, Leibniz typically frames his interest in the Samoyed and other groups in terms of the goal of Christianization alone, and the missionary aim of his ethnoprospecting is, again, evident in his preferred means of collecting samples of local languages. Repeatedly throughout the early years of the eighteenth century, Leibniz recommends to his many correspondents in Russia that they collect samples of languages by writing down these and other familiar texts in each of the indigenous tongues of the Russian Empire.[72] Leibniz's plans for the Russian Empire amount to a concrete application of what he had proposed more generally in the *New Essays*: to move beyond classical texts, as the repositories of human wisdom, and to focus on language itself. As for the Slavic peoples themselves, Leibniz believes that the best source of information will be their own written histories, and he beseeches a number of his Russian correspondents to procure for him a copy of the *Paterikon*, which he believes, sight unseen, to be a comprehensive history of the emergence of the modern Russian state from its ancient Slavonic roots. But he also notes that many of the cultures about which one must learn in order to distinguish the Slavs from the non-Slavs have no historical texts to draw on, or indeed any texts at all, and here all one has to draw on are the fragments of history embedded in spoken language. These are, in Leibniz's view, a source as rich as the most detailed textual history.

What, now, do these considerations on comparative linguistics and political geography have to do with race?

Leibniz not only affirms that language is the best criterion for mapping human diversity; he also explicitly rejects the view that physical appearance can serve as such a criterion. In a 1697 letter to Sparwenfeld, Leibniz writes, "If it is true that the Kalmuks as well as the Moguls [i.e., Mongols] and Tartars [i.e., Tatars] of China depend on the Grand Lama in matters of religion, it is possible that this says something about the relation among their languages and the origin of these peoples. It is simply

[72]Max Müller wrote of Leibniz's contribution to Russian linguistics: "[The linguist Friedrich von] Adelung started out from collections of words that had been compiled under the auspices of the Russian government. But for these collections it is clearly Leibniz who must be thanked. Although Peter the Great had neither time nor inclination for philological studies, his government always kept in view the plan to collect all of the languages of the Russian Empire" (Müller, *Vorlesungen über die Wissenschaft der Sprache*, ed. Carl Böttger, Leipzig: Verlag Gustav Mayer, 1863, 117).

that the size and constitution of their body is so different among them."[73] The Kalmuks (or, more properly, Kalmyks) are, as Leibniz rightly recognizes, Buddhists, and therefore, by his standards, garden-variety pagans in need of conversion (among the non-Abrahamic Asian faiths, Leibniz valued the Chinese state religion, but saw Buddhism, the Vedic religions, and Taoism as idolatry). But the Kalmyks' submission to the Grand Lama at least places them within a known textual tradition, while the great challenge for the creation of Leibniz's ethnolinguistic map of the Russian Empire was posed by peoples who were, in fact or by presumption, nonliterate, such as the "Samoyeds" (today called by their own ethnonym, "Nenets") or the "Siberians" (an umbrella term covering, presumably, a vast number of Mongol-Turkic, Tungusic, and Paleo-Siberian groups). In the early eighteenth century these groups still had a very distinct identity in spite of their political domination by the tsarist regime: the Russian conquest of Siberia and the subordination of indigenous peoples was completed only by the end of the sixteenth century, and the Russians did not reach the Pacific coast until 1639. In his description of the Kalmyks, in any case, Leibniz is (correctly) hypothesizing the existence of a Mongol-Turkic ethnolinguistic group. He is doing so on the basis of shared religion, and he suspects that further investigation would confirm shared origins and common features of language. He cautions that differences in the physical appearance of members of related ethnolinguistic groups can be misleading, and that language is far more important than "race" for determining ancestral relations.

But let us begin to draw together Leibniz's ethnolinguistic project in the Russian Empire with his work as a historian, with which we began this chapter, and finally, in turn, with the properly philosophical basis of his conception of the nature of human diversity. Of particular importance in Leibniz's ethnolinguistic project is the task of relating current ethnic communities to those mentioned in classical texts, to determine, for example, who are the true descendants of the Scythians, and who of the Huns. Thus he mentions to Golovkin the peoples who "earlier emerged from Greater Scythia [and are] now for the must part subject to the Tsar of the Russians, such as the Parthians, Alans, Roxolani, Huns, Bulgars, Avars, Khazars, Hungarians, and Cumans";[74] and to Iavorskiï he speaks of "Great Scythia, the greater part of which belongs to the Russian ruler," and which "is known to have given rise to the ancient Saka, Parthians, Gets, Massagetians, Alans, Huns, Khazars, Bulgarians, Cumans, and the Hungarians themselves."[75] The Dutch explorer Nicolaes Witsen

[73]Leibniz, *Sämtliche Schriften und Briefe*, I, 13, 544.
[74]Leibniz, *Sbornik pisem i memorialov*, 275.
[75]Leibniz, *Sbornik pisem i memorialov*, 278.

had maintained in a 1699 letter to Leibniz,[76] based on linguistic evidence, that the Slavs are the descendants of the Scythians, while at least for a time Leibniz himself believed that it was the Huns who gave rise to the Slavs. Because the Scythians had served since antiquity as a stock figure of brutal and exotic foreigners (as mentioned above, they would also serve as a model or reference point in many early modern accounts of the Native Americans), the identification of any particular modern ethnic group as the descendants of the Scythians was inevitably a highly charged political matter.

Often, it seems that Leibniz alters his account of who is descended from whom in the aim of political expediency. Thus in the 1712 text on the origins of the Slavs, Leibniz identifies them as direct descendants of the Huns, and also believes that the Huns are among the people who in antiquity were called "Scythian." And in a 1699 letter to Witsen, Leibniz mentions the report he has heard of the custom in Russia, wherein authorities of the church are themselves obliged to carry out the execution of criminals, and comments that "this is a custom that still retains something of the Scythian."[77] Yet in the letter to Peter of December 1712, Leibniz characterizes Russian history, as well as Russia in the present, as an ongoing battle of the Russians *against* Scythians ancient and modern. He thus proposes a "Tabor" or wagon fortress (*Wagenburg*) as a strategy for battle against the Turks—now taken by him to be the modern-day Scythians—and notes that it worked well "in ancient times against the Scythian peoples on the flat plains" of the steppe. Who is Scythian and who is not, and thus who is represented as exotic and barbarian and who as civilized, appears more a matter of current geopolitical alliance than something to be determined by historical and linguistic evidence.

A particularly interesting text for understanding Leibniz's historical approach to the study of human diversity, and one that draws us some distance away from the Russian Empire, is the 1697 "On the Origins of the Nations of Transylvania."[78] This region is of particular geopolitical significance from Leibniz's point of view. Since the Middle Ages it had been the home of ethnic German merchants, who were often in danger of political disenfranchisement as a result of being seen as ethnic outsiders. Leibniz notes that the scholarly sources differ as to whether the

[76]Leibniz, *Sbornik pisem i memorialov*, 45.

[77]Leibniz, *Sbornik pisem i memorialov*, 42.

[78]Originally published in G. W. Leibniz, *Viri illvstratis Godefridi Gvil. Leibnitii epistolae ad diversos, theologici, ivridici, medici, philosophici, mathematici, historici et philologici argvmenti, e MSC. avetoris, evm annotationibvs svis primvm divvlgavit Christian. Kortholtvs*, Lipsiae: svmtv B. C. Breitkoptii, 1734, 5–9, republished in an altered form in *Opera Omnia*, II, 206–80. Translated here from the latter edition.

Transylvanian Germans are "new guests," or, rather, "a sort of aboriginal people." The resolution of this question has concrete political relevance:

> And it is a matter of the honor of the Saxons themselves, that they should not be believed to have been given their freedom in the manner of slaves, but that they be called by the promise of the preservation of the liberty that is instilled in them at birth. It would have been worthwhile to know whether the monogram was suspended or imparted by the original privilege of the King. Indeed, the charter of King Andrei calls them not "Saxons," but "Teutons," and the well-known Hermannus, from whom we get "Hermannstadt" [i.e., the German name for Sibiu], is said to have been from Nuremberg.[79]

Using his standard method, here, Leibniz turns to linguistic evidence, in particular, to the origins of ethnonyms: "it is well known besides that very often Germans are all indiscriminately called 'Saxons' by the Italians, the French, and other outsiders, after the appellation 'Franks' begins to be attributed to the French [*Gallis*], and the ensign of the rue, adopted no doubt belatedly by the inhabitants of Hermannstadt, appears to confirm the popular appellation and opinion."[80] And he argues that the best way to determine the precise origins of the Germans in Transylvania is to undertake vastly more systematic linguistic surveys, and to create "dictionaries" that register not only lexical differences between dialects, but also differences of pronunciation:

> Meanwhile I do not wish to deny that the Germans of Transylvania should be specifically referred back to the Saxons, especially if the dialect is known from which it will be more accurately determined whether it is cognate with the dialect of Westphalia or of Lower Saxony. Thus I should like to have some examples of how the words are pronounced by the common people, as well as an index of provincial vocabularies, which, even if they are German, will nevertheless not be known by all Germans. In this way indeed we will be better able to judge as concerns the origins and the dialect.[81]

He notes that "it seems plausible that before the ancient Dacians—also known as Gets (if we are to believe Strabo and others)—there was a Gothic or Germanic people there."[82] This does not mean, however, that

[79]Leibniz, *Opera Omnia*, IV, 2, 206–7.

[80]Leibniz, *Opera Omnia*, IV, 2, 206–7.

[81]Leibniz, *Opera Omnia*, IV, 2, 206–7.

[82]Leibniz, *Opera Omnia*, IV, 2, 206–7. Leibniz is likely drawing on the work of Ogier Ghislain de Busbecq, who in the sixteenth century had established the existence of a variety of Gothic spoken in the Crimean region, leading to speculation that Germanic languages were widespread and ancient in large portions of Southern and Eastern Europe. Crimean

the Germans currently residing in Transylvania may be considered indigenous, since what "was German in that place . . . was long since eradicated by the migrations of peoples."[83] However, the fact that there is an ancient legacy of Germans in this Hungarian territory means that the Saxon merchants who have settled there more recently should not be thought of has having been "given" the right to live there by the king, but rather as being "called" there "by the promise of the preservation of the liberty that is instilled in them at birth."[84]

Key to the project of determining who comes from where in the ethnic *macédoine* of Transylvania is the close, descriptive study of dialectal variations. Leibniz repeats his familiar plea for the composition of more dictionaries, but this time in reference to the multiple dialects of the German language throughout Transylvania, in order to better understand the range and nature of the differences between speakers of German in Germany proper, on the one hand, and on the other those who had migrated into regions where they were in the minority. "It would be most desirable," he writes, "to have a little dictionary of the language of the common people of German Transylvania, and to request that other examples as well be added, which would be genuine, rather than made to fit our own way of speaking."[85] The purpose of dictionaries for Leibniz is plainly descriptive rather than prescriptive. By accurately reflecting local ways of speaking, they will reflect local knowledge, and so too, it may be hoped, they will reflect features of the locality itself. In important respects, Leibniz's call for dictionaries that are "genuine" parallels his ethnographic sensitivity in other domains, among them his siding with the renegade Jesuits in the so-called rites controversy, who held that Catholic ritual in China must be adapted to accommodate elements of traditional ancestor worship, which is an expression of the same concern for the adaptation of universal truth to local contexts that motivates his interest in seeing the Ten Commandments composed in Samoyed.

In the same text Leibniz also shows a remarkable knowledge of the ethnic diversity of Transylvania, and offers sophisticated, if sometimes erroneous, accounts of the origins and connections of the different groups. He wonders whether the Transylvanian Szekers are "truly Hungarian" or rather "Sarmatian," "Slavonic," or "from another Scythian people." He notes that the Wallachian language (which is to say Romanian) "has much in common with Latin," but adds that there may be a Cuman, and

---

Gothic survived into the late eighteenth century. See de Busbecq, *Itinera constantinopolitanum et amasianum*, Antwerp: Ex officina Christophori Plantini, 1581.

[83]Leibniz, *Opera Omnia*, IV, 2, 206–7.

[84]Leibniz, *Opera Omnia*, IV, 2, 206–7.

[85]Leibniz, *Opera Omnia*, IV, 2, 206–7.

thus Turkic, substratum as well.[86] While this is not borne out at a linguistic level, it is worth noting that modern Romanian is in fact the result of the admixture of Latin over a deeper layer of pre-Roman, indigenous languages in the region, particularly Dacian (related to the Thracian of ancient Southeastern Europe). Moreover, recent scholarship has confirmed that Cumans were indeed among the noble families who played an influential role in medieval Transylvania.[87] Leibniz also disputes the view that the Roma people are of Jewish, Near Eastern origin (while also failing to correctly identify their north Indian origin).

In sum, Leibniz shows himself in this text to be unusually well informed about the subtle and complex ethnic patchwork of Southeastern Europe, and also recognizes that the single best method for improving our knowledge of this patchwork is improved, systematic, fieldwork in comparative linguistics. What this work will ultimately yield, at a practical level, is either a justification of the current political order—which ethnic groups have which rights in which regions—or a part of an argument for a change in the political order.

Leibniz's ethnohistorical approach is in clear evidence as well in his narrower work on the heritage of his employer, the Elector of Brunswick. As already mentioned, it was Leibniz's official task, for much of his adult life, to write the history of the House of Brunswick, and to do so in a way that would prove maximally politically advantageous for his employer. An important part of this project involved the establishment of genealogical connections to other noble lineages in different parts of Europe, most notably the House of Este in northern Italy: "Authors remain in agreement that the origins of the Houses of Brunswick and Este are the same, in that they descend from one and the same *tige*, directly along the male line. It is true that very able people have recently called this into question, since those same historians who spoke of it did so almost without basis, and mixed in a number of errors. But I have found very convincing evidence."[88] Thus Leibniz's work as court genealogist required him to do serious research as a medieval historian. Yet he appears constitutionally unable to stay focused on the narrow matter at hand, and evidently seems to believe in earnest that in order to adequately write the narrow history of European noble families that is his official task, he must account for the origins and migrations of European peoples. Indeed, ultimately Leibniz believes that a truly adequate work of history would be one that accounts for the origins of the earth itself, the formation of the

[86]Leibniz, *Opera Omnia*, IV, 2, 206–7.

[87]See in particular Neagu Djuvara, *Thocomerius-Negru Vodă. Un voivod cuman la începuturile* Ţării Româneşti, Bucharest: Humanitas, 2007.

[88]Leibniz, "Lettre de M. Leibniz sur la connexion des Maisons de Brunsvic et d'Este," in *Opera Omnia*, IV, 2, 80–85, 80.

continents and so on: thus we have his important and innovative work on geomorphology from the 1690s, which has subsequently been given the title *Protogaea*. Leibniz's maximally wide-focused approach to history considerably delayed the completion of the narrow for which his employer was waiting; indeed it was unfinished at the time of his death.

Leibniz's method as a historian of medieval noble lineages often involves appeal to the evidence of shared proper names, and when he is writing on relatively distant European dynasties, such as the period of Norman rule in Sicily, he seldom fails to mention, wherever possible, shared names, and evidence of kinship, with the lineage he is officially charged with reconstructing.[89] Not only onomastics, but also ethnonymy, or the study of the names of ethnic groups, helps Leibniz in the task of determining who comes from where, and who is related to whom. He makes much, for example, of the evident connection between the word "German" and the Spanish word for "brother" (*hermano*), both of which appear to be cognate with a number of terms for kinship, for example the notion of *cousin germain* in French.[90] As is typical of Leibniz's approach to ethnonymy, he also speculates that the ultimate origin of the word that gave us the Latin name for Germany, the Spanish word for "brother," a name for a certain type of cousin relation, the adjective "germane," and many other words still, is the proper name of a ruler: "I believe not only that the *Germans* come from *Herminons* or *Hermins*, but also that these people apparently have their name from an ancient prince or hero named *Irmin*, which is the same thing as *Arminius* or *Herman*."[91]

In his 1843 *Critique of Hegel's Philosophy of Right*, Karl Marx wrote scornfully that "the secret of aristocracy is zoology," noting that "this is, of course, why we find such pride in blood and descent, in short, in

[89]Leibniz, "Guillelmi Appuli Historicum poema de rebus Normannorum in Sicilia," in *Opera Omnia*, IV, 23–27, 23. Thus in a text on the Norman dynasty in Sicily, "*Azo* potens Marchio, origine Estensis, ex *Cuniza* veterum Guelforum haerede, pater *Guelfi* Ducis Bavariae, qui stirpem in Germania propagavit, in Brunsvicensibus Principibus adhuc superstitem ex altera conjuge filios habuit *Hugonem* & *Fulconem Hugo* filiam duxit *Roberti Guiscardi*, Normannorum Principis."

[90]Leibniz, "G. G. Leibnitii De origine francorum disquisitio, Annotatiunculis illustrate," in *Opera Omnia*, IV, 2, 146–67, 157. "Nam quemadmodum alibi dudum notavi, Herminones vel Hermunduri, Heermaenner, Hermanni eadem sunt fere vox, quae *Germani*, variata tantum aspiratione initiali; uti idem est *Lotharius*, *Hlotarius*, *Clotarius*; Germani fratres Latinorum iidem cum iis quos Hispani vocant *Hermanos*; denique Gammanus Latinorum idem, qui Hummer Germanorum; ut alia exempla nunc non addam."

[91]Leibniz, "Lettre à M. Nicaise sur l'origine des germains," May 28, 1697, in *Opera Omnia*, IV, 2, 205–6, p. 205. "[J]e crois non-seulement que les *Germains* viennent des *Herminons* ou *Hermins*, mais encore que ces peuples ont apparemment leur nom d'un ancient Prince ou Héros nommé *Irmin*, ce qui est la même chose qu'*Arminius* ou *Herman*."

the life history of the body."[92] Marx meant this half in jest, as a way of exposing the supposedly baser or more lowly underside of aristocratic preoccupation with lineage. But in Leibniz what we see is a very serious confirmation of the converse point: that the study of aristocratic lineages opens up access to vastly broader questions of "zoology," which stands in synecdochically here for the study of the origins, descent, and diversity of natural beings. Leibniz, assigned the task of studying the origins of the House of Brunswick, spins out from this singular generational series an account of the origins of natural order in general. The secret of zoology, so to speak, is in aristocracy.

7.6. Conclusion: Diversity without Race

It may seem a digression for us to focus at such length on the ethnic diversity of the Russian Empire, on Slavs and Finns and Tatars, when we are seeking to tell at least part of the history of the origins of the modern race concept. But to return to some of the conclusions drawn earlier in this book, there is no reason in principle why ethnic difference between, say, Russians and Tatars should not be cognized "racially" in more or less the same way as, say, ethnic difference between African Americans and "white" Americans tends to be cognized in the United States today. "Race," as we are understanding it in this book, is the attribution of essential difference to ethnic others, and such attribution is never based on phenotypic differences between human groups, even if these differences are often invoked in an ad hoc way in order to rationalize the attribution.

Leibniz's own ethnographic interests, while extending to some extent to all regions of the world, are principally focused on Europe and Eurasia, and if we set out looking for his most mature and extensive account of the nature of human diversity, it is in his writings on Eurasian ethnic groups that we are going to find it. But it is not that we are going looking for whatever Leibniz might happen to have to say on the problem of human diversity, and analyzing it for no better reason than that it is Leibniz who said it. On the contrary, Leibniz's general approach to the problem of ethnic diversity in Eurasia, properly understood, provides a useful and important alternative to the usual understanding of the development of the prevailing modern European conception of the nature of human diversity, according to which real subgroups of the human species, more or less rigidly attached to particular geographical regions, may be marked off from one another and taxonomized as if they were in important re-

[92] Karl Marx, *Critique of Hegel's Philosophy of Right*, trans. Joseph O'Malley, Cambridge: Cambridge University Press, 1967, 106.

spects analogous to biological species. What we find is that Leibniz does not at all share in this conception, and quite the contrary stands out in resistance, intentional or no, against it. Neither is he a defender of an explicit theory of racial realism, nor can he properly be described as a liberal racist in Popkin's sense. He does wish for the different cultural groups around Eurasia to be absorbed into the Christian religion, but he does not seem to think that this is incompatible with the preservation of their own cultural form of life, and does not seem to think of it as a matter of bringing these cultures up to a higher form of civilization.

To the extent that Leibniz deploys the term "race," he uses it to describe a temporal lineage or, again, a series of generations. Thus Leibniz writes, for example, of a prince Azon or Albertus of Lombardy, from the end of the tenth century, who lies at the origin of a particular lineage he is seeking to trace. He identifies "Guelph, his eldest son, who died without children; the second, Henri, called 'the Black,' succeeded him, and married Wulfhilde, the daughter of Magnus, the last duke of Saxony of his race."[93] Or again, "If Arminius was of the Gallic race (a very new idea), it is necessary that the Cherusci had been a Gallic colony."[94] Consistently, we see Leibniz using "genus" in Latin texts exactly where we would find "race" in the texts written in French. Often, the Latin term does have a transnational signification, picking out separate but related groups in different emerging nation-states that Leibniz believes to share common linguistic and historical roots. Thus in his *Dissertation on the Origins of the Germans* Leibniz typically refers to "the whole Germanic race [*totum Germanicum genus*]," a category in which he includes Scandinavians and Hollanders.[95]

When Leibniz seeks to account for the basis of ethnic diversity, it is clear that he rejects the idea that there are essential differences between groups. This is so even where phenotypes diverge significantly, as we saw, for example, in his account of the origins of the Kalmyks. For Leibniz, as we have by now seen at length, one must look past outward appearances and instead consider natural language and, wherever possible, historical texts. Over the course of the twentieth century, of course, the science of genetics became increasingly central to the project of uncovering the true

[93]Leibniz, *Opera Omnia*, IV, 2, 81.

[94]Leibniz, "Reponse aux Remarques sur la dissertation précédente, publiées dans le Journal de Trevoux par le P. Tournemine Jésuite," in *Opera Omnia*, IV, 2, 167–73, 173. "Si *Arminius* a été de race Gauloise (sentiment fort nouveau) il faut que les Cherusques ayent été une colonie Gauloise."

[95]Leibniz, "Dissertatio de origine Germanorum," in *Opera Omnia*, IV, 2, 198–205, 202. "Sane hodieque, ubi multa faciliora sunt quam olim itinera, ax centum exteris, qui de citerioribus oris in Scandinaviam venere, . . . vix unum putem accessisse eo itinere, quo totum Germanicum genus venisse volunt sententiae adversae Patroni."

relations between different human populations. Today, genetics does in a more efficient and precise way what historical and comparative linguistics sought to do, most famously in the nineteenth century with the work of the Grimm brothers, Max Müller, and other well-known philologists. Certain questions genetics is now in a position to answer are in an important sense direct descendants of questions first posed in historical linguistics, to the extent that both fields seek to trace out lineages based on a code in need of decipherment, rather than determining fixed and bounded subgroups of human beings based on external features that are supposedly transparently given in perceptual experience.

It may, again, seem a detour to focus at such length on Leibniz's work on the history of European nationalities, when we are principally concerned with the origins of the concept of race in the modern sense of *trans*national, basic human subtypes. But what emerges from an investigation of this work is precisely Leibniz's template for understanding human diversity in general. For him, there simply is no more basic category, other than linguistic communities, into which different groups of people fall.

Leibniz's preoccupation with the origins of European nationalities might be thought to obscure from his view the problem of human diversity at the global scale: he is a "Eurocentrist," one might say, and therefore failed to notice, from his parochial perspective, that there are deeper problems of diversity that arise when we move to the global perspective. But another way of understanding Leibniz's work in this area gives us quite the opposite lesson: that Leibniz did not recognize anything either unique or unifying about "European" as a category, and so the differences that he was highlighting when he separated, say, Turkic Cumans from Latinate Wallachians, or the Sami from the Germanic Scandinavians, were truly differences in the greatest sense in which Leibniz was willing to recognize these within the human species: difference that arises from separate lines of descent, whose distance or proximity to other lines is largely measurable through language.

Because Leibniz was, among other things, a medieval historian, he was well aware of the fact that many of the ethnic groups making up the nations of early modern Europe had Asian origins—indeed, he was even inclined to postulate Eurasian or Pontic origins for groups, such as the Franks, that in fact did not come from there and that we are inclined to think of as, so to speak, aboriginally European. It is also crucial to bear in mind that Leibniz was a German, and was learned in Latin history and literature. As such, he was well aware that his own people were in an important respect "barbarians" par excellence, along with Huns, Scythians, and others who, the standard story went, brought about the decline of classical civilization and precipitated a millennium of darkness. Indeed, he explicitly speaks of the Germans and the Scythians as having parallel histories.[96]

Leibniz's ethnohistorical project may be seen, as we have been arguing, as a pursuit at the more general level of the genealogical project he was official charged with carrying out in the focused project of the history of the House of Brunswick. Ethnohistory, accounting for who came from where, for example, in the Balkans and in the Russian Empire, takes him at least as far as Central Asia and Mongolia. If his project did not extend farther into East Asia to include much work on the origins of the Chinese, this is likely only because, as a matter of historical fact, Chinese noble lineages had played no role in European dynastic successions, which it was his principal charge to study.

In other words, Leibniz's significant work on the origins and relations of European races does not at all indicate a greater interest in the European "race," but indeed the lack of any idea of such a thing. Europeans are Eurasians, for Leibniz, and on his view there is nothing particularly special about this subset of the human species. If he had had the time and the resources, he surely would have set to work, and implored others to set to work, on an ethnolinguistic map of the entire world, which for him would have been an exhaustive record of human diversity. Such a map, again, would have contained nothing about how the different groups of people looked.

Leibniz's ethnohistorical project may also be seen, in turn, as a concrete application of his general, indeed his most basic, philosophical commitment: that the world consists in unity within diversity. This commitment is ordinarily understood to be of interest to Leibniz mostly in accounting for how the world of complex corporeal substances can arise from the basic unities that are for him the ground of all reality. But the idea that diversity or multiplicity must be grounded in unity in order for it to even be possible to speak of the diversity *of* this or that appears to extend as a sort of leitmotif beyond the metaphysics and into the anthropology. Indeed, beyond the problem of ethnic diversity, there is of course also the problem of *individual* diversity: no two human beings are alike; the stock of individual forms nature is capable of producing within a given species appears, in fact, infinite. And yet there is no difficulty at all in recognizing, as one passes from one unique human face to the next, that there is also an identity between them, the identity that comes from shared, equal membership in humanity.

In the modern period, the equality of this membership would come to be questioned, not so much from one individual to the next (though here

[96]See, e.g., Leibniz, "Praefatio accessionibus historicis quibus potissimum continentur scriptores rerum Germanicarum," in *Opera Omnia*, IV, 2, 53–57, 54.

too), as from one human group to the next. The terms in which claims of inequality across groups were generally articulated were the terms of "race": different groups would come to be conceptualized as biogeographically distinct kinds, with different internal capacities that could be placed in comparison with, and measured against, those of other groups.

But Leibniz preserved an older conception of "race" that saw it as a temporal unfolding or a catenary succession: as the successive appearance of links in a chain, each of which retained perfect equality with all other links in virtue of the fact that each is in the end only a moment or an element of a certain generational series. There are indeed subseries, such as the noble families of Europe or the "tribe" or nation of Kalmyks, but in the end these series all join up in the single chain of humanity, in virtue of the fact that all subseries lead back to Adam and Eve as the first parents. Leibniz's conceptualization of race in these terms is rooted in his work both as a historian and as a metaphysician: in his concern to account for the variety of the human species in terms of the unfolding of an original unity.

Chapter 8

Anton Wilhelm Amo

8.1. "The Natural Genius of Africa"

Leibniz was not the only person in Germany to seek to advance his own interests by cultivating connections to Russia. Duke Anton Ulrich of Braunschweig-Wolfenbüttel, to cite another noteworthy case, seems to have hoped to impress Peter the Great through a public demonstration of his own enlightenment, which extended even so far as to copy the Russian tsar's relatively enlightened treatment of his own slave, Abram Petrovich Gannibal, who would eventually be emancipated and rise to a glorious military career in Russia. It is in this geopolitical context that the relatively more humble career of the Afro-German philosopher, Anton Wilhelm Amo, played itself out.

Amo was born at the beginning of the eighteenth century in or near present-day Ghana, and was purchased in Amsterdam for Anton Ulrich (the father of August Wilhelm), to serve as the duke's "chamber moor," at around the age of three.[1] As the East German scholar Burchard Brentjes

[1] A partial list of scholarship on Anton Wilhelm Amo includes the following sources: William E. Abraham, "The Life and Times of Anton Wilhelm Amo, the First African (Black) Philosopher in Europe," in Molefe Kete Asante and Abu S. Abarry (eds.), *African Intellectual Heritage: A Book of Sources*, Philadelphia: Temple University Press, 1996, 424–40; Burchard Brentjes, "Ein afrikanischer Student der Philosophie und Medizin in Halle, Wittenberg und Jena (1727–1747)," in *In memoriam Hermann Boerhaave (1668–1738)*, Halle, 1969; Burchard Brentjes, "Anton Wilhelm Amo, First African Philosopher in European Universities," *Current Anthropology* 16, 3 (1975): 443–44; Johannes Glötzner, *Anton Wilhelm Amo. Ein Philosoph aus Afrika im Deutschland des 18. Jahrhunderts*, Munich: Enhuber, 2002; Paulin J. Hountondji, *Un philosophe africain dans l'Allemagne du XVIIIe siècle*, Paris: Presses Universitaires de France, 1970; Jacob Emmanuel Mabe, *Anton Wilhelm Amo interkulturell gelesen*, Nordhausen: Traugott Bautz, 2007; Simon Mougnol, *Amo Afer. Un Noir, professeur d'université en Allemagne au XVIIIe siècle*, Paris: Harmattan, 2010; Wolfram Suchier, "Ein Mohr als Student und Privatdozent der Philosophie in Halle, Wittenberg und Jena," *Akademische Rundschau* 4 (1916); Wolfram Suchier, "Weiteres über den Mohren Amo," *Altsachsen: Zeitschrift des Altsachsenbundes für Heimatschutz und Heimatkunde* 1, 2 (1918).

argued,[2] the duke appears to have been following the example of Peter the Great in seeking to educate his new charge, in order, in effect, to demonstrate his Enlightenment bona fides and thereby to curry favor with the Russian Empire. Gannibal's fate was much happier than Amo's. The African in Petrine Russia would be legally adopted and raised as a son by the tsar, and would in the end enjoy a glorious career as an officer and engineer. He married into nobility and had generations of descendants, including the father of modern Russian literature, Aleksandr Pushkin.[3]

If the climate for a former slave was better in eighteenth-century Russia than in a small German electorate, the fact remains that in both cases the original idea of providing these men with an education was based on a conviction of the early Enlightenment that some decades later Kant, as a paradigmatic spokesperson of the high Enlightenment, would explicitly contest. The tsar, the duke, and all of the early sources on both Gannibal and Amo certainly never lose sight of these men's origins, but there is no presumption that origins are destiny. To most of those who surrounded him at the Universities of Halle, Wittenberg, and Jena, Amo was, variously, a "Moor," an "Ethiopian," and an African. But these features were not taken generally as markers of inborn limitations. Amo was, for example, addressed by the rector of the University of Wittemberg as a *vir nobilissime et clarissime.* As the abbé Grégoire comments in his *De la littérature des Nègres* of 1808, "the University of Wittenberg did not have, on the basis of a difference of color, the prejudices of so many men who claim to be enlightened."[4]

There are very few traces from Amo's early life that enable us to reconstruct his biography in any detail: a receipt with his name on it here, and there an entry in the church register at Wolfenbüttel's Salzthal chapel. To depart too far from these scattered sources, and to attempt to retell the philosopher's life in richer detail, is to undertake an exercise of the imagination, rather than proper intellectual biography. At the same time however, we do know enough about the life, work, and opinions of many of the people around Amo, whether in the house of Braunschweig-Wolfenbüttel or at the universities of Halle, Wittenberg, and Jena, to be

[2]Burchard Brentjes, *Anton Wilhelm Amo: der schwarze Philosoph in Halle*, Leipzig: Koehler & Amelang, 1976.

[3]Pushkin himself wrote a novel, never published during his lifetime, presenting a fictional account of the life of his own African great-grandfather. The novel's title, *Arap Petrogo Velikogo*, is generally translated as *Peter the Great's Moor*. The Russian term *arap*, "moor" or "slave," is derived from "Arab," thus serving as a vivid reminder of the fluidity of ethnic categories and the importance for their exonymic construction on the forms of exchange and commerce that develop between regions and cultures.

[4]Grégoire, *De la littérature des Nègres*, 200.

able to place his extant works within their context, and to offer some well-grounded speculation about Amo's own intellectual background and motivations.

Amo was baptized at Wolfenbüttel in 1707, and twenty years later was sent to study philosophy and law at the University of Halle. The fact that he was baptized, and thus at least implicitly recognized as a human being with an immortal soul, is not at all unusual for a domestic slave in Europe in the period. A somewhat more unusual life trajectory is revealed, by contrast, from the scattered bits of evidence showing that he was, from an early age, not only highly literate, with elegant handwriting and a cultivated ability for crisp expression, but also entrusted with the responsibility of carrying out financial transactions on behalf of both himself and Anton Ulrich.

After arriving at Halle in 1727, Amo would go on to produce at least three significant philosophical works. He was a defender of a philosophical vision that seems to owe much to the metaphysics of Leibniz, and that stood against the Halle school of medical Pietism as represented, particularly, by Georg Ernst Stahl.[5] Stahlianism had been fiercely contested by Leibniz himself in a bitter and polemical correspondence of 1709 between the Hannover philosopher and the Halle physician. This correspondence was manipulated and published by Stahl in 1720, four years after Leibniz's death, and it is likely that every student at Halle would have known about this work into the 1730s.[6] At issue in the polemic were the opposing views of the two thinkers on the role of the soul in the body and of the constitution of the soul itself: Leibniz defends a mature expression of the doctrine of preestablished harmony, according to which the soul only accompanies the body without causing any of the body's states, while Stahl defends a radical view of the soul's direct implication as the immediate cause of bodily states, as, literally, a motor. A significant implication of this doctrine, for Stahl, is that the soul has a great deal of influence over the conformation and health of the body, and therefore psychological and even moral features of a person can be read directly from the condition of the body. The implications for the question of human equality are not hard to discern.

[5]On the Pietist context of philosophy, and also, importantly, of medicine, in early eighteenth-century Halle and beyond, see, in particular, Johanna Geyer-Kordesch, *Pietismus, Medizin und Aufklärung in Preussen im 18. Jahrhundert*, Tübingen: Niemeyer Verlag, 2000. Pietism was a nebulous movement, and surely more a family resemblance that runs through a number of different, very singular thinkers, rather than a set doctrine to which all these thinkers subscribed.

[6]To appear in English translation as *The Leibniz-Stahl Controversy*, trans. and ed. François Duchesneau and Justin E. H. Smith, New Haven: Yale University Press, forthcoming.

Stahlian conservatism reigned at Halle for much of the first half of he eighteenth century. The great Leibnizian thinker Christian Wolff, previously rector of the University of Halle, was driven out of that city in 1723, having inspired the animosity of Pietist academics such as August Hermann Francke for his apparent defense in his popular philosophy lectures of fatalism and, significantly, for his defense of the integrity and piety of traditional Chinese religion.[7] In the years that followed, Francke assumed the position Wolff had lost, yet in general Pietism failed to provide a viable intellectual program to students for engagement with the main currents of Enlightenment thought.[8] Other, more liberal currents continued on at Halle even after the Pietist coup against Wolff, not least that represented by Christian Thomasius, the son of the teacher of Leibniz, Jakob Thomasius. The younger Thomasius, an ardent defender of what would later be called "human rights," including the right of even atheists not to be tortured, would remain in Halle until his death in 1728.

While we do not have nearly enough textual evidence to chart Amo's course of studies with any precision, we may be confident that he avoided the circles of Francke and other Pietists, and sought out more liberal mentors. Two years after he matriculated at Halle, he defended a thesis titled *De iure Maurorum in Europa* (*On the Right of Moors in Europe*), which has unfortunately been lost. We know from contemporary reports that in this work Amo argues for the freedom and equality of "Moors" in Europe (by which he means black Africans), on the basis of his reading of Roman history and law. In particular, he argues "that the kings of the Moors were enfeoffed by the Roman Emperor," namely, Justinian, and that "every one of them had to obtain a royal patent from him."[9] This meant, in Amo's view, that African kingdoms were all recognized under Roman law, and that therefore all Africans in Europe should have the status of visiting royal subjects with a legal status that precludes their enslavement. Amo's first scholarly work, then, is a remarkable achievement: an argument, made by a slave, against the legitimacy of the institution of slavery, founded in jurisprudence and historical scholarship on Roman law. Nor does this achievement flow automatically from Amo's identity as an African slave in Europe: a near contemporary, the Amsterdam-based Ghanaian minister Jacobus Capitein, would in 1742 write a defense of

[7] Christian Wolff, *Oratio de Sinarum philosophia practica/Rede über die praktische Philosophie der Chinesen*, ed. Michael Albrecht, Hamburg: Felix Meiner Verlag, 1985, 39.

[8] See in particular Martin Brecht, "August Hermann Francke und der Hallische Pietismus," in Martin Brecht (ed.), *Geschichte des Pietismus*, Band I: *Das 17. und frühe 18. Jahrhundert*, Göttingen: Vandenhoeck & Ruprecht, 1993–, 504–7.

[9] See Johann Peter von Ludewig, *Wöchentlichen Hallischen Frage- und Anzeigungs-Nachrichten*, November 28, 1729.

slavery, arguing in particular that baptism as a Christian does not constitute sufficient moral, legal, or theological ground for manumission.[10]

Even if we are unable to read this work today, the fact that Amo wrote it at least shows that he was explicitly interested in the problem of his own status in Europe, in the complicated situation of an African slave among German nobles and philosophers. This interest, however, would retreat from the surface in Amo's subsequent philosophical works, his *On the Impassivity of the Human Mind* of 1734 and his *Tractatus de arte sobrie et accurate philosophandi* (*Treatise on the Art of Soberly and Accurately Philosophizing*) of 1738.[11] There is, finally, a third extant work evidently written by Amo, namely, Johannes Theodosius Meiner's 1734 dissertation at the University of Wittenberg, titled *A Philosophical Disputation Containing a Distinct Idea of Those Things That Belong Either to Our Mind or to Our Living and Organic Body*.[12] In this work, Amo is identified as the president of Meiner's dissertation jury, yet the work itself is written from Amo's own first-person point of view. More than once, we find passages in which the author refers to *On the Impassivity of the Human Mind* as his own, leading inevitably to the conclusion that Amo, and not Meiner, composed the *Disputation*. Thus, for example, the author writes, "Quid mens humana, diximus in diss. nostra. de humanae mentis ἀπαθεία" (We explained what the human mind is in our dissertation on the impassivity of the human mind)."[13]

In contrast with his first treatise on the status of "Moors" in Europe, there is no indication at all of Amo's own biography in his subsequent works. In the *Impassivity*, presented as his inaugural dissertation in philosophy at Halle, the only explicit mentions we have of Amo's African identity are, first of all, in his appellation on the title page as "Antonius Guilielmus Amo, Guinea-Afer," and, second of all, in the dedicatory

[10] See Jacobus Johannes Elisa Capitein, *Dissertatio politico-theologica, de servitute, libertati christianae non contraria; Staatkundig-godgeleerd onderzoekschrift, over de slaverny, als niet strjdig tegen de Christelyke Vryheid . . . ; Uitgewrogte predikatien, zyndende trouwhertige vermaaninge . . .*, Liechtenstein: Kraus Reprints, 1971.

[11] See Antonius Guilelmius Amo, *Tractatus de arte sobrie et accurate philosophandi*, Halle, 1738.

[12] See Johannes Theodosius Meiner, *Disputatio philosophica continens ideam distinctam eorum quae competunt vel menti vel corpori nostro vivo et organico, quam consentiente philosophorum ordine, praeside M. Antonio Guilielmo Amo Guinea-Afro*, Wittenberg: Literis Vidvae Kobersteinianae, May 29, 1734.

[13] Meiner, *Disputatio philosophica*, 4. As the university historian William Clark explains, it was not at all unusual in eighteenth-century Germany for a dissertation to be defended by someone other than its author, and indeed for the author himself to preside over the defense. Thus in this case Meiner was the "author" of the event recorded in the *Disputatio*, while Amo was the author of the text that served as the occasion for the event. See William Clark, *Academic Charisma and the Origins of the Research University*, Chicago: University of Chicago Press, 2007.

epistle from Johannes Gottfried Kraus, the rector of the University of Wittenberg, where Amo matriculated in 1730, and where he was retained as a *Magister legens* in 1734. In the epistle, interestingly, Amo's academic superior effusively praises the "natural genius" of Africa, its "appreciation for learning," and its "inestimable contribution to the knowledge of human affairs" and of "divine things."[14] Kraus places Amo in a lineage that includes a number of classical Latin authors from North Africa, such as Tertullian, Cyprian, Terence, and Augustine, as evidence that Amo comes from a long line of distinguished thinkers. Kraus bemoans the Arabization of North Africa for bringing about the downfall of a great intellectual culture.

The fact that the rector sees Amo as a member, broadly speaking, of the same lineage that produced North African Latin authors, and does not seem to see him as separate from them on "racial" grounds, is noteworthy. Could it be that Kraus was persuaded by Amo's earlier argument in the *On the Right of Moors*, and thinks as a result of this that all Africans, including sub-Saharan Africans, are in effect descendants of Latin antiquity? On such an understanding, one might suppose that the Arabs moved into North Africa and interrupted this lineage, but that the non-Muslim, black Africans to the south of the Sahara preserved it, much as European noble families were thought to have done to the north of the Mediterranean. On this account, the place of Mauretania within Roman antiquity, as the southern- and westernmost reach of the empire into "Moorish" Africa (there is indeed an etymological connection between the "Maur-" in this place-name and the term "Moor"), would be clearly recognized, while that region's subsequent isolation from the Mediterranean world would be seen as a result of the collapse of the political order centered in Rome.

Whatever Kraus's understanding of African history, it is certainly clear from his epistle that he is interested in placing Amo, the deracinated slave, into a history that will be recognized and valued by members of the community at Wittenberg. In the course of this, he characterizes Amo's origins in a way that completely ignores anything we might recognize as "racial," favoring instead the legacies of civilizations and the unity of continents. Kraus would then likely not think of Amo as presenting a counterexample to Hume's racist observation in the decade following the publication of Amo's treatise on the impassivity of the human mind, that no black person has ever accomplished anything of note, since Kraus is not sizing up what Amo has accomplished as representative of the ca-

[14]In Antonius Guilelmius Amo, *Dissertatio inauguralis de humanae mentis apatheia*, Wittenberg, 1734, 19.

pacities of black people. He conceives of Amo as an African, just like Terence and Saint Augustine.

It would be at Wittenberg that Amo would enjoy his greatest success. As Brentjes explains, Stahlianism had made no inroads at all at Wittenberg, and when the young Amo arrived there from Halle he found himself free to pursue his criticism of Stahl's "subjective-idealistic doctrine."[15] At Wittenberg, Amo worked closely with the mechanist physician Martin Gotthelf Loescher, and under his guidance began a close study of medicine. Brentjes plausibly discerns a characterization of Amo's own position vis-à-vis Stahlianism in the 1736 chronicle of a certain Nicholas Gundling, in which "two principal camps" are identified, the "Mechanici" and the "Stahlianer."[16] The former, Gundling explains, "maintain that vital actions in the human body, for the most part, both in a healthy as well as in a sick state, arise and proceed mechanically and by means of the bodily structure." The Stahlians, by contrast, maintain that "it is namely the human soul that is the first mover in the body, and the bodily mechanical structure is only an instrument of the adjoined motor."[17] Again, Gundling does not identify Amo by name, but his description does well pick out the known anti-Stahlians at Wittenberg on whom Amo relies heavily as authoritative sources for his claims in the 1734 treatise, particularly Christian Vater.[18]

Amo's patron and (evidently) friend, Duke August Wilhelm, the son of Anton Ulrich, would die in 1731, leaving Amo to fend for himself in a cultural climate that was growing increasingly intolerant. He would return to Halle as a *Dozent* in 1736, and move to Jena after being hired there in 1739. From surviving correspondences we see that the faculty at Jena is willing, with hopefulness but without much enthusiasm, to give the applicant from Halle a chance, based solely on his merits, without much interest either in his African origins or in the patronage he once enjoyed in Wolfenbüttel. In his initial letter of introduction to the Jena faculty of philosophy, Amo announces that he is indigent, but also that he is very industrious. No mention is made of his previous noble connections in Wolfenbüttel. He gains a particularly strong supporter in the

[15] Brentjes, *Anton Wilhelm Amo*, 41.

[16] Nicholas Gundling, *D. Nicolai Hieronymi Gundlings Vollständige Historie der Gelahrtheit*, Frankfurt, 1736, 4:5236.

[17] Gundling, *Vollständige Historie der Gelahrtheit*, 4:5236–37; Brentjes, *Anton Wilhelm Amo*, 42–43. "[N]un suchen Erstere zu behaupten, dass die Actiones Vitales, in dem menschlichen Cörper, grösten Theils, sowohl in Statu sano, als in Statu morboso, Mechanice und vermittelst der cörperlichen Struktur entstünden und procedirten. . . . Die Herren Stahlianer hingegen statuiren das Gegentheil; Es sei nemlich die menschliche Seele das Primum Movens, in dem Cörper; und die Cörperlich Mechanische Structur nur ein Instrumentum ermeldeten Motoris."

[18] Brentjes, *Anton Wilhelm Amo*, 43.

dean of the faculty, Friedrich Andreas Hallbauer, who writes a note letter to his colleagues on June 29, 1739, presenting various options for the "nostrification," or the transfer of credentials from one university to another, for this impoverished philosopher: "[H]e would either have to be nostrified at no cost, or the cost should be suspended until such time as he gains earnings here; or he should be permitted provisionally to teach, until we can see whether he receives steady applause, in which case he should be allowed to be officially nostrified."[19] Hallbauer concludes, "I will be pleased if you are of the same view," and indeed most of his colleagues are. One professor at Jena, identified only as "Wideburg," presents a number of reasons why Amo's request should be supported, even before those of other applicants: "(1) in his early childhood he was taken from another part of the world; (2) he has turned from paganism to the Christian religion; (3) he has been entirely cut off and abandoned by his family and their associates; and thus (4) possesses nothing other than what he earns through his own industriousness. Since he does not wish to beg, but rather seeks to feed himself in an honest way, we should plainly help him to the extent possible."[20] An arrangement is worked out, and on July 17 Amo presents his first lesson plan for a lecture course in the Michaelmas term, 1739. It includes a curious mixture of topics, such as "[p]arts of the more elegant and curious philosophy; physiognomy; chiromancy; geomancy, commonly known as the art of divination; purely natural astrology, which is opposed to cryptography; dechifratory, or the art of deciphering, which is opposed to the superstitions of the common people and of the ancients, cut down and rejected by all people, and to those things that are the less commended by their ambiguity."[21] Here Amo is evidently attempting to draw in as many students as possible, and to gain their "applausum," which functioned roughly as an eighteenth-century equivalent to high teaching evaluations for nontenured faculty today. Given however that Gallandat's report of his 1752 meeting with Amo in West Africa also mentions the philosopher's interest in astrology and "soothsaying," it is clear that the course advertised here is not simply pandering. He is interested in topics that have fallen off the list of legitimate interests in the intervening centuries, but indeed this is a significant fact about the scope of philosophy in the era that we as historians need

[19] Cited in Brentjes, *Anton Wilhelm Amo*, 63.

[20] Cited in Brentjes, *Anton Wilhelm Amo*, 63.

[21] Cited in Brentjes, *Anton Wilhelm Amo*, 65. "Partes philosophiae elegantioris et curiosae Physiognomiam, Chiromantiam, vulgo Punctir-Kunst, Astrologiam mere naturalem, et quae opponitur Cryptographiae, artem Dechifratoriam, quam Dechifrir-Kunst vocant, Succisis, et reiectis omnibus et Vulgi, et antiquorum Superstitionibus, eisque, quae sua ambiguitate se minus commendant, trimestri temporis Spatio, cum application diligenti, ad vitam in Statu politio prudenter instituendam, perspicue, solide et sufficienter tradam."

to take seriously. In any case, Amo saw his range of interests as eminently reputable. He continues to appeal for his course by adding, "I will be covering these topics clearly, solidly, and exhaustively over the course of the whole term, with diligent application, in the aim of more prudently fostering life in the political state."[22] He signs off on the announcement as "Anton Wilhelm Amo, the African, Master of Philosophy."[23] The struggling philosopher would seem to have arrived at some modest level of success.

After this promising flurry of activity, we unfortunately have no trace at all of Amo's life until the 1747 poem by Johann Ernst Philippi, which we discussed at some length in the introduction, denouncing Amo for his "vile nature." Within the next year, in unknown circumstances (though likely ones involving more than just "melancholy," as Gallandat had conjectured), Amo will return to Africa. Amo, now roughly fifty years old, will be visited there by Gallandat in the early 1750s, and will likely die shortly thereafter. Brentjes claims, based on extensive interviews with people in Amo's native region, that there is a continuous cultural memory of Amo's achievements, passed down through oral tradition from the eighteenth century to the present day.[24]

8.2. Amo's Legacy

After Gallandat, descriptions of Amo's life and work may be divided into a few clear types. Many, especially in the late eighteenth and nineteenth centuries, treat Amo simply as a curiosity, and this is the case whether they are approving or indeed dismissive. A good example of this is J. F. Blumenbach's 1787 contribution to the *Magazine for the Latest News in Physics and Natural History*, an article "On Negroes" in which he describes Amo as "a Negro who was made into a doctor of philosophy [*einen Neger zum Dr. der Weltweisheit creirt*], who showed himself to the best advantage both in his writings as also, later, when he went to Berlin as a Royal Prussian councillor."[25] Blumenbach's claim about Amo's work in Berlin is unsubstantiated, and it is not clear on what he is basing it, though it will subsequently be echoed, on occasion, by other commentators, including Grégoire.

The abolitionist Grégoire writes, in 1808, the longest description to date of Amo's work, and it is highly admiring. But it is almost entirely

[22] Cited in Brentjes, *Anton Wilhelm Amo*, 65.

[23] Cited in Brentjes, *Anton Wilhelm Amo*, 65.

[24] Brentjes, *Anton Wilhelm Amo*, 82.

[25] J. F. Blumenbach, "Abschnitt von den Negern," in *Magazin für das Neueste aus der Physik und der Naturgeschichte*, vol. 4, pt. 3, Gotha, 1787, 9–11.

a pastiche of earlier sources, including Gallandat and Blumenbach. Grégoire does however give some indication of having read at least the *Impassivity*, and he summarizes Amo's philosophical project there with extreme concision, as follows: Amo, he writes, "seeks to establish the differences of phenomena between beings that exist without life, and those that have life. A stone exists, but is not alive."[26] What is important to note at this point is that Grégoire's principal purpose is simply to testify *that* Amo lived and wrote, and therefore that the claim of so many of his contemporaries, that no one of African heritage had ever made any noteworthy intellectual accomplishments, was patently false.

There will be a marked shift in writing on Amo over the course of the twentieth century: from someone who is frequently mentioned, to someone who is, so to speak, used, someone who is conscripted as an early representative of diverse intellectual traditions of importance to the authors invoking Amo's name, not least Marxism, African nationalism, and various hybrids of these. An illustrative example of such an approach can be found in the work of Kwame Nkrumah, the Ghanaian political leader and African nationalist thinker, who published his influential work, *Consciencism*, in 1964 in an attempt to fuse the core doctrines of Marxist-Leninist philosophy with what he saw as some of the basic elements of traditional African thought.[27] Unsurprisingly, Nkrumah goes to some lengths to refute philosophical idealism, "the self-devouring cormorant of philosophy."[28] He distinguishes between two varieties: one that is based in some theory or other of perception, a variety he associates with Berkeley and Leibniz; and one that is motivated by some degree of solipsism, which he associates with Descartes. Nkrumah sees the incipient solipsism contained in Descartes's *cogito* argument as based on the fallacy of supposing that, insofar as one can imagine oneself without any bodily member in particular, one can therefore imagine oneself as entirely nonbodily, and therefore as essentially a thinking thing. Nkrumah appears to believe that Amo, by contrast, rejected the Cartesian account of the mind in favor of a view according to which the mind, in order to accommodate

[26] Abbé Henri Grégoire, *De la littérature des Nègres, ou, recherches sur leurs facultés intellectueles, leurs qualités morales et leur littérature: suivies des notices sur la vie et les ouvrages des Nègres qui se sont distingués dans les sciences, les lettres et les arts*, Paris, 1808, 198–202, 201.

[27] As Nkrumah explains, "*philosophical consciencism* . . . will give the theoretical basis for an ideology whose aim shall be to contain the African experience of Islamic and Euro-Christian presence as well as the experience of the traditional African society, and, by gestation, employ them for the harmonious growth and development of that society." See *Consciencism: Philosophy and Ideology for De-colonization and Development with Particular Reference to the African Revolution*, 2nd ed., New York: Monthly Review, 1970, 70.

[28] Nkrumah, *Consciencism*, 18.

ideas of extended things, must itself be extended, which is to say it must be physically located within at least a portion of the body:

> The eighteenth-century African philosopher from Ghana, Anthony William Amo, who taught in the German Universities of Halle, Jena and Wittenberg, pointed out in his *De Humanae Mentis Apatheia* that idealism was enmeshed in contradictions. The mind, he says, was conceived by idealism as a pure, active, unextended substance. Ideas, the alleged constituents of physical objects, were held to be only in the mind, and to be incapable of existence outside it. Amo's question here was how the ideas, largely those of physical objects, many of which were ideas of extension, could subsist in the mind; since physical objects were actually extended, if they were really ideas, some ideas must be actually extended. And if all ideas must be in the mind, it became hard to resist the conclusion that the mind itself was extended, in order to be a spatial receptacle for its extended ideas.[29]

Subsequently, Nkrumah attributes to Descartes the view that, when the body is harmed, the pain that the mind feels can only ever be accounted for as an intellectual cognizance, and subsequent mental distress, of the fully separate mind. Nkrumah sees Amo by contrast as having explicitly argued against this view in the *Impassivity*:

> Descartes . . . tried to solve the mind-body problem by resorting to a kind of parallelism. He instituted parallel occurrences, and thus explained pain as that grief which the soul felt at the damage to its body. On this point, as on several others, Descartes was assailed by the critical acumen of the Ghanaian philosopher Anthony William Amo. According to Amo, all that the soul could do on Descartes' terms is to take cognizance of the fact that there is a hole in its body or a contusion on it, and unless knowledge is itself painful, the mind could not be said to grieve thereat.[30]

Nkrumah has, then, attributed to Amo two anti-Cartesian views: first, that the mind must be extended in order to accommodate ideas of extended things; and, second, that the mind must somehow be more integrated with the body than Descartes is able to admit, in order for it to properly be said to feel pain when the body is injured, rather than simply to take cognizance of the injury.[31]

[29]Nkrumah, *Consciencism*, 18–19.

[30]Nkrumah, *Consciencism*, 87.

[31]William E. Abraham, like Nkrumah, sees Amo's *Apatheia* as principally a critique of Cartesian dualism, and he confirms Nkrumah's account of Amo's criticism of Descartes on the experience of pain. "Amo claimed confusion," Abraham writes, "in Descartes' presentation of the thesis that it was the function of an organ to receive sensible forms (e.g., by feeling) while to judge forms when received (e.g., by taking cognizance of what is felt) was the function of the mind. Yet taking cognizance of bodily pain or contact should not require the

Amo cites or discusses Descartes on five distinct occasions in the *Impassivity*. The first occurrence is for corroboration of his own view that the soul cannot undergo passion through contact, since whatever touches or is touched is a body.[32] The second occurrence also invokes Descartes approvingly, in order to draw a distinction between the way ideas are formed in the mind of God and of other thinking substances that lack a "very tight bond and commerce with the body." Amo denies here that there could be any representation in God's mind, "since representation supposes the absence of the thing to be represented." Instead, God's nonsensory thoughts about created substances are ones, presumably, that concern the concept of these substances directly, as fully present to God's mind. The third occurrence appears to be an invocation of Descartes, again, in order to clarify the notion of "internal senses," defining these as "passions or affections of the soul." However, subsequently Amo will set up the difference between his own view and Descartes's precisely on this point: he denies that there can be passions of the soul at all, since all sensation occurs only in "the living and organic body." Amo cites an important letter to Princess Elisabeth, in which Descartes explains that "there are two things in the human soul on which all the cognition that we are able to have of its nature depends, one of which is that it thinks, the other that, united to a body, it is able *to act* and *to suffer* together with it."[33] Here Amo states his opposition starkly: "In reply to these words we caution and dissent: we concede that the mind acts together with the body by the mediation of a mutual union. But we deny that it suffers together with the body." In his final reference to Descartes, Amo again criticizes him, not so much for holding the wrong view of whether or not the soul may experience passions, but rather for contradicting himself on this matter: by his own lights, Amo thinks, Descartes is in truth compelled to share Amo's own view that the soul, to the extent that it is defined as a thinking

---

mind itself to *feel* pain or contact, or *sense* anything at all. A faculty of sense is not an apposite feature of minds. Hence, Amo denied that the mind could *feel*, urging that sense organs were only a medium, but not an instrument, in a theoretical conception of the occurrence of sensing. In this theory, without sense organs, there would be no sensing; and the entity with the faculty of sense should be the entity comprising living organs, namely the body." See William E. Abraham, "Anton Wilhelm Amo," in Kwasi Wiredu (ed.), *A Companion to African Philosophy*, London: Blackwell, 2004, 190–99, 195. See also William E. Abraham, "The Life and Times of Anton Wilhelm Amo, the First African (Black) Philosopher in Europe," in Molefi K. Asante and Abu S. Abarry (eds.), *African Intellectual Heritage: A Book of Sources*, Philadelphia: Temple University Press, 1996, 424–40.

[32] He refers to René Descartes, *Renati Descartes Epistolae, Partim Latino sermone conscriptae, partim e Gallico in Latinum versae*, pt. 3, Amsterdam: Typographia Blaviana, 1683, 420.

[33] Descartes, *Epistolae*, pt. 1, letter 29, 59.

thing, cannot undergo passions, since thinking is an action of the mind, not a passion.

In sum, Amo does indeed criticize Descartes in the *Impassivity*, but not at all for the reasons Nkrumah and others have held. Amo does not criticize Descartes for conceiving the mind as excessively distinct from the body, but rather as not nearly distinct enough. Far from rejecting Cartesian dualism, on the contrary Amo offers a radicalized version of it. For him, all sensation is "suffering" in living beings, which is to say undergoing passion. But if the mind can do nothing but think, then it follows that it can undergo no passions at all. It follows, in turn, that the sensation of pain is something that occurs entirely within the body, while if the mind is involved at all this will be through a simple cognizance of the pain the body is feeling. In other words, the objection that Nkrumah believes Amo is leveling against Descartes, that the mind cannot feel pain, is one that could more rightly be raised against Amo himself.

How can we make sense of this perceived need to set Amo up against Descartes? There has been considerable debate over the past century as to what precisely constitutes a contribution to or an instance of African philosophy. On a certain definition, Amo's work cannot be considered such a contribution, since he plainly had his intellectual formation within the context of the European intellectual tradition. Thus the Ghanaian scholar Kwame Gyekye writes: "The cultural or social basis (or relevance) of the philosophical enterprise seems to indicate that if a philosophy produced by a modern African has no basis in the culture and experience of African peoples, then it cannot appropriately claim to be an *African* philosophy, even though it was created by an African philosopher. Thus, the philosophical works of the eminent Ghanaian thinker Anton Wilhelm Amo, who disntinguished himself by his philosophical acumen in Germany in the eighteenth century, cannot be regarded as *African* philosophy."[34] Other scholars, however, have with varying degrees of explicitness attempted to identify distinctively African contours. Thus Kwasi Wiredu represents Amo's contribution to philosophy as principally a rejection and critique of Descartes's dualistic ontology, arguing further that Amo's strength lies in his points of disagreement with Descartes, and his weakness in his points of agreement. Cartesian dualism is "a conceptual inconsistency dear to much Western metaphysics," while Amo's critique of it, Wiredu speculates, may come from residual commitments that he absorbed from his early life surrounded by fellow members of the Akan culture. "May it not be," Wiredu asks, "that some recess of Amo's

[34]Kwame Gyekye, *An Essay on African Philosophical Thought: The Akan Conceptual Scheme*, Cambridge: Cambridge University Press, 1995 [1987], 33–34.

consciousness was impregnated by the concept of mind implicit in the language and thought of the Akans?" He continues,

> [I]n the Akan conceptual framework, insofar as this can be determined from the Akan language and corpus of communal beliefs,[35] the feeling of a sensation does not fall within the domain of the mental, if by "mental" we mean "having to do with the mind." Mind is intellectual not sensate. This is obvious even at the pre-analytical level of Akan discourse. The Akan word for mind is *adwene*, and I would be most surprised to meet an Akan who thinks one feels a sensation—a pain, for instance—with his or her *adwene*. No! You feel a pain with your *honam* (flesh), not with your *adwene*.[36]

This is a very intriguing speculation, but what if a more proximate source of Amo's particular philosophical commitments can be located? In fact, as we have already seen, properly understood Amo's position is not really a critique of Cartesian dualism at all. Rather, it is a critique of a philosophical position that was very influential in the precise context in which Amo worked in the 1730s in Halle and Wittenberg: namely, the vitalism of Stahl and his commitment to a psychosomatic medical philosophy of health and illness.

Does this account, if correct, mean that Amo's African identity played no role in the philosophical commitments he took up? By no means. In fact, Amo's anti-Stahlianism took shape within a very specific context in Halle, in which the liberal Leibnizian-Wolffian philosophy took on a political significance in a sectarian battle against the conservative Stahlians and other Pietists. The somewhat unorthodox popularizer of G. W. Leibniz's philosophy, Christian Wolff, had been ousted from his chair at the University of Halle in 1723, after which his adversary Joachim Lange imposed a version of Pietism as the dominant philosophical current at the university. However subterranean currents of Wolffianism endured at Halle, particularly among students. Amo, who arrived there four years after Wolff's ouster, appears to have been one such student.

Amo's particular version of dualism, as spelled out in the *Impassivity*, makes perfect sense in light of his parti pris for the Leibnizian-Wolffian

[35] Among the declarations in the influential resolution made by the Commission on Philosophy at the Second Congress of Negro Writers and Artists, held in Rome in 1959, is the idea "that the African philosopher must base his inquiries upon the fundamental certainty that the Western philosophic approach is not the only possible one; and therefore . . . that the African philosopher should learn from the traditions, tales, myths, and proverbs of his people, so as to draw from them the laws of a true African wisdom complementary to the other forms of human wisdom to bring out the specific categories of African thought." Here, Wiredu is implementing this very approach.

[36] Kwasi Wiredu, "Amo's Critique of Descartes' Philosophy of Mind," in Kwasi Wiredu (ed.), *A Companion to African Philosophy*, Blackwell Publishing, 2004, 200–206, 204.

camp against the Stahlians. Moreover, Amo's affiliation with this camp would have made particular sense for an African philosopher working in Germany in the early eighteenth century, insofar as it provides the resources for a properly egalitarian and antiracist philosophical anthropology. We do not need to go back to the intellectual context of Akan culture in West Africa, and to oppose it to Western rationalist metaphysics, in order to make sense of Amo's philosophy. It is enough to understand the much more local divisions between different philosophical positions to which Amo was exposed in the learned world of Lower Saxony.

When we do so it becomes clear, in particular, that it is not to refute Cartesian dualism that Amo presents his own account of the mind-body relationship. It is, rather, to position himself against the Stahlian tradition of medical philosophy, one of the central convictions of which had been that it is the soul that is in the end responsible not just for the motions of the body, but also for its conformation and appearance. Amo instead inserted his own views within a broadly Leibnizian theory according to which mind and body harmoniously run on two distinct tracks, so to speak, while all of the states of the body unfold from entirely mechanical causes. If we properly understand the context and the stakes of these two positions, we also see why, in the period, the former would have been understood as lending philosophical support to the racist view that black Africans are in some way morally degenerate or inferior, while the latter, Leibnizian view was understood, by contrast, as showing the way to a properly egalitarian philosophical anthropology.

8.3. The Impassivity of the Human Mind

The 1734 *Impassivity* is Amo's major extant philosophical work, and it is written in the typical style of an early eighteenth-century academic philosophy dissertation. It consists in rigorous definitions, theses, and proofs, all extremely concise and generally lacking in contextualizing information. The principal thesis, contained already in the full title, is that the human mind is entirely incapable of sensation, that sensation is a capacity of the body alone, while the mind does nothing but think. These theses are established by various means, including straightforward deduction from definitions, but also appeals to authority, and here not least to the Bible. Significantly, Amo also cites a wide variety of contemporary medical and physiological literature, demonstrating a willingness to supplement his philosophical arguments with a deep knowledge of relevant empirical data.

Amo understands "life" as a property of certain bodies, namely, organically structured bodies, such as those of humans and animals. For him, to live and to exist are not the same thing, and the mortal life of the body is not the same thing as the eternal existence of the soul. Life, in this reduced sense, is simply a temporary condition of the body, which, arranged in a certain way, is able to sense. "If therefore the body is killed and can be killed," Amo argues, "it follows that it lives; if it lives, it senses."[37] Amo insists that there can be no intermediate principle, such as blood or breath, that binds, or mediates between, body and soul. Thus in considering the passage in Leviticus 17, which maintains that "all life is in the blood,"[38] Amo argues that this is not incompatible with his own strict dualism, since life is in any case only a property of the body.

It is important to emphasize here, moreover, that Amo's dualism is in one central regard deeply opposed to Cartesian dualism: Descartes, while arguing that ontologically speaking the mind and body have nothing in common, nonetheless goes to great lengths to explain how the soul has its own "seat" within the body, is intimately wrapped up with the body, and can therefore be affected in diverse ways by the changing succession of states of the body. Here is not the place to determine whether Descartes's theory is coherent or not. It is enough for us to note that Amo believes it most certainly is not. The soul can undergo no passion through contact, as he argues against the Cartesian view, "[f]or whatever touches and is touched is a body."[39] Amo also excludes the possibility that the soul could be affected by "communication" or "penetration," concluding instead that the succession of states of the soul must unfold from within itself: "Every spirit operates spontaneously," Amo writes, "i.e., intrinsically. It determines its operations toward a goal that is to be pursued, and operates absolutely without any external influence."[40]

Amo's work reveals a firm commitment to a clear philosophical position, backed up by solid argumentation. His principal concern is to remove the mind (*mens*), as a variety of spirit (*spiritus*), from the functioning of the body. In order to do this, he believes it is necessary to argue that the mind is not capable of passions of any sort. Spirit is by definition active, and therefore to say that such an entity undergoes passion is to utter a straightforward contradiction. Spirit operates not through outside influence, either from another spirit or from a material body, but rather through intentions, that is, "from a precognition of a thing that should arise, and of an end that it intends to pursue through its operation."[41]

A human mind, for Amo, is not just any subvariety of spirit; it is, in particular, "a purely active and immaterial substance, having commerce

[37] Amo, *De humanae mentis apatheia*, 16.
[38] Amo, *De humanae mentis apatheia*, 17–18.
[39] Amo, *De humanae mentis apatheia*, 6.
[40] Amo, *De humanae mentis apatheia*, 7.
[41] Amo, *De humanae mentis apatheia*, 7.

with the living and organic body, having knowledge and operating from intention according to a determinate end of which it is conscious."[42] Now, it might seem that passions would be an unavoidable consequence of such commerce, but Amo quickly goes on to explain this notion as holding simply "that the body is made use of on behalf of the subject inhering in it," and "on behalf of the instrument of its operation and the medium."[43] The body, in turn, is described as "most elegant, first made by the creator from diverse vital and animal organs, and propagated from there through generation."[44] This claim is evidently continuous with the theory of organic structure developed in much greater detail in the Meiner dissertation, and also plainly evident in Leibniz's model of organic bodies or "divine machines," which consist in individual organic beings nested within the structure of other organic beings.[45] For Amo, evidently, as for Leibniz, to be an organic individual is to be elegantly structured by God at the creation, to consist in mutually subordinating and dominating organic systems, which are transmitted and activated through the chain of reproduction, but not initially brought into being through generation.

Amo cites the above-mentioned Wolffian physician Christian Vater's 1712 work, *Physiologia experimentalis*, as his authority,[46] yet, again, it is clear that his conception of spirit, body, and the relationship between them is broadly Leibnizian. For both philosophers the succession of states of the mind unfolds not as a result of changes in the body, but rather as a result of consciousness, or perhaps also subconscious perception, of ends. Leibniz describes this succession (at least in certain periods of his career) as the unfolding of an individual substance's complete concept, whereas Amo accounts for it simply in terms of "precognition." But the common heritage of both accounts is clear: both are looking for a way to account for the succession of the states of the soul without appeal to bodily change. Amo never mentions Leibniz (or Wolff, for that matter) in his 1734 dissertation. Yet, again, the position he takes up echoes the Leibnizian side of the debate between Leibniz and Stahl. In broad outline, these positions remained vital options at Halle into the 1730s, as students in philosophy there were practically forced to take sides between the liberal, Enlightenment-oriented Wolffian school on the one hand, and the conservative Pietist school on the other.

Pietism was a multifaceted movement, and many thinkers, including Stahl, who were loosely affiliated with it did not necessarily subscribe to all of its core tenets. But again, one enduring theme in Pietist medicine, a

[42] Amo, *De humanae mentis apatheia*, 8.
[43] Amo, *De humanae mentis apatheia*, 8.
[44] Amo, *De humanae mentis apatheia*, 9.
[45] See Smith, *Divine Machines*, esp. chaps. 3 and 4.
[46] See Christian Vater, *Physiologia experimentalis*, Wittenberg, 1712.

theme that is vigorously promoted by Stahl in his polemic against Leibniz, is the idea that medicine and morality cannot be separated, that the well-being or detriment of the body is largely or even principally traceable to states of the soul. For example, Stahl's general account of the causes of birth defects attributes the most important role to the imagination of the mother, and counsels, for the avoidance of birth defects, that women avoid strong passions that might be communicated to the developing fetus. This narrow concern fits into a broader theory of mind-body interaction on which, Stahl believes, the soul just is the direct cause of motions in the body, and changes in the body from without are directly experienced by the soul. Leibniz would accuse Stahl, in effect, of inadvertently making the soul play the role of intermediary once attributed to animal spirits, plastic natures, and other principles that were conceived as straddling the boundary between the mental and the physical. But Stahl believes that the boundary must be crossed by the soul itself, since otherwise there would be no way of retaining a role for the soul in the responsibility for the health and well-being of the body, and if this were to happen, wanton immorality would be the eventual result. An unspoken implication of this broad way of thinking is that we are all, in the end, responsible for the condition of our bodies, and if dark skin, for example, is implicitly or explicitly judged inferior to light skin, it follows on the Pietist line of thinking that the African is in the end guilty of some sort of moral failure that caused him to have the complexion he has.

Now it would be simplistic to suppose that Amo adopted a broadly dualist and harmonist philosophy *because* he was African. What can be affirmed with a high degree of certainty however is that it was easier for an African student of philosophy to find a home in the Leibnizian philosophy than in the Pietistic school that sought to ground the conformation of the body in a person's moral character. Mind-body dualism, as we have been arguing throughout this book, for all its inadequacies, effectively helped to prevent the fragmentation of the human species into different races. As long as humanity is rooted in something nonphysical, the observation of physical differences between different human groups cannot lead to the conclusion that there are real, essential boundaries between these groups. The unity of the human species is ensured, so long as the specific differentium of humanity is placed beyond the scope of naturalistic study of human variety. Yet the nondualism that threatened the unity and well-boundedness of humanity came not only from antitraditional empiricist thinkers such as Locke, but also from relatively conservative thinkers such as Stahl. Such thinkers did not believe that races are bounded off from one another as permanent, fixed natural kinds, but they did tend to believe that the state of any individual's body is a result of the way that individual conducts himself or herself. Thus birth defects are a result of the improper passions that a pregnant woman has allowed herself to undergo, for example, and features such as complexion or relative hirsuteness are a result of the way one or one's immediate ancestors have led their lives.

The connection between conduct and phenotype is very explicit in a thinker such as Bulwer, as we saw earlier, and rather more subtle in Stahl. It is nonetheless very evident that Stahl sees Leibniz's philosophy of preestablished harmony as threatening first and foremost because it removes the soul from any responsibility for the condition of the body, and most importantly for the health of the body. In a context in which the concept of race had not been entirely separated from questions of medicine and environmental health, we find an African student of philosophy gravitating toward a philosophical theory on which the true self, the soul, is held to unfold in its states entirely independently of the series of states of the body. This is probably not a coincidence. Amo's mentors are effusive in their praise for Amo *as* an African philosopher, while Amo himself prefers not to mention *who* he is in relation to *what* he believes. But we are nonetheless justified in hypothesizing a connection between Amo's identity as an African thinker in Europe, on the one hand, and on the other at least some of the philosophical commitments he takes up, particularly the commitment to the unity of the human species. Amo sees humanity not as a matter of gradation, measurable in the states of the body, but as a yes/no matter, a question of threshold, determined by the possession of a human soul, which is an entity that itself cannot be either healthy or ill, able or disabled, and cannot in any meaningful sense be said to have a race.

Amo's particular formulation of the core doctrine of his philosophy is in terms of the "apathy" or "impassivity" of the human mind. This is to say that, for him, the changes in the body brought about by the senses cannot lead to any modifications of the mind, since the only thing the mind does is think. Amo states this point as follows: "Every spirit is beyond any passion." He explains, "No parts, properties, or effects of another entity can be made present in spirit by a certain mediating act. Otherwise the spirit would contain in its essence and substance something other than what it should contain. Likewise, to contain and to be contained are material concepts, nor can they be truly predicated of spirit. Therefore spirit does not sense through communication, i.e., in such a way that the parts, properties and effects of the material entity should be made present in that same certain mediating act."[47] This is certainly going well beyond Cartesian dualism, which was supplemented or modified by the French philosopher so as to make room for the possibility of passions

[47]Amo, *De humanae mentis apatheia*, 5.

of the soul. For Descartes, soul and body are ontologically completely distinct, yet there must be a way for each to influence the other. In Leibniz, though the possibility of mind-body causation is strictly speaking impossible, nonetheless there remains a perfect tracking of the states of the one in the states of the other, with the result that, at the phenomenal level, Leibniz is perfectly comfortable speaking of ordinary instances of the passions having an effect on body and soul simultaneously, for example in the influence of the maternal imagination on the developing fetus. Amo does not address many of the instances of apparent psychophysical interaction that a direct confrontation with Stahl, for example, would have forced him to address, but it is clear enough in his short treatise that, for him, the separation of body and spirit must be total. Spirit can undergo no passions, and none of the parts, properties, or effects of the material entity with which it is associated can be appropriately attributed to it. Again, it follows from this that for Amo a spirit cannot have a race.

The most obvious antecedent for this line of thinking is Leibniz's argument, against Locke, that the outward conformation of a human body can have no relevance to the determination of that person's humanity. Humanity, as Leibniz says, consists in rationality alone, and this is something that is entirely nonbodily. What is more, for Leibniz the state of the body, including the conditions of its upbringing, could very well be such that the rationality of the embodied human cannot be perceived from outside. Leibniz was only tangentially concerned with race here (Locke, again, had insinuated that it is in Africa that the human species blends gradually with the apes), yet the implication of his argument for the idea of race is clear: for Leibniz, insofar as membership in the human species is a question with only two possible answers, determined by the presence or absence of a rational soul, there can be no meaningful distinction between human races.

In a rare instance of unclarity, Burchard Brentjes seeks in his study to identify Amo's philosophy with that of John Locke, and even identifies Locke's "grounding of human freedom in natural law" as "the first philosophical grounding of the equality of all races."[48] As we have seen in the previous chapter, this is a dubitable characterization of Locke's philosophical project, and indeed given that the position Amo takes up is clearly in line with the general account of the mind-body relation Leibniz offers in the *New Essays Concerning Human Understanding*, written as a respectful polemic against Locke's philosophy, it is simply not possible that here Amo would be following in a Leibnizian and a Lockean vein at once. Brentjes has trouble understanding the core project of Leibniz's philosophy, which is precisely to show how a commitment to a thoroughgoing

[48]Brentjes, *Anton Wilhelm Amo*, 43.

mechanism can be perfectly compatible with antimaterialism. This difficulty on the part of Amo's greatest interpreter no doubt arises from his education in communist East Germany, and it prevents the author from grasping perhaps the most important feature of Amo's position in the history of German philosophy: if there is any respect in which the African thinker's commitments reveal a concern for grounding racial equality in a philosophical account of the human being, this account depends precisely on a rejection of Lockean materialism in favor of the view that human nature is entirely independent of the material conformation of the body, residing instead in an immaterial soul that harmoniously accompanies the body but has no real connection to it.

In 1737 Amo would give a lecture course at Halle titled "De harmonia, seu concordia rerum," which as Brentjes acknowledges appears to have been devoted to a defense of Leibnizian preestablished harmony.[49] If Amo agreed with the philosopher whose work he taught that year, then we can conclude that he was indeed a mechanist, and not a materialist, and that, as had been the case for Leibniz himself, his belief in human equality was directly rooted in the philosophical view that a person is not a physical body at all, but a soul that accompanies the body in perfect harmony.

8.4. Conclusion: From Philippi to Kant

In the 1764 *Observations on the Feeling of the Beautiful and Sublime*, Kant tells the story of a "Negro carpenter" who, though he appeared to have said something of interest (and something misogynistic, but that's another story) about the nature of marriage, nonetheless "was quite black from head to foot, a clear proof that he was stupid."[50] A somewhat less well-known claim occurs a few sentences earlier in the same text, where Kant speaks of the "fundamental difference" between the black and white races, which "appears to be as great in regard to mental capacities as in color."[51] He describes the "religion of fetishes so widespread among them," which is "perhaps a sort of idolatry that sinks as deeply into the trifling as appears to be possible to human nature. A bird's feather, a cow's horn, a conch shell, or any other common object, as soon as it becomes consecrated by a few words, is an object of veneration and of invocation in swearing oaths."[52] There is, here, no trace of a thought that knowledge and reason might be advanced by an attempt to understand

[49]Brentjes, *Anton Wilhelm Amo*, 50.

[50]Immanuel Kant, *Beobachtungen über das Gefühl des Schönen und Erhabenen*, Riga: Friedrich Hartknoch, 1764, 102–3.

[51]Kant, *Beobachtungen*, 103.

[52]Kant, *Beobachtungen*, 103.

why this sort of conch shell and not another is venerated, that there could be an internal logic and foundation for "idolatrous" beliefs and practices. To this extent, Kant is repeating a pattern we have seen several times before, in numerous authors in the centuries preceding the Königsberg philosopher's own observations on human diversity: he is simply failing to take an interest in native knowledge systems, presuming that whatever is not European is for that same reason not rational and not worthy of attention.

But what is relatively new here is that Kant seems to directly peg the absence of reason to physical traits that are supposed to be markers of racial difference. It is not just that dark-skinned people are cut off, for contingent biographical reasons, from the proper training and deployment of reason, but that they are cut off from reason *because* of their dark skin. The racialization of the distinction between European reason and native ignorance is new, but the presumption that there is nothing to learn from non-European peoples is not. When Kant says that the black skin is a sure sign of stupidity, he has in effect found a convenient outward marker of a stupidity that was presumed to be there in native peoples long before race came to be conceived as the biological or essential basis for differing intelligences in different human groups.

It is hard to pinpoint the moment at which thinking about non-European peoples became explicitly racialized in this way. Certainly, at any given time in the early modern period there was a great variety of viewpoints. In the era of Anton Wilhelm Amo, in particular, at least in the microcultures of Wolfenbüttel, Halle, Wittenberg, and Jena, there was no single prevailing viewpoint on the natural aptitudes of sub-Saharan Africans. What is certain, however, is that within a few decades after Amo's flight from Germany at the end of the 1740s, the racist view of the natural inferiority of blacks would be predominant, whereas it had not been when Amo began his study in the 1720s. One very significant lexical transition that can be dated to around 1750, is the shift from "Moor" to "Negro" in the way Europeans speak of sub-Saharan Africans.[53] Amo inhabits a world in which he is conceptualized as a Moor. This is a quasi-racial designation, but it is also rooted in classical antiquity, connotes the Roman province of Mauretania, and overlaps conceptually with the figure of the Muslim in the European imagination. "Negro," by contrast, marks a rupture with the past and with thinking of human diversity in terms of religion and history. The people Amo's colleagues at Jena thought of as Moors would be described just a few decades later by Kant, in nearby Königsberg, as Negroes. Writing in the *Observations* of 1764,

[53]On this transition, see in particular Miriam Claude Meijer, *Race and Aesthetics in the Anthropology of Petrus Camper (1722–1789*, Amsterdam: Rodopi, 1999.

and plainly echoing Hume, Kant asserts that it is impossible to cite a single example in which a Negro has shown talents, and that "among the hundreds of thousands of blacks who are transported elsewhere from their countries, although many of them have even been set free, still not a single one was ever found who presented anything great in art or science or any other praiseworthy quality, even though among the whites some continually rise aloft from the lowest rabble, and through superior gifts earn respect in the world."[54] It is hard not to suspect that Kant has Amo in particular in mind here, that he has an interest in denying what he has heard of the Halle philosopher, even if we know, in fact, that this line of argumentation is highly derivative in Kant, and is almost certainly being unreflectively borrowed from some other textual source, without much thought given to whether any actual Africans fit the description.

As we have seen, Amo had been a philosophical adversary of the Pietists, and a defender of a broadly Leibnizian philosophical outlook. It seems, moreover, to have been a shift in German academic culture in favor of Pietism in the 1740s that caused Amo's life in Germany to be plagued by racist exclusion in a way that it had not been in the beginning. As Ursula Goldenbaum has argued,[55] the early Kant's views took shape in a broadly Pietist context, and it would not be at all surprising if the history of the unfortunate "Moorish" philosopher driven out of Germany by like-minded xenophobes was still lingering in Kant's memory when he parroted the widespread prejudice about the unimprovability of black people. Whether he has Amo in mind or not, Kant's view here marks a significant departure from the sort of Enlightenment thinking that brought the duke of Braunschweig-Wolfenbüttel to send his *Kammermohr* to study at university.

What changed, now, between the Enlightenment as understood by Duke Anton Ulrich and the Enlightenment promoted by Kant? There is no short answer, of course; the full account would be multifactorial, and would extend over several volumes. The greater part of the answer surely has to do with social and economic history, in particular the continuing growth of a global economy based on the forced labor of Africans, and a corresponding rationalization of this system by appeal to an imagined inborn inferiority of the exploited group. But at a more local and idea-historical level, we may simply say that the universalism of the early Enlightenment gave way to a fragmented view of humanity, on which "barbarous" peoples lay beyond the pale of rationality and morality not just as a contingent consequence of their place of origins, but as a result of who

[54]Kant, *Beobachtungen*, 103.

[55]See for example Ursula Goldenbaum, *Appell an das Publikum: Die öffentliche Debatte in der deutschen Aufklärung, 1687–1796*, Berlin: Akademie Verlag, 2004.

they, to speak with Philippi, were in their "natures." These two competing visions of the Enlightenment would remain in tension throughout the eighteenth century, and while neither would ever completely recede, the Kantian would be plainly predominant by the latter part of the century.

A third vision, by contrast, will remain largely unattested between the work of Leibniz and the Counter-Enlightenment philosophical anthropology of J. G. Herder. On this view, non-European cultures are neither essentially irrational, nor potentially rational but circumstantially deprived of the cultural *Bildung* that could awaken their faculty of reason. Rather, on this view, non-European cultures are already on an equal footing with European cultures; as a result of a universal, equally distributed human essence, every cultural form is held to be equal to every other. This conviction, as we have already seen, is present in Leibniz's promotion of the study of language and folklore as being no less valuable than the study of textual traditions such as those of Chinese or Western philosophy. This third tradition will be treated in the following chapter.

One thing the dualist and nonracist strain of modern philosophy brings into clear focus is the basic non sequitur at the heart of racism: the completely unjustified move from the observation of measurable physical differences between different human populations—an observation that is not in itself misguided—to the claim that these different bodies *must* be underlain by minds or souls with a different set of nonphysical capacities that somehow match up with the physical differences. Kant's inference from blackness to stupidity is perhaps the most vivid demonstration of this fallacy, but it is only a relatively more extreme expression of a broad tendency in thinking about racial difference that we see in a wide variety of authors, all of whom share in some way or other in the nondualist view that the state of the body is somehow a report upon the state of the soul. Add to this the further conviction, shared by most of these authors, that a dark-skinned body is somehow degenerated from the original ideal type of the species, and we begin to understand the sort of reasoning that could make Kant's inference make sense.

This is by no means an attempt at a philosophical defense of dualism; it is only an attempt to describe how the naturalization of the human being, while opening up the possibility of scientific study of humanity, at the same time made possible a certain kind of thinking about human difference that could not have developed as long as a philosophical anthropology that rooted human unity in a nonphysical criterion held sway.

Chapter 9

Race and Its Discontents in the Enlightenment

9.1. Introduction

In his classic study, *The History of the Race Idea, from Ray to Carus*,[1] Eric Voegelin argues that the insertion of human beings into nature at the beginning of the modern period was not in itself problematic, so long as a body-soul dualism preserved the true determinant of membership in the human species from the naturalizing effect of taxonomy: these two "contradictory constructions," the natural order in which the body is placed and the supernatural order in which the soul is preserved, Voegelin writes, "are more or less joined in that at one time the inclusion of man in the animal kingdom is presented as a conventional classification that does not affect the true ontic status of man as a rational soul substance."[2] Voegelin sees eighteenth-century thinking about human diversity as dominated by two distinct possibilities: "We can either let the soul substance dominate and declare the diversity of races irrelevant compared with many forms of the whole, the spiritual unity, or we can shift the emphasis to the causal-scientific explanation of the physical diversity and completely ignore the problem of the spirit and its connection to the body or completely separate the spirit from the body and treat its problems as independent of the physical diversity."[3] He notes, however, that a third position is carved out by Johann Gottfried Herder, one that takes reason as fundamental to the understanding of what human beings are, yet nonetheless takes reason as manifested in human diversity, or, conversely, takes human diversity as itself a reflection of a universal underlying reason:

> Neither the Cartesian dichotomy of body and *anima* or *ratio* nor the Kantian division into sensory nature and reason is satisfactory for Herder—he

[1] See Eric Voegelin, *Die Rassenidee in der Geistesgeschichte von Ray bis Carus*, Berlin: Junker und Dünnhaupt Verlag, 1933; published in English as *The History of the Race Idea, from Ray to Carus*, in Klaus Vondung (ed.), *The Collected Works of Eric Voegelin*, vol. 3, trans. Ruth Hein, Columbia: University of Missouri Press, 1989.

[2] Voegelin, *History of the Race Idea*, 66.

[3] Voegelin, *History of the Race Idea*, 66.

> sees more. As a philosopher of the Enlightenment, Herder, like Linnaeus and Buffon, also sees man as a rational person, raised above the animal kingdom because of his reason. But for Herder reason is not a substance that can be isolated and, freed of all connections to the senses, added on to the body or soma as a differential element. Instead, he sees reason as an autonomous, psychic-lawful essence, as the spiritual unity of man.[4]

For Herder, in Voegelin's view, a human being is fully identified neither with a nonphysical transcendental reason, nor with a certain category of natural being inscribed in the order of nature, but rather with a reason that is manifested in the various ways in which human beings inhabit nature, or in which they adapt nature to their distinct human needs. This manifestation would come to be familiar under the concept of "culture," which, again, for Herder, is something that is distributed evenly throughout the many diverse expressions of human social life. Kant, by contrast, would flatly dismiss the suggestion that human groups, living beyond the pale of European history, might be living *for* something, might be living lives of value. Thus in the *Critique of the Power of Judgment* he describes the existence of the order of nature as having a fitting and proper finality to it only insofar as it contributes to the sustenance of human beings, but adds that if the human beings in question are, for example, the primitive New Hollanders (i.e., Australians), or the indigenous people of Tierra del Fuego, it is in turn hard to answer the question why human beings should exist at all, and thus indeed why anything should exist at all.[5]

On Voegelin's interpretation, it is precisely Herder's commitment to the essential embodiment of reason that saves him from sharing in Kant's "paltry image of man." On Herder's understanding, the notion of embodiment may be extended well beyond the individual organic body, to include embeddedness within a culture. And here, for Voegelin, there can be no question of better or worse, but only suitedness to local exigencies. Thus Herder declaims, "What and wherever thou art born, O man, there thou art what thou shouldst be!"[6] This even if one is a New Hollander.

Here, as before with Leibniz, we may seem to have wandered, again, into questions of history and culture, and thus to be far from the question

[4]Voegelin, *History of the Race Idea*, 67.

[5]Kant, *Kritik der Urtheilskraft*, AA V, 378. "Geht man aber davon ab und sieht nur auf den Gebrauch, den andere Naturwesen davon machen, verlässt also die Betrachtung der innern Organisation und sieht nur auf äussere zweckmässige Beziehungen, wie das Gras dem Vieh, wie dieses dem Menschen als Mittel zu seiner Existenz nöthig sei; und man sieht nicht, warum es denn nöthig sei, dass Menschen existiren (welches, wenn man etwa die Neuholländer oder Feuerländer in Gedanken hat, so leicht nicht zu beantworten sein möchte)."

[6]Johann Gottfried Herder, *Ideen zur Philosophie der Geschichte der Menschheit*, pt. 2, Riga: Johann Friedrich Hartknoch, 1786, 262. "Wo und wer du gebohren bist, o Mensch, da bist du, der du seyn solltest."

of race. But it is possible to work our way back in a few simple steps. It is not simply that Herder's philosophy of history and his commitment to cultural parity cause him to downplay the perception of the sort of cultural differences that a thinker such as Kant would see as requiring an explanation in terms of essential racial differences. It is, much more strongly, that an a priori or philosophical commitment to cultural parity reduces or even eliminates the perception of the salience of empirical data about racial difference. To return to a point already made: racial differences must not be understood as simply phenomenally given. Rather, it would be more correct to affirm that such differences result from the ad hoc and a posteriori way one makes sense of a prior belief in the unbridgeability of the gap between two cultures or ethnicities. And one will not perceive the need to go looking for these racial differences unless one has a prior commitment to deeper differences, a commitment grounded ultimately in the history and politics of relations between different social groups. Again, as the abbé Henri Grégoire put it so lucidly, "those who have wanted to disinherit the Negroes have used anatomy to their advantage."[7] The desire to disinherit comes first, the identification of salient anatomical differences second.

It might be supposed, then, that much talk of race in the eighteenth century involved little more than a transposition into the description of human phenotypic diversity of preoccupations with real or imagined cultural differences, and that these differences, in turn, were supposed to result, at the individual level, in a hierarchy of intellectual capacities. How else do we account for such an obvious non sequitur from Kant, for example, when he infers from the blackness of a person's skin to the stupidity of what that person is saying? For Kant, even an African living in a traditional African society is not "where he should be," to adapt Herder's phrase, as the only place human beings should be in Kant's view is within the fold of history as it unfolds from Europe. The African's phenotypic traits, such as skin color, come to stand in for this supposed cultural-historical backwardness, even if Kant himself fails to notice that he is making such a substitution. And this is where the Herderian injunction to be where one is appears to falter, since inevitably groups of people are not only, or even primarily, located at different points in a Cartesian plane, but always also at different points in a supposed vertical hierarchy.

Kant is a liberal racist, in Popkin's sense. He does not in fact believe that Africans and Europeans are essentially different, that they have separate origins or separate internal natures. Kant is a monogenist, and his criterion for the unity of the human species, like Leibniz's before him, is based on the fact that this species constitutes a sort of generational chain

[7] Grégoire, *De la littérature des Nègres*, 14.

extending back to the first parents. As Kant writes in *On the Different Races of Men* of 1775, "In the animal kingdom, the natural classification into genera and species is based on the common law of reproduction, and the unity of the genera is nothing other than a unity of the generative power."[8] Notwithstanding this unity, however, Kant believes that Africans are culturally backward, and his substitution of skin color as a marker of culture is nothing more than a failure on his part to be consistent. That there was no price to pay for such inconsistency, that one could say such things and get away with them, shows, perhaps, how little the "liberality" of liberal racism helped to soften it. One did not need an explicit theory of the essential inferiority of Africans in order to dismiss anything an African might say as stupid. In fact one did not even have to worry about the fact that one did not have such a theory, and could not justify a claim about an African's stupidity if called on to do so.

There was of course at least a pretence of empiricism in many Enlightenment-era claims about the inferiority of Africans, and an implicit or explicit avowal of a preparedness to change one's views about Africans should new evidence go against these views. Thus Hume, as we saw in the introduction, mentions that he has heard "talk of one Negro as a man of parts and learning," and while he dismisses the report as an exaggeration, he also intends to signal that he would be prepared to acknowledge the accomplishments of a "Negro" were he to come across any solid evidence for them.[9] Unlike Amo's peers, who saw him not as a "Negro" but as a "Moor," and who took his accomplishments as a reconfirmation of "the natural genius of Africa" rather than as the falsification of a reigning theory of Africa's lack of genius, Hume would pretend to be ever on the lookout for exceptional Africans, ever ready to be proved wrong: a pretence of falsificationism lending an air of scientific respectability to a prejudice.

Herder, for his part, would seek to disrupt this contrived search, by calling into question the very idea that "genius" must be measured in relation to the sort of accomplishments Europeans value. For him, it does not take an Anton Wilhelm Amo or a Joseph Bologne de Saint-George (a French-Caribbean composer of the late eighteenth century) to prove that Africans are talented, because it is not only the European traditions of philosophy or musical composition that afford people the opportunity to cultivate their talent. It is this rejection of European civilizational superiority, and the corollary underlining of the reality and legitimacy

[8] Kant, *Von den verschiedenen Racen der Menschen*, AA 2, 429. "Im Thierreiche gründet sich die Natureintheilung in Gattungen und Arten auf das gemeinschaftliche Gesetz der Fortpflanzung, und die Einheit der Gattungen ist nichts anders, als die Einheit der zeugenden Kraft."

[9] Hume, *Philosophical Works*, 3:228.

of cultural diversity, that at the same time seem to diminish for him the reality and importance of racial diversity.

In response to Herder's general line of thinking, Kant returns to his worry about the evident meaninglessness of lives spent beyond the pale of history: "Does the author [Herder] really mean that, if the happy inhabitants of Tahiti, never visited by more civilized nations, were destined to live in their peaceful indolence for thousands of centuries, it would be possible to give a satisfactory answer to the question why they should exist at all?"[10] This is exactly what Herder means, and his commitment to cultural parity goes together with a rejection of the idea that culture is underlain by race. In the end, again, different cultural profiles are only local expressions or inflections of "one and the same great portrait" of humanity.[11]

9.2. The Significance of Skin Color

As already mentioned, Kant is a defender of monogenesis. He is also an enemy of evolution.[12] These two facts together entail that, for him, all human beings share in the same nature, and that any differences between different human groups can be explained as a result of contingent circumstances. It is in this respect that Kant's dismissal of the African carpenter's observation, on the basis of skin color alone, serves as such a vivid example of Popkin's notion of liberal racism: it is not supported by Kant's own theoretical commitments. Kant was not alone in this particular failure, and indeed much of his prejudice seems to be borrowed from French sources. Buffon and Pieter Camper,[13] two centrally important sources for Kant's own anthropological views, along with those of many others in the eighteenth century, would attempt to enumerate the different basic subtypes of human being, implicitly or explicitly on the model of the project of zoological and botanical taxonomy, while at the same

[10]Immanuel Kant, "Review of Herder's *Ideas on the Philosophy of the History of Mankind*," in *Political Writings*, ed. H. S. Reiss, Cambridge: Cambridge University Press, 1970, 201–20, 219–20.

[11]Herder, *Ideen*, pt. 2, 94.

[12]See in particular Catherine Wilson, "Kant and the Speculative Sciences of Origins," in Justin E. H. Smith (ed.), *The Problem of Animal Generation in Early Modern Philosophy*, Cambridge: Cambridge University Press, 2006, 375–401.

[13]Camper remains rooted in what we might call the "physiognomical" tradition: his central concern is to discern traits of character from perceptible physical traits, whether these physical traits are associated with an entire population or attached to a single individual. See Camper, *The Works of the Late Professor Camper: On the Connexion between the Science of Anatomy and the Arts of Drawing, Painting, Statuary, &c.*, trans. T. Cogan, London: C. Dilly, 1794.

time all would explicitly defend the view that (1) human beings are all descended from the same two ancestors; (2) there is no speciation over time; (3) differences between human "races" can be attributed largely, perhaps entirely, to differences of diet, custom, and environment.

These three points together signal that all of these authors were well aware that a human "race" is not biologically comparable to an animal species. Yet this awareness did not slow down the project of racial science at all. The history of racial science was not based on the presumption of essential differences between different races. While, as we have seen, polygenesis served as an important undercurrent in sixteenth- and seventeenth-century libertine thought, in figures such as Lucilio Vanini, for whom it constituted one part of a radical defense of a thoroughgoing naturalistic account of human origins, by the eighteenth century this was a doctrine that had taken on distinctly racist overtones. Of all the major Enlightenment thinkers, Voltaire was virtually alone in maintaining unequivocally that "the race of Negroes is a species of men different from ours."[14]

Unlike the sixteenth-century libertines who conflated Africans, Native Americans, and mythical nymphs, eighteenth-century polygenists such as Voltaire knew very well the unfantastic reality of the day-to-day lives of the Africans whose *fraternité* with Europeans they sought to deny. In a text of 1733 on "The Negro," Voltaire, seeking to emphasize the distance between Europeans and Africans, falls back on the ambiguity of the terms "race" and "species," and in so doing foils his own comparison: "The Negro race," he writes, "is a species of men as different from ours as the breed of spaniels is from that of greyhounds."[15] Of course, spaniels are a different "race," in the French sense, from greyhounds, but they are not a different "espèce." It is at least possible that Voltaire is using the term "species" here in a way that is synonymous with "breed," in which case this text alone, however racist, would not win for him the title of polygenist. In another text, from 1756, Voltaire uses "race" in a way that perhaps suggests our own sense of "species," writing that "it is only allowed to a blind person to doubt that whites, negroes, albinos, Hottentots, the Chinese, the Americans, are from entirely different races."[16] We might suppose that in general when Voltaire speaks of races he means "entirely different races," and that what makes them entirely different is that they do not have shared ancestry. This is still surprising, however, since no contemporary theory of dog-breeding held that dissimilar breeds had separate origins, and in any case Voltaire knew perfectly well, just

[14]Voltaire, "Essai sur les moeurs," in *The Works of Voltaire*, ed. and trans. Tobias Smollett, New York: E. R. DuMont, 1901, 39:240–41.

[15]Voltaire, "Le Nègre," in *Works of Voltaire*, vol. 19.

[16]Voltaire, *Essai sur les moeurs et l'esprit des nations*, in *Oeuvres complètes de Voltaire*, Paris: Renouard, 1819 [1756], 1:6.

as every racist knows, that positing separate ancestries does not cause the reality of interfertility to disappear. Voltaire leaves this question unaddressed to make a series of exaggerated claims about physiological differences between Europeans and Africans: that the latter have "wool" instead of hair, that "their eyes are not formed like ours," and so on. He relates with relish an anecdote about Frederik Ruysch's anatomical study of a Black African, whose skin, the author writes, was then sold to Peter the Great (the same man who had adopted his own African slave as a son). "This membrane is black," Voltaire relates, "and it is this that communicates to Negroes this intrinsic blackness that they only lose through illness."[17] Unsurprisingly, he soon moves on from supposed physical to supposed mental attributes, running together "their round eyes, their flat nose, their lips that are always big, the wool of their heads, even the measure of their intelligence,"[18] as if the last of these flowed directly from the series preceding it.

Andrew S. Curran has compellingly argued that Voltaire's commitment to polygenesis should not be dismissed as mere scandal-mongering, and that, although it was a minority view in his period, particularly in the wake of Buffon's gradationism, nonetheless Voltaire contributed significantly to one of the major transformations that the figure of the African underwent in European thought during the eighteenth century: "[F]rom a barbaric heathen (a moral and religious category) who could be redeemed through slavery, to a subhuman (racial category) for whom human bondage seemed the logical but regrettable extension of the race's many shortcomings."[19] Curran is certainly right: Voltaire's views, which were shocking in the early eighteenth century, would by the nineteenth century be fairly commonplace. But this does not make them any more well grounded at the time he presents them. As Curran notes, Voltaire placed much of his confidence in Ruysch's supposed discovery that the reticulum mucosum of Africans, more commonly known as the stratum spinosum or Malpighian layer, is black like "gauze" or "ink." Voltaire hoped to make this feature into the specific differentium of a particular sort of natural kind. But even if it were as black as ink, this would only push the explanation of the natural difference between blacks and whites a few epidermal layers deeper, and would in no way provide a basis for any claim of essential difference.

It is useful to dwell on Voltaire here only because he gives such clear voice to a widespread Enlightenment view of race that would have somewhat softened echoes in the German authors of principal interest to us in this chapter. In fact, it is not hard to discern the derivative character of

[17] Voltaire, *Essai sur les moeurs*, 7.
[18] Voltaire, *Essai sur les moeurs*, 7.
[19] Curran, *Anatomy of Blackness*, 148.

the most notorious claims about race in the most important eighteenth-century philosophers, and indeed typically the direction of influence is from France outward to other parts of Europe. Kant, as we have seen, is a monogenist who believes that reason is a distinguishing mark of humanity, and yet he is still prepared to take the color of an African's skin as "a clear proof that that he was stupid."[20] We see almost exactly the same non sequitur, expressed in the same language, in Montesquieu's *Spirit of the Laws* of 1750, though here it is an inference from physiology to absence of sympathy, rather than from physiology to absence of a shared rational faculty: "Those concerned are black from head to toe," Montesquieu writes, "and they have such a flat nose that it is almost impossible to feel sorry for them."[21] The author goes on to assert that skin color, much more than the shape of the nose, is so fundamental to the way in which we perceive shared humanity that it is in effect psychologically impossible to acknowledge the humanity of sub-Saharan Africans. "It is so natural to think that color constitutes the essence of humanity. . . . It is impossible for us to suppose that those people are men."[22] Montesquieu is not in any sense a "racial scientist," and here he is not addressing the question of whether a taxonomical distinction between different human races carves nature at its joints, or fails to do so. He is, rather, describing what he takes to be an innate disposition, at least among Europeans, to perceive physical differences between Europeans and Africans as essential differences.[23]

Those authors who are generally acknowledged as architects of modern racial realism typically emphasize the unity of the human species, portraying racial difference as a contingent and reversible effect of migration, diet, and custom. Thus Christoph Meiners, often cited as one of the eighteenth century's staunchest racial realists, writes in his 1785 *Outline of a History of Mankind*,

[20]Kant, *Beobachtungen*, 102–3. For some insightful discussion of the problem of placing this sort of claim in relation to Kant's, let us say, more venerable contributions to the history of philosophy, see in particular Robert Bernasconi, "Will the Real Kant Please Stand Up: The Challenge of Enlightenment Racism to the Study of the History of Philosophy," *Radical Philosophy* 117 (2003): 13–22.

[21]Montesquieu [Charles-Louis de Secondat, Baron de la Brède], *De l'esprit des lois, nouvelle edition, avec les dernieres corrections et illustrations de l'auteur*, 2 vols., Edinburgh, 1750, 1:343.

[22]Montesquieu, *De l'esprit des lois*, 1:343.

[23]For a general overview of the historical development of the question of the significance of skin color, see Renato Mazzolini, "Skin Color and the Origin of Physical Anthropology (1640–1850)," in Susanne Lettow (ed.), *Reproduction, Race and Gender in Philosophy and Early Life Sciences*, Albany: State University of New York Press, 2014.

> All the peoples of the earth constitute a single genus [*Geschlecht*], or a single species of creatures. In this single human kind [*Menschen-Geschlecht*] we must suppose that there are two completely distinct lineages [*Stämme*], in each lineage several races, (a) in each race countless varieties [*Varietäten*], (b) and finally a great multiplicity of subvarieties [*Spielarten*], that have arisen from the mixture of people from different lineages and races. It seems peculiar to me that color has been chosen as the exclusive or primary trait on the basis of which the similarity and difference of peoples is defined.[24]

For Meiners, there are in the end only two basic color types of the human species, the white and the brown, which are associated respectively with the "Caucasian" and the "Mongolian" races. But there is nothing fundamental about color difference for the classification of races. Color, in Meiners's view, is "much more changeable than the other traits . . . by which the peoples of the earth are distinguished from one another. . . . The principal cause of this is climate, and, when this has exercised its influence, the descent [*Abstammung*], and much less the diet."[25]

For the most part, the question whether skin color could reveal anything truly significant about different human groups retreated into the background in the mid-eighteenth century, upstaged by a frenzy of research into the physical causes of these differences. This research took for granted that the difference is significant, even if it failed to give an explicit account of the nature of this significance. There was no shortage of treatises incorporating the latest discoveries from Newtonian physics and optics into the quest for an answer to this natural enigma; an early contribution to this genre was Pierre Barrère's anonymously published *Dissertation sur la cause physique de la couleur des Nègres* (*Dissertation on the Physical Cause of the Color of Negroes*) of 1741.[26] Barrère is intent to show that blackness pertains essentially to black people, to such an extent that no amount of treatment of the skin after death can remove it. Thus he relates that "if, after long maceration of the skin of a Negro in water, one detaches the epidermis or outer layer of skin, it will be found to be black, very thin, and to appear transparent when it is held up to the light." From this, Barrère concludes, "it is evidently proven that the color of Negroes is not, so to speak, picked up [n'est pas, pour ainsi parler, une

[24] Christoph Meiners, *Grundriß der Geschichte der Menschheit*, Lemgo: Meyersche Buchhandlung, 1785, 17.

[25] Meiners, *Grundriß*, 45.

[26] [Pierre Barrère], *Dissertation sur la cause physique de la couleur des Nègres, de la qualité de leurs cheveux, et de la dégénération de l'un et de l'autre*, Paris: Pierre-Guillaume Simon, 1741.

couleur d'emprunt]."[27] Blackness, works like this show, was held to be in need of scientific explanation; it was held to be an exceptional property. In the *Physical Geography* of 1802, Kant himself would compose a list of no fewer than seven "curiosities about the color of black people," consisting mostly in fantastic stories told by unreliable observers, as for instance that African babies are born white, and turn black only over the course of their first few months.[28]

One problem that occupied the minds of many thinkers, including Kant, was albinism, which seemed to suggest that skin color may in some cases have nothing at all to do with either lineage or geography. The seventh item in Kant's list of curiosities concerns "white Moors, or Albinos, who are conceived by black parents." Kant relates that "they are Moorish in appearance, have curly, snow-white, woolly hair, are pale, and can only see in the moonlight."[29] Again, here, Kant's observations are largely derivative of French sources. In his *Dissertation physique à l'occasion du Nègre blanc* [*Physical Dissertation on the Occasion of a White Negro*], published anonymously in 1744, Pierre-Louis Moreau de Maupertuis for his part had used the appearance of an albino African in Paris, who had generated much talk in the salons and academies, as an opportunity to present his own theory of the inheritance of traits, according to which albinism amounts to something akin to a genetic mutation.[30] For Maupertuis, the seeds of both parents form the fetus, carrying a sort of "memory" of the parts of the parental body from which they came, and thus there is no need to resort to any preformed being to explain reproduction. The fact that a case of apparent racial difference between parent and child could serve as an occasion to present an antipreformationist account of generation shows the extent to which race had come to be seen as constituting an essential difference between kinds. In view of this essentialism, the production of one race by another was interpreted as a sort of rupture in the order of a nature supposed to be governed by a general law according to which like begets like: this in spite of the obvious point of disanalogy between races and species, that there are no impediments at all to cross-fertility between members of different human groups.

Such a rupture will be perceived only where differences of skin color are held to be a relevant criterion in distinguishing like from unlike, and the mid-eighteenth century seems to have witnessed a steep rise in interest

[27] Barrère, *Dissertation*, 4.

[28] Kant, *Physische Geographie, auf Verlangen des Verfassers*, Königsberg: Göbbels und Unzer, 1802, pt. 2, 1.2, "Einige Merkwürdigkeiten von der schwarzen Farbe der Menschen," 4–5.

[29] Kant, *Physische Geographie*, 5.

[30] [Pierre-Louis Moreau de Maupertuis], *Dissertation physique à l'occasion du Nègre blanc*, Leiden, 1744.

in skin color as a marker of essential difference. This interest seems moreover to have been just as strong among people, such as Kant, whose theoretical commitments concerning the unity of the human species would have required them, on reflection, to acknowledge the relative superficiality of differences of skin color. In other words, skin color was held to be important even when those who held it to be so could offer no coherent explanation of why it was important.

9.3. Kant: From Non Sequitur to Critique?

There is ongoing scholarly debate about the development of Kant's thinking on race over the course of his career. Pauline Kleingeld argues, for example, that by the 1790s the philosopher will come to downplay the importance of racial distinctions, in favor of a more truly egalitarian and cosmopolitan understanding of the unity of the human species.[31] What is perhaps Kant's most significant contribution to his era's debates about race comes, however, several years earlier. Kant's *On the Different Races of Men* of 1775 constitutes one of the most formidable attempts in the modern period to treat the concept of race within the framework of a broader philosophical project. We have already mentioned Kant's observation in this work that natural division into genera and species is based on "the unity of the generative power." This is significant, it was argued, and indeed an important echo of the theory earlier articulated by Leibniz, to the extent that it grounds the unity of a biological kind in its history, or in the reproductive chain extending back to the first parents. It follows from this, inter alia, that for Kant wherever any two individual beings share the same first parents, they therefore belong to the same kind.

Kant makes a sharp distinction between the grouping of entities based on the unity of the generative power, on the one hand, and, on the other, artificial division, which is "based upon classes and divides things up according to similarities."[32] Kant is very clear that only the latter sort of division can be appropriate for humankind. "[A]ll human beings anywhere on earth," he writes, "belong to the same natural genus, because they always produce fertile children with one another even if we find

[31] Pauline Kleingeld, "Kant's Second Thoughts on Race," *Philosophical Quarterly* 57, 229 (2007): 573–92. See also Raphaël Lagier, *Les Races humaines selon Kant*, Paris: Presses Universitaires de France, 2004. For another very compelling account of the development of Kant's views on race, and their relationship to his philosophy, at different stages of his career, see Robert Bernasconi, "Kant's Third Thoughts on Race," in Stuart Elden and Eduardo Mendieta (eds.), *Reading Kant's Geography*, Albany: State University of New York Press, 2011, 291–318.

[32] Kant, *Von den verschiedenen Racen der Menschen*, AA 2, 429.

great dissimilarities in their form."[33] The unity of the natural genus is for Kant identical with the unity of the reproductive power. Since plainly there are no natural impediments to reproduction between members of two different races, it follows that any division of the human species into races can only be artificial. In accounting for the unity of the human kind as a natural genus, Kant believes that we must presume that "all human beings belong to one line of descent from which, regardless of their dissimilarities, they emerged, or from which they at least might possibly have emerged."[34] The polygenetic alternative, Kant thinks, is "a view that needlessly multiplies the number of causes."[35]

Kant goes on to distinguish between "deviations" and "degenerations": the former are heritable differences between populations; the latter are deviations that "can no longer produce the original formation of the line."[36] Races, Kant believes, "are deviations that are constantly preserved over many generations," and that come about as a result of migration or interbreeding.[37] By these criteria, Kant concludes, "Negroes and whites are clearly not different species of human beings (since they presumably belong to one line of descent), but they do comprise two different races."[38]

The litmus test for determining whether two individuals belong to different species is by consideration of the nature of the offspring of two individuals: Kant believes that a child of black and white parents is "necessarily" a "half-breed," unlike the child of, say, a blond parent and a brown-haired parent. Recall, now, Mill's articulation of the notion of natural kind. For Mill, "brown-haired people" and "blond-haired people" cannot possibly constitute natural kinds, since nothing beyond the principle of classifying them in this way is entailed in the classification. There are no common properties beyond the one that is picked out. For Mill, by contrast, "an hundred generations have not exhausted the common properties of animals or of plants."[39] Mill does not mean, here, that these properties are reproduced through one hundred generations, but rather that over so many generations of *human* investigation of these plants and animals, we continue to discover new common properties. This is clear from the fact that he adds "sulphur and phosphorus" to the list, that is, natural kinds that plainly do not reproduce themselves in the way plants and animals do. Here, by contrast, Kant follows Leibniz (and

[33] Kant, *Von den verschiedenen Racen der Menschen*, AA 2, 429.
[34] Kant, *Von den verschiedenen Racen der Menschen*, AA 2, 430.
[35] Kant, *Von den verschiedenen Racen der Menschen*, AA 2, 430.
[36] Kant, *Von den verschiedenen Racen der Menschen*, AA 2, 430.
[37] Kant, *Von den verschiedenen Racen der Menschen*, AA 2, 430.
[38] Kant, *Von den verschiedenen Racen der Menschen*, AA 2, 430.
[39] Mill, *System of Logic*, 156.

Aristotle, one might add) in identifying the kindhood of living entities as a different matter than that of other sorts of entity in nature, to the extent that it is precisely reproduction, in the sense of "begetting" that provides the criteria for kind membership. In this respect, we might suspect that Mill is running together ontological domains that had been traditionally kept quite separate. One might also suggest, in passing, that work on the history of the philosophy of biology has been distorted in the post-Millian era: we tend to overlook the important distinctions pre-Millian philosophers made between physical species and the species of plants and animals, allowing the criteria for the former to overshadow the analysis of the latter. This, as we already saw in the treatment of Leibniz in chapter 7, is problematic for our understanding of the history not just of the concept of biological species, but also of race.

Kant is far closer to Leibniz here than to Mill. For him, race is not like hair color, not in view of the depth of common properties beyond those guiding our classification, but in virtue of the power to transmit properties across generations. Variations within a race, such as blondness among white people, are not necessarily preserved across generations, while racial traits always are, and this over many generations. This is why designations such as "mulatto," "quadroon," and so on pick out real entities for Kant: these are indeed artificial divisions, to the extent that the only meaningful "natural" division is the one imposed by reproductive boundaries between creatures, but at the same time Kant does not understand "natural" here as a synonym of "real." Rather, artificial divisions do pick out something real for Kant, even if they do not necessarily pick out something significant; they amount to one way of carving up nature at some of its many joints, even if this is not a way that nature itself requires us to notice. In this respect, the question whether races are "natural kinds," if posed to Kant, would be difficult to answer, for his distinction between the natural and the artificial places race outside the usual understanding that would develop in the Millian tradition of the notion of natural kinds. Yet there is also certainly nothing about race that is constructed or illusory for Kant.

With the theoretical reflections completed, Kant moves on to list the divisions of the races. His system differs to some extent, though not radically, from those of his predecessors. It includes "(1) the white race; (2) the Negro race; (3) the Hun race (Mongol or Kalmuck); and (4) the Hindu or Hindustani race."[40] He maintains that "it is possible to derive all of the other hereditary characters of peoples from these four races either as mixed races or as races that originate from them."[41] In his account

[40]Kant, *Von den verschiedenen Racen der Menschen*, AA 2, 432.
[41]Kant, *Von den verschiedenen Racen der Menschen*, AA 2, 432.

"of the immediate causes of the origin of these different races," Kant distinguishes between "seeds" (*Keime*) and "natural predispositions" (*natürliche Anlagen*).[42] The former have to do with particular parts of an organic being, while the latter concern "the size or the relationship of the parts to one another." Kant gives as an example of the first sort of cause the "seeds" in birds that yield new feathers when these birds move to a cooler climate. Adaptation to new environments can never lead to true speciation for Kant, since "apparent new species are really nothing other than deviations and races of the same genus whose seeds and natural predispositions have only occasionally developed in different ways in the long course of time."[43]

The ability to adapt to new climates, Kant thinks, cannot be explained by chance or "general mechanistic laws."[44] Instead, as Kleingeld explains, the natural predispositions of living beings can only be thought of as preformed, and supposed to be envisioned by the creator in anticipation of all future environmental circumstances in which the members of a given species might find themselves: "Any possible change," Kant writes, "with the potential for replicating itself must instead have already been present in the reproductive power, so that chance development appropriate to the circumstances might take place according to a previously determined plan."[45] As for human beings, Kant thinks that they were created in order to be able to inhabit every climate on earth: a topic to which he often returns in contemplating, with marvel, the existence of Arctic peoples. Consequently, "numerous seeds and natural predispositions must lie ready in human beings either to be developed or held back in such a way that we might become fitted to a particular place in the world."[46] Thus, human racial variation for Kant is of a pair with human biogeographical domination of the earth.

Whether or not Kleingeld is correct in holding that Kant's view on race underwent a transformation between the 1770s and the 1790s, it is clear that the sort of biogeographical flexibility of the human species that Kant describes in 1775 would serve as a very compelling basis for the peculiar sort of universalism he will defend much later in the *Critique of the Power of Judgment* and elsewhere. Now, the key concern is not with a human population's racial attributes, but rather with the question of where they stand in relation to the fold of history, that is, the progress of reason and sensibility that begins in Europe, that eventually engulfs

[42]Kant, *Von den verschiedenen Racen der Menschen*, AA 2, 434.
[43]Kant, *Von den verschiedenen Racen der Menschen*, AA 2, 434.
[44]Kant, *Von den verschiedenen Racen der Menschen*, AA 2, 435.
[45]Kant, *Von den verschiedenen Racen der Menschen*, AA 2, 435.
[46]Kant, *Von den verschiedenen Racen der Menschen*, AA 2, 435.

large sections of the world, and that, Kant hopes, may someday cover the entire globe.

In a remarkable passage of the third *Critique*, the philosopher explains that people in extreme corners of the globe are indeed well adapted to their environments, and that this adaptation indicates the existence of a "previously determined plan":

> In cold lands the snow protects the seeds from frost; it facilitates communication among humans (by means of sleds); the Laplanders find animals there that bring about this communication (reindeer), which find adequate nourishment in a sparse moss, which they must even scrape out from under the snow, and yet are easily tamed and readily deprived of the freedom in which they could otherwise maintain themselves quite well. For other peoples in the same icy regions the sea contains a rich supply of animals which, even beyond the nourishment and clothing that they provide and the wood which the sea as it were washes up for them there for houses, also supplies them with fuel for warming their huts.[47]

Thus, Kant thinks, there is an "admirable confluence of so many relations of nature for one end," that end being, namely, "the Greenlander, the Lapp, the Samoyed, the Yakut, etc."[48] And yet, Kant goes on, "one does not see why human beings have to live there at all."[49] This whole confluence, it turns out, is literally for nothing, and thus "to say that moisture falls from the air in the form of snow, that the sea has its currents which float the wood that has grown in warmer lands there, and that great sea animals filled with oil exist **because** the cause that produces all these natural products is grounded in the idea of an advantage for certain miserable creatures would be a very bold and arbitrary judgment."[50] Here Kant is speaking only of adaptations of Arctic human life that have to do with cultural practice, but he may as well have mentioned more directly "racial" adaptations, such as the supposed fact, reported in *On the Different Races of Men*, that "Greenlanders (are) not only far smaller in stature than Europeans," but also that "the natural heat of their bodies [is] noticeably greater."[51] The important point here is that human adaptation to environments, whether racial or cultural, cannot be conceived as an end in itself, since in Kant's view a life spent merely satisfying animal needs of warmth, food, and sexual gratification—which he supposes to be all there is to the lives of indigenous peoples—is, again, for him, a life not worth living. Kant will go on in the third *Critique* to say explicitly

[47] Kant, *Kritik der Urtheilskraft*, AA V, 369.
[48] Kant, *Kritik der Urtheilskraft*, AA V, 369.
[49] Kant, *Kritik der Urtheilskraft*, AA V, 369.
[50] Kant, *Kritik der Urtheilskraft*, AA V, 369.
[51] Kant, *Von den verschiedenen Racen der Menschen*, AA 2, 436.

that grass exists to nourish herbivores, herbivores to nourish carnivores, and that these all in turn exist only "[f]or the human being, for the diverse uses which his understanding teaches him to make of all these creatures; and he is the ultimate end of the creation here on earth, because he is the only being on earth who forms a concept of ends for himself and who by means of his reason can make a system of ends out of an aggregate of purposively formed things."[52] To the extent that the lives of indigenous peoples do not, from Kant's perspective, yield such a system of ends, Kant believes, again, that the question why human beings exist at all, "if one thinks about the New Hollanders or the Fuegians, might not be so easy to answer."[53] Elsewhere, as we have already seen, in his 1785 review of Herder's *Ideas for the Philosophy of the History of Mankind*, Kant describes the lives of South Sea Islanders as no more meaningful than the lives of the sheep they tend, while the meaninglessness of sheep's lives, in turn, has to do with the fact that lack a share in reason, and cannot participate in the kingdom of ends.[54] The importance of race is indeed diminished in the *Critique of the Power of Judgment*, relative to what we find in early writings such as the *Observations on the Feeling of the Beautiful and the Sublime* of 1764, where Kant seems to take a person's racial identity as necessarily determining that person's capacity for aesthetic experience. In the critical work, by contrast, Kant will have little interest in race as an innate and unmalleable marker of difference between groups of people, and instead will prefer to distinguish between people on civilizational grounds: he dispenses with race, in order to return to a more classical division between the civilized and the barbarian.

In Kant, however, the criterion for membership within the fray of civilization is no longer Christian baptism. Instead, the Enlightenment philosopher effectively replaces the shibboleth of religion by the shibboleth of historicity: a life is properly human, and not merely animal, if it is lived in accordance with a system of ends, and the greatest system of ends is the one that has set European civilization upon the path of progress. To live outside of this system of ends becomes tantamount to living as an animal, or simply living for the sake of living, which for Kant is the same as living for nothing at all. In this respect, Kleingeld is right that Kant takes a turn toward a more properly universalist and egalitarian conception of humanity. His version of the Enlightenment is no longer "equality among equals," with the implicit proviso that not all humans are equal. It is now *potential* equality among all humans, where this potentiality can

[52] Kant, *Kritik der Urtheilskraft*, AA V, 427.
[53] Kant, *Kritik der Urtheilskraft*, AA V, 378.
[54] Kant, "Recension," AA 8, 65.

be fulfilled only through an absorption of non-European peoples into the unfolding of European history.

Into his mature, critical period, Kant continues to try to engage with racial difference, as opposed to differences of culture and history, in a rigorous philosophical manner. In a remarkable passage of the review of Herder's *Ideas*, Kant seems practically to satirize his own use of antinomies in his major critical works, arguing that "one may prove, if one wishes, from numerous descriptions of various countries, that Americans, Tibetans, and other genuine Mongolian peoples are beardless—but also, if one prefers, that they are naturally bearded and merely pluck their hair out."[55] From this seemingly trivial and erroneous example—surely, one supposes, the matter of whether Mongolian peoples pluck their facial hair or not could be resolved without recourse to philosophical "proof"—Kant moves on to a somewhat more weighty example, at least to the extent that beliefs about it do not seem to be so easily susceptible to change in the face of empirical evidence: "Or one may prove that Americans and Negroes are races which have sunk below the level of other members of the species in terms of intellectual abilities—or, alternatively, on the evidence of no less plausible accounts, that they should be regarded as equal in natural ability to all the other inhabitants of the world."[56] Next, Kant describes the situation of the European theorist of human difference in remarkable terms, as a dilemma between two unsatisfying horns: the philosopher is at liberty to choose, Kant writes, "whether he wishes to assume natural differences or to judge everything by the principle *tout comme chez nous*, with the result that all the systems he constructs on such unstable foundations must take on the appearance of ramshackle hypotheses."[57]

There is much that would warrant unpacking here. It is hard to imagine that Kant really believes that the question of the fundamental unity or diversity of the human species is akin to the question of the hirsuteness of Mongolians. The latter question concerns only *what* the particular physical differences between different groups of people are, while the former concerns *whether* physical differences between groups mark out deep, essential boundaries between them. Surely Kant understands the difference in kind between these two sorts of question, and the presumption that he does understand it encourages the conclusion that he is being somewhat facetious. But what stands out most of all in his description of the two camps is a single phrase that, whether left in French or rendered in translation, could easily go unnoticed: *tout comme chez nous* and, the

[55]Kant, "Recension," AA 8, 62.
[56]Kant, "Recension," AA 8, 62.
[57]Kant, "Recension," AA 8, 62.

alternative, *tout comme ici*, are variations on what is sometimes called "the Harlequin principle," coming from a popular early modern pantomime character, Harlequin, the Emperor of the Moon. This principle was most strongly associated with the philosophy of Leibniz. The latter would write to Sophie Charlotte in 1794 for example, endorsing the fictional emperor's view, "always and everywhere, everything is the same as here."[58] This motto means many different things for Leibniz, and the way it applies to his complicated philosophical views certainly extends well beyond his philosophical anthropology. Nonetheless, it does most certainly capture Leibniz's beliefs about the unity of the human species, which we investigated in detail in chapter 7, and in particular about the way this unity underlies all appearances of difference. The fact that Kant picks this quintessentially Leibnizian motto to describe Herder's anthropology underlines a continuity that to us, today, is largely forgotten: in his anthropology, Herder is not simply going against his age by supporting the unity of the human species, he is also picking back up a broadly Rationalist, and particularly Leibnizian, vision of the world as characterized by unity behind multiplicity.

9.4. J. G. Herder: The Expectation of Brotherhood

Herder is virtually alone among his contemporaries in recognizing the fundamental provincialism of racial taxonomies. He writes that "some have for example ventured to call 'races' four or five divisions of [the human species]," but for his part insists that "there is no cause for this name." He continues,

> Race has to do with a difference of origin, which here is either not at all to be found, or in each of these bands of the world among each of these colors includes the most different races. For every people [*Volk*] is a people [*Volk*]: it has its own national character [*Nationalbildung*], as in its language. . . . In short, neither four or five races, nor exclusive varieties are to be found in the world. Colors blend into one another; the ideas [*Bildungen*] serve the genetic character; and in the whole it is all in the end only a shading of one and the same great tableau, which extends throughout all of the spaces and periods of the earth. It thus belongs not so much to systematic natural history, as to the physical-geographical history of humanity.[59]

For Herder, the human species is a unity; successive generations of it are linked as in a chain. Variety within the human species is cultural variety,

[58]Leibniz, *Die philosophischen Schriften*, 3:343.
[59]Herder, *Ideen*,, 94.

rooted particularly in language, and it is the task of the philosopher to identify the unity behind this variety. This variety is not the outward sign of any essential difference, since shared essence just is shared membership in a single chain.

Herder wishes to deny that human racial differences can be explained in terms of greater or lesser proximity on a continuum to apes, that in effect one group's greater apelikeness always proves on analysis to be illusory:

> Most of these apelike people are in countries in which there never were apes, as in the backward-slanting skull of the Kalmuks and the Moluccans, the jutting ears of the Pevas and the Amikuanes . . . etc., show. These things are also, once one is over the initial trickery of the eye, so little apelike, that indeed Kalmucks and Negroes remain fully human in the conformation of the face as well, and the Moluccans exhibits abilities, that many other nations do not have. In truth ape and man were never of one and the same family, and I would like for every last trace of the story to be corrected, according to which somewhere in the world they live together in common, fruit-bearing community. Every species [*Geschlecht*] has been done well enough by nature, and given its own heritage. She divided the apes into so many families [*Gattungen*] and types [*Spielarten*] and spread these as widely as she was able; but you, man, honor yourself. Neither the pongo nor the *longimanus* is your brother, but indeed the American and the Negro are. You should therefore not oppress them, nor kill them, nor steal from them: for they are men as you are. With the ape you should expect no brotherhood.[60]

The unity of the human species thus requires radical discontinuity between it and neighboring rungs on the ladder. Difference within the human species, in turn, endures at the level of culture: a difference so apparently natural or species-like that it takes a philosopher intent on finding the unity behind the multiplicity to see past it. In the end, the unity and the apparent difference both are rooted for Herder in culture, the distinguishing feature of which is that it is something that governs the reproductive patterns of its members, even if it does not define the natural boundaries of interfertility. For this reason, in Herder's view the attempt to assimilate so-called races to species amounts to a simple failure to carve nature at its joints.[61]

[60]Herder, *Ideen*, 92–93.

[61]For an important and very detailed study of the interactions between Kant and Herder that led to the development of their respective conceptions of philosophical anthropology, see John Zammito, *Kant, Herder, and the Birth of Anthropology*, Chicago: University of Chicago Press, 2002.

Racial divisions for Herder must be understood principally as "an aide for our memory."[62] The human species, on the final analysis, "appears on the earth in so many different forms; yet it is everywhere one and the same human family [*Menschengattung*]."[63] This principle is to be upheld, Herder believes, on fundamentally Leibnizian grounds: since "human understanding seeks unity in all multiplicity," and since the divine understanding "everywhere weds unity with the most innumerable variety."[64] This variety, for Herder, is to be found entirely *within* the human species, rather than *between* humanity and its nearest neighbors. "I would never wish," he writes, "for the bordering of human beings on the apes, also, to be driven so far that in looking for a ladder of things, we forget the shoots and the intermediate spaces, without which there is no ladder."[65] A ladder is possible only if there are discontinuities, and thus for him as for Leibniz (as we saw in chapter 7) the existence of a scale of being does not at all imply a continuum of beings such that there are no real divisions between them. There must be spaces between the rungs, or else it is not a ladder at all. This commitment to discontinuity also carries with it a firm opposition to evolution.

For Herder, as again for Leibniz, "every man is in the end a world."[66] This is to say, inter alia, that each individual member of a species is irreducibly different from each other member. It is precisely this difference, moreover, that ensures the unity of a species: given that there is endless variety within the chain of generations, no amount of further variation can be expected to cause a new chain to split off from an old one. The cycle of generation preserves just the right amount of difference: a constant outpouring of variety, rooted in an unchanging succession of the same. This sameness is, as Aristotle would have put it, a sameness in kind but not in number.[67]

As has already been noted, behind the problem of human racial diversity, there is a deeper and more pervasive philosophical problem of individual diversity: how it is, namely, that each person is identical to every other in respect of humanity, yet irreducibly singular as a person.

[62]Herder, *Ideen*, 85.

[63]Herder, *Ideen*, 85.

[64]Herder, *Ideen*, 87.

[65]Herder, *Ideen*, 91–92.

[66]Herder, *Ideen*, 86. For a very insightful account of the Leibnizian elements of Herder's thought, see Nigel DeSouza, "Leibniz in the Eighteenth-Century: Herder's Critical Reflections on the *Principles of Nature and Grace*," *British Journal for the History of Philosophy* 20, 4 (July 2012): 773–95. See also Beate Monika Dreike, *Herders Naturauffassung in ihrer Beeinflussung durch Leibniz' Philosophie*, Wiesbaden: Steiner, 1973; Günter Arnold, "'. . . der grösste Mann den Deutschland in den neuern Zeiten gehabt': Herders Verhältnis zu Leibniz," *Studia Leibnitiana* 37, 2 (2005): 161–85.

[67]Aristotle, *On the Generation of Animals*, 2.4 738b28.

Augustine understands the problem of individual diversity and that of group diversity as two instances of the same general problem, and resolves them both by appeal to the inherence of an immortal rational soul; thus for him the "same account which is given of monstrous births can be given of monstrous races."[68] Kant notes for his part in the *Anthropology* that "[n]ature has sufficient supply at her disposal so that she does not have to send, for want of forms in reserve, a person into the world who has already been there."[69] Yet Kant, unlike Herder, is not fully ready to recognize that this very same account of human individuality also provides a key to thinking about racial diversity, that there is indeed an antiracist implication to this observation: to wit, that observable physical differences between different human groups tell us nothing about real subdivisions of the human species.

Over the course of the modern period, as we have been seeing, there is a strong association between opposition to scientific racism, on the one hand, and to the evolution of species, on the other. Those who did not or could not see species as separating off from each other also could not see different human groups as on their way to separating off from humanity, or as closer to the apes than other human groups. But as we have seen, at the same time the scientific racism facilitated by evolutionary thought is very obviously in tension with the conception of human racial differences as a static grid, which at least intimates polygenesis, and takes the differences between different groups as fixed and unchanging.

One difficult lesson here, and indeed one that we have been learning throughout this book, is that modern racism must be seen as in part a consequence of the abandonment of a conception of humanity as bound together in "a golden chain" that includes all and only the generations of human beings, and its replacement by a conception of humanity as branching off from a tree that includes all other animal species. There was, at the moment the arboreal model was beginning to appear attractive, no prima facie reason why there should not have been human kinds intermediate between ourselves and apes. But the assumption that these kinds might best be sought among non-European men and women—already recognized throughout history *as* men and women—rested on two great mistakes, one conceptual and one empirical. The conceptual mistake was to presume that there was any ground for believing that there was anything more in need of explanation about the appearance or nature of, say, Africans, than there was about Europeans; to speak in structuralist

[68]Augustine, *City of God*, vol. 2, book. XVI, chap. 8, 117.

[69]Kant, *Anthropologie in pragmatischer Hinsicht*, AA 7, 321. ". . . und die Natur hat Vorrath genug in sich, um nicht der Armuth ihrer vorräthigen Formen halber einen Menschen in die Welt zu schicken, der schon ehemals drin gewesen ist."

terms, there was no good conceptual reason, as Herder plainly saw, why "black" is the marked category and "white" the unmarked one. Once the arbitrariness of this marking is apparent, one quickly sees that there is no good conceptual basis for supposing that Africans are any more intermediate between the human and the ape than, say, Europeans are. And this is well understood by Herder virtually alone, among Enlightenment European thinkers, who challenges his readers to identify anything that is truly more simian about non-European groups.

It would take a few more centuries for science to work out exactly why a commitment to evolution need not carry with it a commitment to the biological significance for the human species of the category of race. But where antievolutionist, antiracist thinkers such as Herder and before him Leibniz were perhaps more prescient was not in biology but in cultural anthropology: in their understanding that apparently natural differences between human groups are in fact rooted in culture, and that cultural differences are, in the end, local variations on universal human potentialities, local inflections of a global order: diversity compensated by unity.

This tradition of thinking about human diversity by no means solves all problems, particularly in the political sphere. To adapt a Leibnizian notion, it conceptualizes human cultures as "windowless," as "worlds apart," which can have no meaningful or desirable overlap, but which instead remain eternally rooted to fixed geographical regions. This overlooks the reality of human history, which in fact witnesses a constant, vast web of cultural and material exchange. In the end what Herder is calling for is little different from what, say, white nationalist separatists in the United States would find amenable, and what has on occasion brought them to peaceful summits with representatives of black nationalist parties: each in his own place, and there will be no cause for antagonism. But such a call can have different valences, as is often noted today, depending on who is making it, and on whether that person belongs to an historically oppressed group or not. For this reason, any serious analyst would have to give a different reading of, say, the Black Panthers' participation in such a summit than the participation of the American Nazi Party. Though the contours of Herder's own context have faded with time, it is important to recall that he imagines himself to be writing from what today would be called a "subaltern" position, speaking up against the specter of French political and cultural domination across Europe.

9.5. J. F. Blumenbach: Variety without Plurality

Recall Anténor Firmin's observation, cited as an epigraph at the beginning of this book, of a deep tension between his own commitment to science and his certainty of his own equality with the members of white race, whom the science of the day was holding up as superior. If Kant and Herder had come to understand the problem of human diversity as fundamentally rooted in culture and history, many of their contemporaries, not least J. F. Blumenbach, were actively seeking to comprehend the problem of diversity in purely quantitative and synchronic terms. This is precisely the approach that would come to dominate by the time Firmin wrote in the nineteenth century.

Blumenbach's influential work on racial difference is riddled with the same fundamental inconsistency we have by now seen in a number of different authors: at the height of the modern taxonomic project, he continues this project at the subspecies level in the case of human beings, thereby implying that there are real or natural boundaries between subspecies kinds. Yet at the same time he denies that the sort of features that make interspecies boundaries real arise at the subspecies level. Blumenbach picks out, and names, the "races of men," even as he deprives them of the features that he himself explicitly cites as making biological kinds real. It is as if we were seeing in Blumenbach an illustration of the thesis that the tendency of the human mind to carve nature up into kinds is so great that it cannot leave off where it knows nature has left off, but must keep on picking out kinds at the subkind level. Or at least that it must do so with respect to that particular kind that is of most interest to it, namely, its own kind: the human kind.

In his 1775 work *De generis humani varietate nativa* (*On the Innate Variety of Mankind*, also more commonly, though less correctly, translated as *On the Natural Varieties of Mankind*), Blumenbach divides the human species into three basic races—the Caucasian, Mongolian, and Ethiopian—and posits an additional two—the American and Malay—that he sees as intermediaries between pairs formed from the first two. Blumenbach is insistent however that all of these races belong to the same species, insofar as shared species membership may be confirmed whenever different populations "agree so well in form and constitution, that those things in which they do differ may have arisen from degeneration."[70] For Blumenbach, the only other possibility in accounting for racial variety would be to hold to a polygenetic account of human origins, that is, the view that the different races all have separate creations.

According to Blumenbach, there are three causes that change the form of animals within the same species: climate, mode of life, and hybridism. The first two can, for him as for most of his contemporaries, bring about

[70]Blumenbach, *On the Natural Varieties of Mankind*, in *The Anthropological Treatises of Johann Friedrich Blumenbach*, trans. and ed. Thomas Bendyshe, London: Anthropological Society, 1865, 188.

only relatively slight changes. But what about the third? Blumenbach believes that there is a natural variety of humankind, yet this is, as he puts it, variety within unity. There are "no human hybrids," though he explicitly acknowledges "mulatto" as a meaningful scientific classification. These two beliefs, taken together, are worthy of some pause. The very term "mulatto," as we have seen, derives from the Spanish word for "mule," which is perhaps the most familiar hybrid of the animal kingdom, invoked continuously since Aristotle as an example of nature's wisdom in rendering interspecies offspring sterile, and thus nipping the proliferation of monstrosities in the bud. For Blumenbach, as for Aristotle, the infertility of the offspring of hybrid copulations stems from "the providence of the Supreme Being, lest new species should be multiplied indefinitely."[71] Blumenbach thus takes the term "mulatto" as, strictly speaking, a misnomer: a mulatto is not, biologically, comparable to a mule, for from a mulatto, as he goes on to note, one can derive "quateroons," "octoroons," and so on.

On the other hand, the possibility of real human hybrids of the sort that Locke had imagined, with other animal species, is one that Blumenbach absolutely denies: "That men have very wickedly had connexion with beasts seems to be proved by several passages both in ancient and modern writers. That however such a monstrous connexion has anywhere ever been fruitful there is no well-established instance to prove. . . . And even if it be granted that the lascivious male apes attack women, any idea of progeny resulting cannot be entertained for a moment, since . . . travellers relate that the women perish miserably in the brutal embraces of their ravishers."[72] There are no hybrids between humans and apes, and therefore every higher primate is, as for Tyson and Leibniz, as we have seen, either one or the other. Thus, in order to ensure a safe taxonomic distance away from the purported *Homo sylvestris*, Blumenbach writes that he is "induced to consider even that famous animal the orang-utan as a quadruped."[73] In contrast with Locke and others, Blumenbach denies the possibility of hybrids issuing from the pairing of a human and an animal of another species, on the grounds that "the Supreme Being foresaw these disgusting kind of unions and took care to render them futile."[74] The impossibility of hybrids is for Blumenbach part of a conception of the human species that rejects gradation between humanity and animality, and instead sees humanity as something one either possesses or does not possess, and refuses to locate the anthropological difference

[71] Blumenbach, *On the Natural Varieties of Mankind*, 73.
[72] Blumenbach, *On the Natural Varieties of Mankind*, 80–81.
[73] Blumenbach, *On the Natural Varieties of Mankind*, 86.
[74] Blumenbach, *On the Natural Varieties of Mankind*, 80–81.

in anatomical features.[75] Just as for Tyson, the difference lies in speech, which is not simply a side effect of the possession of a certain bodily conformation, but rather a sign of the inward possession of reason. Thus all human beings possess speech, and, in contrast with the view of, for example, Monboddo, no apes possess it. Blumenbach writes,

> It is indeed beyond all doubt that the fiercest nations, the Californians, the inhabitants of the Cape of Good Hope, &c. have a peculiar sort of speech, and plenty of definite words, and that animals on the contrary, whether they be like man in structure, as the famous orangutan is, or approach man in intelligence, to use the words of Pliny about the elephant, are destitute of speech, and can only emit a few and those equivocal sounds. That speech is the work of reason alone, appears from this, that other animals, although they have nearly the same organs of voice as man, are entirely destitute of it.[76]

Because he has such a clear commitment to the discreteness of the human species, for Blumenbach primatology can only be a distraction from the problem of diversity. "Enough then has been said about the Troglodyte and the Satyr," he writes hastily, "now we must come more closely to the principal argument of our dissertation, which is concerned with the question; *Are men, and have the men of all times and of every race been of one and the same, or clearly of more than one species?*"[77]

Blumenbach observes, without naming any names, though almost certainly thinking of Voltaire, that a certain love of novelty has caused some people to adopt the polygenist view, and he goes on to explicitly associate this theory with freethinking and impiety: "The idea of the plurality of human species has found particular favour with those who made it their business to throw doubt on the accuracy of Scripture. For on the first discovery of the Ethiopians, or the beardless inhabitants of America, it was much easier to pronounce them different species than to inquire into the structure of the human body, . . . to compare parallel examples from the universal circuit of natural history, and then at last to come to an opinion, and investigate the causes of the variety."[78] In fact, for Blumenbach, the differences between different human subgroups are so gradual that "[o]ne variety of mankind does so sensibly pass into the other, that you cannot mark out the limits between them."[79] Blumenbach believes that

[75]We are borrowing this apt phrase from Markus Wild's *Die anthropologische Differenz: der Geist der Tiere in der frühen Neuzeit bei Montaigne, Descartes und Hume*, Berlin: De Gruyter, 2006.

[76]Blumenbach, *On the Natural Varieties of Mankind*, 83–84.

[77]Blumenbach, *On the Natural Varieties of Mankind*, 98.

[78]Blumenbach, *On the Natural Varieties of Mankind*, 98.

[79]Blumenbach, *On the Natural Varieties of Mankind*, 98–99.

the "whole bodily constitution, the stature, and the colour, are owing almost entirely to climate alone."[80] Unlike Bernier, who had contrasted the "natural blackness" of Africans with the circumstantial climatic blackness of southern Indians, Blumenbach associates *all* pigmentation in all parts of the world entirely with climate, as well as with lifestyle, thus explicitly drawing together, like so many of his predecessors—including Evelyn, Acosta, Burton, and many other early theorists—the questions of race and class identity: "Anatomists not unfrequently fall in with the corpses of the lowest sort of men," he writes, "whose reticulum comes much nearer to the blackness of the Ethiopians than to the brilliancy of the higher class of European."[81] Thus, Blumenbach concludes that color is an "adventitious and easily changeable thing, and can never constitute a diversity of species."[82]

Upon reflection, one might suppose that *any* racialist thinker in a preevolutionary context who is committed to monogenesis must of necessity be committed to a sort of variety without plurality. There cannot be real or essential divisions between races, since the human species was created as one, and all descend from the first two progenitors, connected to them, as Leibniz figuratively expressed it, in "an eternal golden chain." The project of racial typology went ahead, explicitly on the model of species taxonomy, even as its principal contributors insisted that there could be no species-like or essential divisions within the human species, to the extent that this species consists entirely in descendants of the same original ancestors, and to the extent that speciation was not yet a consideration.

For Blumenbach, perhaps the most famous of racial typologists, all races descend from the same progenitors and commitment to their existence presupposes a theory of eternal species fixism. The races pass imperceptibly from one into the other, so that you cannot mark out the boundaries between them; attempts to enumerate the number of races are "arbitrary indeed," even including his own attempt. Whatever we might ascribe to racial difference can be accounted for almost exclusively in terms of climate, and color in particular is so superficial that it can easily change over the course of an individual's life, either through changes in diet or climate, or simply through internal changes in the life cycle, of the same sort as we see in the graying of hair. For a reifier of race, we see here, Blumenbach certainly cedes quite a bit to what we today would call the constructionist camp.

[80] Blumenbach, *On the Natural Varieties of Mankind*, 101.
[81] Blumenbach, *On the Natural Varieties of Mankind*, 108.
[82] Blumenbach, *On the Natural Varieties of Mankind*, 113.

This much comes out even more clearly in the 1795 edition and expansion of the work, in which he has now abandoned Linnaeus's four-part division of the races in favor of his own distinctive, and certainly more influential, five-part division: "Five principal varieties of mankind may be reckoned. As, however, even among these arbitrary kinds of divisions, one is said to be better and preferable to another; after a long and attentive consideration, all mankind, as far as it is at present known to us, seems to me as if it may best, according to natural truth, be divided into the following varieties . . . Caucasian, Mongolian, Ethiopian, American, and Malay."[83] Blumenbach continues, prefixing a "double warning" to his enumeration of the different characteristics of the five races, "First, that on account of the multifarious diversity of the characters, according to their degrees, one or two alone are not sufficient, but we must take several joined together; and then that this union of characters is not so constant but what is liable to innumerable exceptions in all and singular of these varieties. Still this enumeration is so conceived as to give a sufficiently plain and perspicuous notion of them in general."[84] There seems to be an odd vacillation in the first part of the passage: how can an "arbitrary kind of division" be made "according to natural truth"? In his simultaneous desire to carve up human subgroups, even as he concedes that these do not exist in nature, we might detect in a racial typologist such as Blumenbach a grasping for language that he does not yet have, and that would not be available until the advent of the cognitivist account of race described in chapter 1. The best that he can do is to speak of "variety" in terms of "preferability"; what he cannot do is explain *why* it should be preferable, given that he also acknowledges that racial difference marks out no real difference, that any divisions are necessarily arbitrary, and so on.

Like many other naturalists of the eighteenth and nineteenth centuries, Blumenbach saw the problem of the variety of kinds as ultimately to be resolved by reference to what we have already seen Kant calling "the unity of the generative power." Thus Blumenbach explains: "As I am going to write about the natural variety of mankind I think it worth while to begin from the beginning, that is, with the process of generation itself."[85] He is committed to the "like begets like" principle along broadly Aristotelian lines, but this principle fails to adequately account for diversity within a kind. Aristotle himself had been able to account for diversity rather easily, in terms of his broader metaphysics of matter and form, which posited that perfect offspring would be literally a re-production of

[83]Blumenbach, *On the Natural Varieties of Mankind*, 264.

[84]Blumenbach, *On the Natural Varieties of Mankind*, 265.

[85]Blumenbach, *On the Natural Varieties of Mankind*, 60.

the father, as the contributor of form, but that such reproduction never happens, in consequence of the limitations of the mother's material cause. In the absence of this explanatorily satisfying schema, modern generation theorists were often at a loss to account for the multiplicity within the unity of a species, to explain why, in spite of the fact that all members of a species are in a sense identical with respect to kind, they are nonetheless each distinct as individuals, and often distinct as populations within the same species as well. The latter sort of distinctness, for Blumenbach as for many of his contemporaries, is generally conceptualized in terms of deviation relative to the original makeup of the species.

In an important sense, in fact, the very existence of variety within a species is itself a consequence of the failure of the "like begets like" principle to fully hold. Thus Blumenbach writes: "The offspring at last brought to light, and in the process of time become adult, can produce like with the other sex of its species, whose posterity ought to go on forever like their first parents. What then are the causes of the contrary event? What is it which changes the course of generation, and now produces a worse and now a better progeny, at all events widely different from its original progenitors?"[86] Here, evidently, commitment to real racial divisions involves a recourse to explanations that are both premature and impious: "The idea of the plurality of human species has found particular favour with those who made it their business to throw doubt on the accuracy of Scripture."[87] On the first discovery of the Ethiopians, he writes, "it was much easier to pronounce them different species than to inquire into the structure of the human body."[88] When this structure is investigated, we find no markers of unbridgeable difference, whether in skeletal structure, skin color, or elsewhere; instead, different groups "do so run into one another, and that one variety of mankind does so sensibly pass into the other, that you cannot mark out the limits between them."[89] Blumenbach writes of an "almost insensible and indefinable transition" from "the pure white skin of the German lady through the yellow, the red, and the dark nations, to the Ethiopian of the very deepest black."[90] The "insensible transition" characterizes not only the entire species as one scans from one individual to another, but also each single individual as it passes from one stage of life to another: "All of us are born nearly red, and at last in progress of time the skin of the Ethiopian infants turns to black, and ours to white, whereas in the American the primitive red colour remains, excepting so far as that by change of cli-

[86] Blumenbach, *On the Natural Varieties of Mankind*, 70–71.
[87] Blumenbach, *On the Natural Varieties of Mankind*, 98.
[88] Blumenbach, *On the Natural Varieties of Mankind*, 98–99.
[89] Blumenbach, *On the Natural Varieties of Mankind*, 98–99.
[90] Blumenbach, *On the Natural Varieties of Mankind*, 107.

mate and the effects of their mode of life those colours sensibly change, and as it were degenerate."[91] Blumenbach's treatise also shows an enduring preoccupation with the same sort of cultural practices of bodily modification that had so interested Bulwer more than a century earlier. The German naturalist reports of the "Hottentot" practice of removing a testicle, the Chinese and Native American practices of altering the shape of the head, Turkish and Tahitian methods of permanent hair removal.[92] He mentions tattooing and scarification in California and New Holland, foot binding in China, and so on, as examples of "deformities." He disputes Linnaeus's view that the "custom of making women thin by a particular diet" counts as a deformity, on the grounds that, as he puts it, this practice "is very ancient, and has prevailed amongst the most refined nations."[93] Here we see a problem that also goes back to Bulwer's denunciatory anthropology, as for example in his distinction between cutting the nails and shaving the beard: one culture's deformation is, so to speak, another culture's culture.

From what we have seen, Blumenbach seems like an odd figure to pick out as one of the important founders of racial science, to be responsible in part for the paradoxical situation diagnosed by Firmin. In fact, though, what Blumenbach shows us is that racial science did not necessarily imply racial *realism*, any more than, say, Robert Boyle's experiments with the air pump in the seventeenth century involved a metaphysical commitment to the existence of a real vacuum.[94] Racial science, for Blumenbach, is mostly a descriptive project, while racial realism by contrast is a philosophical theory based on an interpretation of what that descriptive project yields. Blumenbach contributes to the rise of racial realism not by defending the philosophical theory—in fact, he explicitly rejects the theory—but rather, principally, by providing a method for investigating racial difference quantitatively. Blumenbach contributes nothing new to the study of skin color, and in fact runs rather hastily through the prominent theories of pigmentation in order to come to the data that he believes to be most revealing about human difference: namely, those that measure variations in the size of crania.

Craniometry provides Blumenbach with a quantitative basis for the study of human difference that allows for what we might think of as "statistical racial realism," that is, a commitment to real differences between different human groups that is based on measurable average variation, rather than on some key feature that every member of a given

[91] Blumenbach, *On the Natural Varieties of Mankind*, 108.

[92] Blumenbach, *On the Natural Varieties of Mankind*, 127.

[93] Blumenbach, *On the Natural Varieties of Mankind*, 128.

[94] See in particular Michael Hunter, *Boyle: Between God and Science*, New Haven: Yale University Press, 2009.

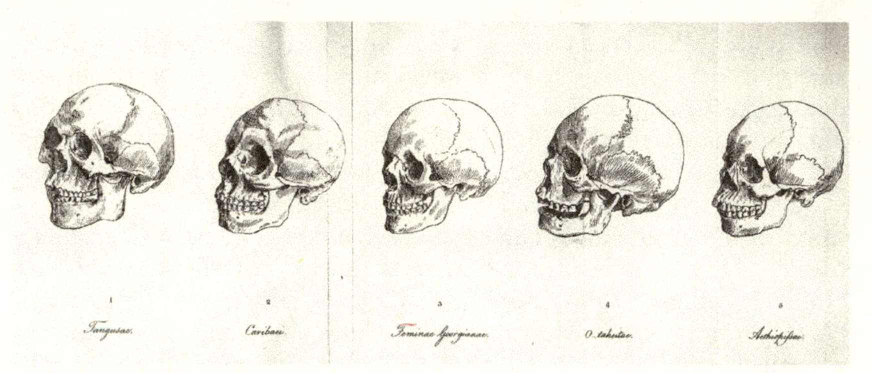

Figure 4. "Racial craniometry," from J. F. Blumenbach's *De generis humani varietate nativa* (1795).

group must have. In other words, craniometry made the study of human diversity scientific in a way that it could not be so long as this study was rooted in the qualitative description of visible features of the body ("woolly hair," "flat nose," and so on). At least since Leibniz's *Directions Pertaining to the Institution of Medicine* of 1671, researchers had been concerned to root their claims about health and sickness, about physiological causes of temperament, and about variety within or between different populations, in relatively stable, fixed features of the body. There is nothing more stable in the body than the skeleton; it is, one might say, the substratum of the human corporeal self, staying the same, more or less, across the life cycle, indeed even resisting decomposition after death.

What Blumenbach adds to this conviction is a methodological commitment to establishing claims about physical differences between different groups in relation to the measurement of large samples. Again, the establishment of measurable skeletal differences between different human groups entails no commitment to essential differences between them; but it does locate variation somewhere that cannot be easily accounted for by appeal to the three sources of variation—mode of life, environment, and hybridity—that Blumenbach identifies at the outset as the sources of human diversity. This in turn makes it fairly easy for others to fill in the ontological commitment to real racial differences that Blumenbach himself leaves out.

The categories that are of more interest to Blumenbach than those based on skin color are, again, those based on the relatively more stable and more readily quantifiable physical features of human populations, in particular skeletal features. Acknowledging the precedent set by Linnaeus, Buffon, and other authors,[95] Blumenbach divides the human species into five distinct races, though he insists that he is doing so principally on the basis of convenience and repeats throughout his account that these races blend "insensibly" into one another. The Caucasian race is held to be the most primordial, while the American and Malay are held to be "intermediate" races: the American between the Caucasian and the Mongolian, and the Malay between the Caucasian and the Ethiopian. The account of the Mongolian race, to cite just one example, begins with color ("yellow"), and then moves on to hair ("black, stiff, straight, and scanty") and shape of the head ("almost square"), face ("broad, at the same time flat and depressed"), glabella ("flat, very broad"), as well as nose, cheeks, eyelids, and chin. As to the Ethiopian race, Blumenbach offers a similar description: "head narrow, compressed at the sides; forehead knotty, uneven; malar bones protruding outwards."[96] He notes that some people, such as Voltaire ("witty, but badly instructed in physiology"),[97] have argued that the Ethiopian variety is "a peculiar species of mankind," principally on the ground that "it is so different in colour from our own."[98]

Blumenbach has little patience for the exaggerated claims of the polygenists. He notes of the assertion that Ethiopians are closer than other races to the apes, that "I willingly allow [it] so far as this, that it is in the same way that the solid-hoofed variety of the domestic sow may be said to come nearer to the horse than other sows."[99] There is no respect at all in which Blumenbach supports any actual relations of descent between horses and pigs. However ungrounded his perception of a greater physical resemblance between Ethiopians and apes than between Caucasians and apes, his purpose in bringing up the similarity between the hooves of horses and a certain kind of pig is precisely to highlight the difference between his own view and that of someone like Voltaire. Blumenbach thinks that the supposedly similar appearance between two species emerges out of the fact that two different species encounter similar environmental circumstances. This is an account something like what we today would call "convergent evolution," but which can for Blumenbach himself be only convergent degeneration.

Although he himself does not believe any real taxonomic difference exists between people of different races, nonetheless he considers it worthwhile to recite the conventional names of the products of so-called mixed-race unions. He calls these children "hybrid offspring,"

[95]Blumenbach, *On the Natural Varieties of Mankind*, 266–67.
[96]Blumenbach, *On the Natural Varieties of Mankind*, 266.
[97]Blumenbach, *On the Natural Varieties of Mankind*, 270.
[98]Blumenbach, *On the Natural Varieties of Mankind*, 270.
[99]Blumenbach, *On the Natural Varieties of Mankind*, 271.

but then adds, parenthetically, "if we may use that word."[100] Of course, by his own standards Blumenbach knows full well that we may not use that word. For him, there are only insensible gradations between human populations, while there is an absolute boundary between the human species and animal species, a boundary that the supreme being protects by ensuring that the couplings of humans with animals cannot yield offspring. Yet none of this stops Blumenbach from telling us that the offspring of a black man and a white woman "is called Mulatto, Mollaka, Melatta," and that of a Native American man and a European woman a "Mameluck." A European man and a Mulatto woman give us a "Terceron, Castiço," the offspring of two Mulattoes a "Casque," and of a black person and a Mulatto, a "Griff." A Terceron and a European yield a "quateroon" or "postiço," while a quateroon male and a white female result in a "quinteroon," called by the Spanish a "Puchuela." And so on. Blumenbach concludes from this florid proliferation of names that "it is plain therefore that the traces of blackness are propagated to great-grandchildren," yet he quickly adds that "from all these cases, this is clearly proved, which I have been endeavouring by what has been said to demonstrate, that colour, whatever be its cause, be it bile, or the influence of the sun, the air, or the climate, is, at all events, an adventitious and easily changeable thing, and can never constitute the diversity of species."[101] Blumenbach is evidently doing two things simultaneously: repeating the folk-taxonomical designations that encourage realist thinking about race, while at the same time denying that these designations name anything real or fixed. To the extent that he stands as a representative of the emerging "racial science," which would at least implicitly take racial categories as picking out something real, and would remain the predominant account of human variety in physical anthropology until the mid-twentieth century, one possible conclusion to draw is that the use of names, or of a naming scheme, strongly encourages the view that there are real kinds corresponding to those names. Blumenbach denies that quinteroons are a real variety of being in nature, on a par with elephants or aluminum, but the naming scheme already carries with it a reality-generating commitment that the explicit denial does little to eliminate.

By the time Blumenbach is writing, there have already been a great number of similar divisions into a handful of basic human subtypes. Blumenbach himself mentions Linnaeus's division of the human species into American, European, Asiatic, and Negro; Buffon's into Lapp, Tartar, South Asian, European, Ethiopian, and American; and Meiners's

[100] Blumenbach, *On the Natural Varieties of Mankind*, 111.
[101] Blumenbach, *On the Natural Varieties of Mankind*, 112–13.

minimalist division of the human species into two races: "(1) handsome, 2) ugly; the first white, the latter dark."[102] Arguably, it is the last of these that distills all the others down to their essence: a binary distinction—us and them—grounded not in a nature knowable by science, but in a fairly provincial aesthetic sensibility.[103]

[102] Blumenbach, *On the Natural Varieties of Mankind*, 268.

[103] Throughout this book, we have avoided discussion of the complex role of aesthetics, ideas about beauty, sensibility, and taste, in the construction of racial categories and judgments. On this vast topic, see, to begin, David Bindman, *Ape to Apollo: Aesthetics and Race in the 18th Century*, Ithaca, NY: Cornell University Press, 2002. From an art-historical perspective, see particularly Jean Devisse and Michel Mollat, *L'image du noir dans l'art occidental*, Paris: Bibliothèque des Arts, 1979; David Bindman, Henry Louis Gates Jr., et al., *The Image of the Black in Western Art, Volume I: From the Pharaohs to the Fall of the Roman Empire*, Cambridge, MA: Belknap, 2010; followed by *The Image of the Black in Western Art, Volume II: From the Early Christian Era to the "Age of Discovery," Part 1: From the Demonic Threat to the Incarnation of Sainthood*, Cambridge, MA: Belknap, 2010; *The Image of the Black in Western Art, Volume II: From the Early Christian Era to the "Age of Discovery," Part 2: Africans in the Christian Ordinance of the New World*, Cambridge, MA: Belknap, 2010; *The Image of the Black in Western Art, Volume III: From the "Age of Discovery" to the Age of Abolition, Part 1: Artists of the Renaissance and Baroque*, Cambridge, MA: Belknap, 2011; *The Image of the Black in Western Art, Volume III: From the "Age of Discovery" to the Age of Abolition, Part 3: The Eighteenth Century*, Cambridge, MA: Belknap, 2011.

Conclusion

THAT THINKING ABOUT HUMAN DIVERSITY comes to be expressed specifically in *global* typologies of human subgroups only toward the end of the seventeenth century is easy to understand. This is the first time in history that we see great progress in attempts at global knowledge of any sort at all: increasingly accurate world maps, the measurement of longitude, of magnetic variation, and so on. That these global typologies should in turn reproduce the language, and some of the presuppositions of biological taxonomy, flows in turn from the tremendous recent progress of the taxonomic project, combined—which we may see only in retrospect—with what may be an innate tendency to conceptualize ethnic difference along the same essentialistic lines by which we cognize differences of kind throughout the biological world. This appears to have been the case, as we have seen, even when the explicit scientific theory of ethnic or racial difference did not involve a commitment to any species-like or biologically significant essential differences between races. The modern race concept was constructed by practices of naming and ordering, much more than it was by overt expressions of racial realism. It owes much more to the systematic project of Blumenbach than to the juvenile taunts of Voltaire.

Some recent cognitive literature suggests that a basic division of human society into essentialized subtypes has been with us all along, even if there is by no means an explicit theory of race in all places and times. One way of understanding the difficulty of eliminating racial thinking, on the supposition that this literature has identified a real feature of the human mind, is to suppose that it is not divisions in nature that remain stubbornly fixed and ineliminable, but rather certain patterns of thought. Such a supposition long predates the development of any explicitly cognitive approach. To cite one of many examples we have seen, in the eighteenth century Montesquieu argues not that there is an essential difference between Europeans and Africans, but only that Europeans are innately disposed to see the difference of skin color as marking out an essential difference. This is not, to be sure, what the current cognitive literature on race says, since no serious scientist claims today that the perception of salient difference between Europeans and Africans is fixed and insurmountable. We know from comparative and historical data that there are cases of the perception of what might seem to be less significant phenotypic differences than those that often obtain between Europeans and Africans as being no less salient, or revealing of supposed essential differences, for that reason. Nonetheless, Montesquieu does place the sources of racial difference in, so to speak, the mind of the beholder, rather than in the joints of nature. To some extent, this is also what Kant, Blumenbach, and so many other exponents of modern racial science do, when they distinguish, to use Kant's terminology, between artificial and natural divisions. They identify the entire human species as the only true natural division, in virtue of the unity of its reproductive power, while identifying racial divisions as ultimately grounded only in the classificatory interests of the people doing the classification.

At the same time, the distinction between natural and artificial divisions, which is made explicit by Kant but pervades the work of all three of these authors, does not appear to constitute an explicit argument against the view that race is a natural kind. Indeed, it would be hard to position the views of most of the authors we have investigated in relation to the Millian conception of such kinds. For Mill, in contrast to Leibniz, Kant, Blumenbach, and others, the principle of natural unity of "Kinds" in general arises from the fact that they share properties beyond the properties in virtue of which they are classified; for Kant and the others on whom we have focused, the principle of the natural unity of *living* kinds in particular arises from their power to transmit kind membership across generations. Here, then, the criteria for shared kindhood for a group of living beings, such as "Negroes," cannot be the same as the ones Mill would set up, for example, for phosphorus. Rather, for Kant and others, racial divisions are artificial to the extent that they are not grounded in a principle of reproductive unity, but real to the extent that they pick out "deviations" that are actually there. These deviations, however, are in the end much like the admixtures of tin that turn copper to bronze: it is up to the name-giver to decide where the one kind of thing leaves off and the other begins. Race is not a natural kind for the authors we have studied—even, or particularly, the authors who were so influential in the rise of racial science—but this is in large part because these authors reject the idea, pace Mill, that there can be a general account of natural kinds that runs races, species, and the elements of the periodic table together indiscriminately. Rather, living kinds deserve a distinct philosophical analysis, and on this analysis one finds no basis for the identification of races as natural. The importance of this point can be seen only if we work our way back into a non-Millian frame of mind, part of which involves taking seriously the central significance of what we now call the "philosophy of biology" throughout so much of the history of philosophy.

It was in large part the systematization of nature, in often avowedly artificial classifications, that led to the emergence of racial realism in the modern period. It was quite enough to devise complicated schemata or groupings of all natural beings, including human beings and their purported subtypes, to reify the categories of race that so many modern people have taken for granted. It was not necessary, in addition, to produce philosophical arguments in defense of racial essentialism. Race, then, as an entity on a par with phlogiston, cosmic radiation, or gluons, is not invented in the modern period. What *is* invented is a system of racial typology, which in turn promotes a new way of talking about human diversity—a discourse, if you will—and which supervenes on the prior and parallel project of biological taxonomy, even as it explicitly and repeatedly denies that the divisions it is making are actually given in nature. This new typology, finally, may be said to be the result of an increasing concern in the modern period to understand the human being as a thoroughly natural being, as exhaustively comprehensible within the terms of a system of nature that also includes primates, quadrupeds, molluscs, and plants. The insertion of the human being into such a system of nature, as we have attempted to show here, had profound implications for philosophical anthropology in the widest sense: for the understanding of human nature, and of the nature of human difference.

There is a certain responsibility, in addressing a subject as contemporary and unresolved as race, to not treat it from a dusty and antiquarian point of view, but to bring something to current discussions that might help to lessen this idea's harmful effects. We began this book with an epigraph from W.E.B. DuBois, identifying "color prejudice" in the southern United States as a "curious kink of the human mind," and then proceeded to investigate the concept of race as it unfolds from the Spanish Renaissance to the German Enlightenment, thus in a chapter of history that plays out mostly before the institutions of prejudice that interest DuBois had taken shape. This may seem an avoidance of the pressing matter at hand, but the approach here has been motivated by the conviction that these curious kinks of U.S. history, which in the end is the history of utmost concern to the present author, may best be seen as a local and recent inflection of a deep global history. This history must be uncovered and analyzed in order for the seemingly intractable local pathologies, the "kinks" to which DuBois refers, to be properly diagnosed and remediated. There can be no easy division between the antiquarian and the contemporary, since the way we talk about race is in large part an accrual or a distillation of history. There may be transhistorical and innate predispositions to divide human society into a fixed number of essentialized subgroups, but it would be extremely hasty to suppose that these kinks of the human mind are somehow fixed in the human brain. Between any possible predisposition and the actual modern history of thinking about race, there is a tremendous amount of room for conceptualizing alternative paths our deep-seated propensities for thinking about human diversity might have taken, and could still yet take.

Recent work in the "philosophy of race," particularly in the Anglo-American tradition, has provided remarkable insight, borrowing much from empirical psychology, into the way implicit bias functions to heighten and perpetuate racial prejudice in society.[1] This is valuable work, but it has not yet offered much in the way of positive prescriptions for correcting those false beliefs we evidently harbor unknowingly. One possible path toward correction might be discovered in the project of improved historical awareness. Our perception of social reality, our implicit biases, and our explicit beliefs are all historically conditioned. For this reason, the categories that come into play in much of our effort to make sense of social reality are much better understood not as natural kinds, or even as candidates for natural kindhood, but as historical kinds, to be questioned and challenged not only in clinical experiment and conceptual analysis, but also in the archives: the open record of our wrongs, conceptual and moral at once.

[1] In addition to the titles cited in chapter 1, see Lawrence Blum, "Stereotypes and Stereotyping: A Moral Analysis," *Philosophical Papers* 33, 3 (2004): 251–90; Daniel Kelly and Erica Roedder, "Racial Cognition and the Ethics of Implicit Bias," *Philosophy Compass* 3, 3 (2008): 522–40; Daniel Kelly, Luc Faucher, and Edouard Machery, "Getting Rid of Racism: Assessing Three Proposals in Light of Empirical Evidence," *Journal of Social Philosophy* 41 (2010): 293–322; Eric Schwitzgebel, "Acting Contrary to Our Professed Beliefs; or, The Gulf between Occurrent Judgment and Dispositional Belief," *Pacific Philosophical Quarterly* 91, 4 (2010): 531–53; Nilanjana Dasgupta, "Implicit Attitudes and Beliefs Adapt to Situations: A Decade of Research on the Malleability of Implicit Prejudice, Stereotypes, and the Self-Concept," *Advances in Experimental Social Psychology* 47 (2013): 233–79.

Biographical Notes

THIS IS A PARTIAL LIST of names of figures who play a role, large or small, in the history of the concept of race in the early modern period in Europe.

José de Acosta (1539–1600), a Spanish Jesuit missionary and naturalist in Latin America, and the author of an important study of the cultural and natural diversity of the New World.

Anton Wilhelm Amo (ca. 1703–53), a West African legal theorist and philosopher, originally sent to Germany as a slave of Duke Anton Ulrich of Braunschweig-Wolfenbüttel, and later active at the Universities of Halle, Wittenberg, and Jena.

Francis Bacon (1561–1826), a British philosopher and statesman. One of the founders of the modern scientific method, and the author of an influential work of utopian fiction.

Pierre Barrère (1690–1755), a French physician and naturalist, and the author of an anonymous treatise on the causes of skin pigmentation in Africans.

François Bernier (1625–88), a French physician and philosopher. A disciple of Pierre Gassendi, and attending physician to the Grand Moghul in Delhi, he traveled extensively in Asia, and is widely considered to be one of the most influential founders of the modern race concept.

Johann Friedrich Blumenbach (1752–1840), a German philosopher, anthropologist, and anatomist. A dialog partner to Kant, he is widely considered to be one of the founders of modern quantitative "racial science" and of physical anthropology.

Jacobus Bontius (1591–1631), a Dutch physician, traveler in the East Indies, and founder of the discipline of tropical medicine. He is credited among Europeans with first identifying and describing the orangutan.

Giordano Bruno (1548–1600), an Italian philosopher and mathematician. Executed by the Inquisition for heresy, he partially defended the libertine view that not all human beings are descended from Adam.

Georges-Louis Leclerc, Comte de Buffon (1707–88), a French naturalist and philosopher. A towering figure in Enlightenment natural history, and the author of the influential multivolume work, *Histoire naturelle*.

John Bulwer (1606–56), an English Baconian natural philosopher, and the author of several works on sign language, gesture, bodily modification, and related topics.

Robert Burton (1577–1640), an English author, whose masterwork is the 1621 *Anatomy of Melancholy*.

Pieter Camper (1722–89), a Dutch physician, naturalist, and anatomist. He wrote extensively on human physical variety, and was an important influence on Kant's views in this area.

Jacobus Johannes Elisa Capitein (ca. 1717–47), a Dutch minister of Ghanaian origin. Sold into slavery, he was also a defender of the institution of slavery within Europe.

Andrea Cesalpino (1519–1603), an Italian naturalist, influential in the development of modern systematic taxonomy.

Charles Marie de la Condamine (1701–74), a French naturalist and explorer, he made important contributions to the botanical surveying of Latin America.

Gabriel Daniel (1649–1728), a French Jesuit historian and satirist, author of the anti-Cartesian work, *Le voiage du monde de Descartes* of 1690.

Erasmus Darwin (1731–1802), an English naturalist and evolutionist, author of the 1794 *Zoonomia*, and grandfather of Charles Darwin.

René Descartes (1596–1650), a French philosopher and mathematician. One of the most influential figures of modern philosophy.

Denis Diderot (1713–84), a French Enlightenment thinker, a materialist and atheist, and one of the principal authors of the *Encyclopédie*.

Olaudah Equiano (1745–97), also known as Gustavus Vassa, a Nigerian-born author and abolitionist.

John Evelyn (1620–1706), an English author and environmental thinker associated with the Royal Society of London.

Abram Petrovich Gannibal (1696–1781), an African-born Russian statesman and general. Born into slavery, he was adopted by Peter the Great, and would be hailed as a symbol of Enlightenment ideas of human equality.

Pierre Gassendi (1592–1655), an influential French materialist philosopher and adversary of Descartes.

Henri Grégoire (1750–1831), a French Catholic priest and abolitionist, and the author of an influential treatise on the cultural attainments of Africans.

Hugo Grotius (1583–1645), a Dutch author and legal theorist, made important contributions to international law and also to the study of geography.

Matthew Hale (1609–76), an English jurist and legal theorist, made important contributions to geography and to the study of human migration.

Thomas Herbert (1606–82), an English traveler and author, whose reports from Africa and Asia were very influential in seventeenth-century Europe.

Johann Gottfried Herder (1744–1803), a German philosopher and author, argued for the equality of all human cultures.

Georg Horn (1620–70), a German author, theorized about the origins of the Native Americans.

David Hume (1711–76), an important Scottish Enlightenment philosopher.

Immanuel Kant (1724–1804), an important German philosopher, wrote extensively on anthropology and geography.

Jean-Baptiste Lamarck (1744–1829), a French naturalist, developed an influential theory of the mechanisms of biological heredity.

Isaac La Peyrère (1596–1676), a French theologian and the most prominent defender of the pre-Adamist theory.

Gottfried Wilhelm Leibniz (1646–1716), an important German philosopher and polymath, wrote extensively on geography, linguistics, and anthropology.

Carolus Linnaeus (1707–78), a Swedish naturalist and the founder of modern biological taxonomy.

John Locke (1632–1704), an important English philosopher and an influential theorist on the nature and structure of Native American societies.

Edward Long (1734–1813), a British colonial administrator in Jamaica, he was an influential historian of the island and a defender of slavery.

Hiob Ludolf (1624–1704), an important German founder of the scholarly discipline of Orientalism, wrote extensively on the languages and cultures of Ethiopia.

Olaus Magnus (1490–1557), a Swedish Catholic priest, wrote extensively on the history and cultures of Scandinavia.

Nicolas Malebranche (1638–1715), a French Oratorian priest and an influential philosopher, author of the 1676 masterwork, *The Search After Truth*.

Pierre Louis Moreau de Maupertuis (1698–1759), a French philosopher and mathematician, wrote on the causes of albinism.

Christoph Meiners (1747–1810), a German polygenist thinker and one of the most influential founders of modern "racial science."

James Burnet, Lord Monboddo (1714–99), a Scottish philosopher, jurist, and linguist, speculated on possible kinship relations between human beings and apes.

Michel de Montaigne (1533–92), an important French writer of the Renaissance, defended a variety of moral relativism in an essay on "cannibals."

Montesquieu [Charles Louis de Secondat de la Brède] (1689–1755), a French author and political thinker, wrote extensively on legal institutions and their possible basis in custom and morality.

Paracelsus [Philippus Aureolus Theophrastus Bombastus von Hohenheim] (1493–1541), a Swiss physician, astrologer, and mystic, defended the doctrine of polygenesis.

Willem Piso (1611–78), a Dutch physician and naturalist, author of the influential 1648 work, *Natural History of Brazil.*

John Ray (1627–1705), an influential English naturalist and taxonomist.

Jean-Jacques Rousseau (1712–78), a Swiss philosopher and writer, one of the most influential political theorists of the modern period.

Johan Gabriel Sparwenfeld (1655–1727), a Swedish statesman and linguist, one of the founders of the comparative study of Slavic languages.

Baruch Spinoza (1632–77), a Dutch philosopher, one of the most important figures of early modern Rationalist philosophy.

Georg Ernst Stahl (1659–1734), a German physician, chemist, and philosopher, loosely affiliated with the school of Pietist thought.

Edward Tyson (1651–1708), an English physician and anatomist, author of an influential anatomical study of a chimpanzee.

Lucilio Vanini (1589–1619), an Italian Renaissance freethinker and defender of polygenesis.

Garcilaso de la Vega (1539–1616), an Incan chronicler and political theorist.

Voltaire [François-Marie Arouet] (1694–1778), a French Enlightenment thinker and author, a vociferous opponent of equality between human races.

John Wallis (1616–1703), an English mathematician and philosopher, corresponded with Edward Tyson on questions relating to physiology and the evolution of human anatomy.

Bibliography

Works Originally Published Prior to 1900

Acosta, José de. *The Naturall and Morall Historie of the East and West Indies Intreating of the Remarkable Things of Heaven, of the Elements, Mettalls, Plants and Beasts Which Are Proper to That Country: Together with the Manners, Ceremonies, Lawes, Governments, and Warres of the Indians. Written in Spanish by the R.F. Ioseph Acosta, and Translated into English by E.G.*, London: Printed by Val: Sims for Edward Blount and William Aspley, 1604.

Amo, Antonius Guilelmius. *Dissertatio inauguralis de humanae mentis apatheia*, Wittenberg, 1734.

———. *Tractatus de arte sobrie et accurate philosophandi*, Halle, 1738.

Augustine, Aurelius. *The City of God*, trans. Marcus Dodds, Edinburgh: T&T Clark, 1871.

Bacon, Francis. *Nova Atlantis. Mundus alter et idem, sive Terra australis antehac semper incognita: longis itineribus peregrini Academici nuperrime lustrata AUthore Mercurio Britanico . . .*, Utrecht: Apud Joannem a Waesberge, 1643 [1624].

———. *The Works of Francis Bacon, Baron of Verulam, Viscount St. Alban, and High Chancellor of England*, London: J. Walthoe et al., 1740.

———. *The Essays, or Counsels Civil and Moral*, ed. Samuel Harvey Reynolds, Oxford: Clarendon, 1890.

[Barrère, Pierre]. *Dissertation sur la cause physique de la couleur des Nègres, de la qualité de leurs cheveux, et de la dégénération de l'un et de l'autre*, Paris: Pierre-Guillaume Simon, 1741.

Bernier, François. *Abregé de la philosophie de M. Gassendi*, Paris: Jacques Langlois, 1671.

———. *The History of the Late Revolution of the Empire of the Great Mogol: Together with the Most Considerable Passages for 5 Years Following in That Empire: To Which Is Added a Letter to the Lord Colbert Touching the Extent of Indostan . . . and the Principal Cause of the Decay of the States of Asia*, London: Moses Pitt, 1671.

———. "Nouvelle division de la terre," *Journal des Sçavans* (April 24, 1684): 133–40.

———. *Un libertin dans l'Inde Moghole. Les voyages de François Bernier (1656–1669)*, ed. Frédéric Tinguely, Paris: Éditions Chandeigne, 2008.

Blumenbach, Johann Friedrich. "Abschnitt von den Negern," in *Magazin für das Neueste aus der Physik und der Naturgeschichte*, vol. 4, pt. 3, Gotha, 1787, 9–11.

———. *De generis humani varietate nativa*, 3rd ed., Göttingen: Vandenhoek and Ruprecht, 1795.

———. *The Anthropological Treatises of Johann Friedrich Blumenbach*, trans. and ed. Thomas Bendyshe, London: Anthropological Society, 1865.

Brunnhofer, Hermann. *Giordano Bruno's Weltanschauung und Verhängniss*, Leipzig: Fues's Verlag, 1882.

Bruno, Giordano. *Opera Latine conscripta*, ed. F. Fiorentino et al., Naples, 1879–91.

Buffon, George-Louis Leclerc, Comte de. *Histoire naturelle, générale et particulière, avec la description du cabinet du Roy*, Paris, 1749.

———. *Oeuvres complètes de M. le Cte.de Buffon*, vol. 3, Paris: Impriméerie Royale, 1775.

———. *A Natural History, General and Particular, Containing the History and Theory of the Earth, &c.*, 8 vols., trans. William Smellie, London: T. Kelly, 1781.

Bulwer, John. *Anthropometamorphosis: Man Transform'd, or, The Artificial Changeling*, London, 1650.

Burton, Robert. *The Anatomy of Melancholy*, New York: New York Review Books Classics, 2001.

Busbecq, Ogier Ghislain de. *Itinera constantinopolitanum et amasianum*, Antwerp: Ex officina Christophori Plantini, 1581.

Camper, Petrus. *The Works of the Late Professor Camper: On the Connexion between the Science of Anatomy and the Arts of Drawing, Painting, Statuary, &c.*, trans. T. Cogan, London: C. Dilly, 1794.

Capitein, Jacobus Johannes Elisa. *Dissertatio politico-theologica, de servitute, libertati christianae non contraria; Staatkundig-godgeleerd onderzoekschrift, over de slaverny, als niet strjdig tegen de Christelyke vryheid . . . ; Uitgewrogte predikatien, zyndende trouwhertige vermaaninge . . .* , Liechtenstein: Kraus Reprints, 1971.

Condamine, M. de la. "Sur l'arbre du Quinquina," in *Histoire de l'Académie Royale des Sciences, Année MDCCXXXVIII*, Paris: Imprimérie Royale, 1740, 226–43.

Cyrano de Bergerac, Hercule Savinien. *Histoire comique des états et empires de la Lune et du Soleil*, ed. P. L. Jacob, Paris: Adolphe Delahays, 1858 [1655].

Danckaerts, Jasper. *Journal of Jasper Danckaerts, 1679–80*, ed. Bartlett Burleigh James and J. Franklin Jameson, New York: Charles Scribner's Sons, 1913.

Daniel, Gabriel. *Le voiage du monde de Descartes*, Paris: Simon Bénard, 1691.

Darwin, Erasmus. *Zoonomia: Or, the Laws of Organic Life*, vol. 1, 2nd ed., London: J. Johnson, 1796 [1794].

Descartes, René. *Renati Descartes Epistolae, Partim Latino sermone conscriptae, partim e Gallico in Latinum versae*, pt. 3, Amsterdam: Typographia Blaviana, 1683.

———. *Discours de la méthode*, in *Oeuvres de Descartes*, ed. Charles Adam and Paul Tannery, Paris: Léopold Cerf, 1902.

Diderot, Denis, et al. *Encyclopédie ou Dictionnaire raisonné des sciences, des arts et des métiers*, Paris, 1765.

———. *Oeuvres complètes*, ed. J. Fabre, J. Dieckmann, J. Proust, and J. Varloot, Paris, 1975–.

Equiano, Olaudah. *The Interesting Narrative of the Life of Olaudah Equiano, or Gustavus Vassa, the African, Written by Himself*, ed. Werner Sollors, New York: Norton, 2001 [1789].

Evelyn, John. *Fumifugium, or The Inconveniencie of the Aer and Smoak of London Dissipated. Together with Some Remedies Humbly Proposed by J. E. Esq*, London: W. Godbid for Gabriel Bedel, and Thomas Collins, 1661.

Fage, Robert. *A Description of the Whole World, with Some General Rules Touching the Use of the Globe*, London: J. Owsley, 1658.

Firmin, Anténor. *De l'égalité des races (anthropologie positive)*, Paris: Librairie Cotillon, 1885.

Forbes, John, Alexander Tweedie, and John Conolly (eds.). *Cyclopaedia of Practical Medicine, Comprising Treatises on the Nature and Treatment of Diseases, Materia Medica and Therapeutics, Medical Jurisprudence, Etc.*, vol. 3, London: Sherwood, Gilbert, and Piper, and Baldwen and Craddock, 1834.

Gassendi, Pierre. *The Mirrour of True Nobility and Gentility: Being the Life of the Renowned Nicolaus Claudius Fabricius, Lord of Pieresk, Senator of the Parliament at Aix*, trans. William Rand, London: Humphrey Moseley, 1657.

Ger'e, V. I. *Leïbnits i ego vek*, Saint Petersburg: V. Golovin, 1868.

Grégoire, Henri. *De la littérature des Nègres, ou, recherches sur leurs facultés intellectuelles, leurs qualités morales et leur littérature: suivies des notices sur la vie et les ouvrages des Nègres qui se sont distingués dans les sciences, les lettres et les arts*, Paris, 1808.

Grotius, Hugo. *De origine gentium americanarum dissertatio*, Paris, 1642.

———. *Hugonis Grotii de Origine gentium armericanarum dissertatio altera, adversus obtrectatorem*, Paris: S. Cramoisy, 1643.

Gundling, Nicholas Hieronymus. *D. Nicolai Hieronymi Gundlings Vollständige Historie der Gelahrtheit*, vol. 4, Frankfurt, 1736.

Hale, Matthew. *The Primitive Origination of Mankind, Considered and Examined according to the Light of Nature*, London: William Godbid, 1677.

Hall, Thomas. *The Loathsomnesse of Long Haire*, London: J. G. for Nathanael Webb and William Grantham, 1654.

Herbert, Thomas. *Some Yeares Travels into Africa & Asia the Great. Especially Describing the Famous Empires of Persia and Industant. As Also Divers Other Kingdoms in the Orientall Indies and Iles Adjacent*, London: Jacob Blome and Richard Bishop, 1638.

Herder, Johann Gottfried. *Ideen zur Philosophie der Geschichte der Menschheit*, pt. 2, Riga: Johann Friedrich Hartknoch, 1786.

Herodotus. *The Histories*, trans. Robin Waterfield, Oxford: Oxford University Press, 1998.

Hilpert, Johann. *Disquisitio de praeadamitis, anonymo Exercitationis & Systematis theologici auctori opposita*, Amsterdam: Apud Johannem Janssonium juniorem, 1656.

Hippocrates. *Upon Air, Water, and Situation; upon Epidemical Diseases; and upon Prognosticks, in Acute Cases Especially . . .*, trans. and ed. Francis Clifton, London: J. Wattes, 1734.

Horn, Georg. *De originibus Americanis, libri quatuor*, Halberstadt: Sumptibus Ioannis Mülleri, 1669.

Hulsius, Antonius. *Non ens prae-adamiticum, sive confutatio vani et socinizantis cujusdam somnii, quo S. Scripturae praetextu in cautioribus nuper imponere conatus est anonimus fingens ante Adamum primum homines fuisse in mundo*, Leiden: Apud Johannem Elsevirium, 1656.

Hume, David. *The Philosophical Works of David Hume*, Bristol: Thoemmes Press, 1996.

Kant, Immanuel. *Beobachtungen über das Gefühl des Schönen und Erhabenen*, Riga: Friedrich Hartknoch, 1764.

———. *Physische Geographie, auf Verlangen des Verfassers*, Königsberg: Göbbels und Unzer, 1802.

———. *Kants Werke*, Akademie Textausgabe, 8 vols., Berlin: Walter de Gruyter, 1968.

———. *Political Writings*, ed. H. S. Reiss, Cambridge: Cambridge University Press, 1970.

Kircher, Athanasius. *China monumentis*, Amsterdam: Apud Jacobum à Meurs, 1667 [repr., Frankfurt: Minerva, 1966].

Lafitau, Joseph-François. *Moeurs des sauvages amériquains, comparées aux moeurs des premiers temps*, Paris: Saugrain l'aîné, Charles Estienne Hochereau, 1724.

Lamarck, Jean-Baptiste. *La philosophie zoologique*, Paris: Dentu, 1809.

La Peyrère, Isaac. *Relation de l'Islande*, Paris: Louis Billaine, 1663 [1644].

———. *Prae-Adamitae, sive Exercitatio super versibus duodecimo, decimotertio, & decimoquarto, capitis quinti Epistolæ d. Pauli ad Romanos. Quibus inducuntur primi homines ante Adamum conditi*, [Amsterdam], 1655.

———. *Men before Adam, or, A discourse upon the Twelfth, Thirteenth, and Fourteenth Verses of the Fifth Chapter of the Epistle of the Apostle Paul to the Romans: by Which Are Prov'd That the First Men Were Created before Adam*, London, 1656.

Las Casas, Bartolomé de. *Apologética historia sumaria*, ed. Edmundo O'Gorman, Mexico City: Universidad Nacional Autónoma de México, Instituto de Investigaciones Históricas, 1967 [1551].

———. *In Defense of the Indians*, trans. and ed. Stafford Poole, DeKalb: Northern Illinois University Press, 1992 [1552].

Leibniz, Gottfried Wilhelm. *Otium hanoveranum, sive Miscellanea, ex ore & schedis illustris viri, piae memoriae, Godofr. Guilielmi Leibnitii*, ed. Joachim Friedrich Feller, Leipzig: Impensis Joann. Christiani Martini, 1718.

———. *Viri illvstratis Godefridi Gvil. Leibnitii epistolae ad diversos, theologici, ivridici, medici, philosophici, mathematici, historici et philologici argvmenti, e MSC. avetoris, evm annotationibvs svis primvm divvlgavit Christian. Kortholtvs*, Lipsiae: svmtv B. C. Breitkoptii, 1734.

———. *G. G. Leibnitii Opera Omnia*, ed. Louis Dutens, Geneva: Fratres De Tournes, 1768.

———. *Sbornik pisem i memorialov Leïbnitsa otnosyashchikhsya k Rossii i Petru Velikomu*, ed. V. I. Ger'e, Saint Petersburg, 1873.

———. *Die philosophischen Schriften von G. W. Leibniz*, ed. C. I. Gerhardt, Berlin, 1875–90.

———. *Mittheilungen aus Leibnizens ungedruckten Schriften*, ed. Georg Mollat, Leipzig, 1893.

———. *Sämtliche Schriften und Briefe*, Berlin: Berlin-Brandenburgische Akademie der Wissenschaften, 1923–.

Le Prieur, Philippe. *Animadversiones in librum Prae-Adamitarum: in quibus confutatur nuperus scriptor, et primum omnium hominum fuisse Adamum defenditur*, Paris: Apud Ioan. Billaine, 1656.

Linnaeus, Carolus. *The Linnaean Correspondence*, http://linnaeus.c18.net/Letters/.

Locke, John. *The Works of John Locke Esq; In Three Volumes*, London: John Churchill, 1714.

———. *The Second Treatise of Government*, New York: Macmillan, 1987.

Long, Edward. *The History of Jamaica*, 3 vols., London, 1774.

Ludewig, Johann Peter von. *Wöchentlichen Hallischen Frage- und Anzeigungs-Nachrichten*, November 28, 1729.

Ludolf, Hiob. *Historia Aethiopica*, Frankfurt, 1681.

Magnus, Olaus. *Historia de gentibus septentrionalibus, earumque diversis statibus, conditionibus, moribus, ritibus, superstitionibus, disciplinis, exercitiis, regimine, victu, bellis, structuris, instrumentis, ac mineris metaliicis, & rebus mirabilibus* . . . , Rome, 1555.

Malebranche, Nicolas. *The Search After Truth*, trans. Thomas M. Lennon and Paul J. Olscamp, Cambridge: Cambridge University Press, 1997.

Marx, Karl. *Critique of Hegel's Philosophy of Right*, trans. Joseph O'Malley, Cambridge: Cambridge University Press, 1967.

Marx, Karl, and Friedrich Engels. *Marx/Engels Collected Works*, Moscow: Progress, 1975–2005.

[Maupertuis, Pierre-Louis Moreau de]. *Dissertation physique à l'occasion du Nègre blanc*, Leiden, 1744.

Megiser, Hieronymus. *Wahrhafftige gründliche und aussführliche so wol Historische alss Chorographische Beschreibung der überauss reichen mechtigen und weitberhumbten Insul Madagascar*, Altenburg, 1609.

Meiner, Johannes Theodosius. *Disputatio philosophica continens ideam distinctam eorum quae competunt vel menti vel corpori nostro vivo et organico, quam consentiente philosophorum ordine, praeside M. Antonio Guilielmo Amo Guinea-Afro*, Wittenberg: Literis Vidvae Kobersteinianae, May 29, 1734.

Meiners, Christoph. *Grundriß der Geschichte der Menschheit*, Lemgo: Meyersche Buchhandlung, 1785.

Mill, John Stuart. *System of Logic: Ratiocinative and Inductive*, in John M. Robson (ed.), *The Collected Works of John Stuart Mill*, vol. 7, Toronto: University of Toronto Press, 1974.

Monboddo, James Burnet, Lord. *Of the Origin and Progress of Language*, Edinburgh, 1773.

Montesquieu, Charles-Louis de Secondat, Baron de la Brède. *De l'esprit des lois, nouvelle edition, avec les dernieres corrections et illustrations de l'auteur*, 2 vols., Edinburgh, 1750.

More, Henry. *An Antidote against Atheism, or, An Appeal to the Naturall Faculties of the Minde of Man, Whether There Be Not a God*, London: J. Flesher, 1655 [1653].

More, Thomas. *De optimo rei publicae statu deque nova insula Utopia*, Leuven, 1516.

Newton, Isaac. *Opticks: or, A Treatise of the Reflections, Refractions, Inflections and Colours of Light*, 4th ed., London: William Innys, 1730 [1704].

Nietzsche, Friedrich. *Der Antichrist*, in *Digital Critical Edition of the Complete Works and Letters*, ed. Paolo D'Iorio, based on the critical text by G. Colli and M. Montinari, Berlin: de Gruyter 1967–.

Ortelius, Abraham. *Thesaurus geographicus*, Antwerp: Ex officina Christophori Plantini, 1587.

Oviedo y Valdés, Gonzalo Fernández de. *La historia general y natural de las Indias*, ed. José Amador de Los Ríos, Madrid: Imprenta de la Real Academia de la Historia, 1851 [1535].

Paracelsus [Philippus Aureolus Theophrastus Bombastus von Hohenheim]. *Astronomia Magna: oder Die gantze Philosophia sagax der grossen und kleinen Welt des von Gott hocherleuchten erfahrnen und bewerten teutschen Philosophi und Medici Philippi Theophrasti Bombast, genannt Paracelsi magni*, Frankfurt: Sigismund Feyerabend, 1571 [1537–38].

Philippi, Johann Ernst. *Belustigende Poetische Schaubühne, und auf derselben I. Ein Poßirlicher Student, Hanß Dümchen aus Norden, nebst Zwölf seiner lustigen Cameraden. II. Die Academische Scheinjungfer, als ein Muster aller Cocketten. III. Herrn M. Amo, eines gelehrten Mohren, galanter Liebes-Antrag an eine schöne Brünette, Madem. Astrine. IV. Der Mademoiselle Astrine, Parodische Antwort auf solchen Antrag eines verliebten Mohren*, Cöthen: in der Cörnerischen Buchhandlung, 1747.

Piso, Willem, and Georg Markgraf. *Historia naturalis Brasiliae auspicio et beneficio illustriss I. Mauritii Com. Nassau . . . , In qua non tantum plantae et animalia, sed et indigenarum morbi, ingenia et mores describuntur et iconibus supra quingentas illustrantur*, Amsterdam: Apud Lud. Elzevirium, 1648.

Plinius Secundus, Caius. *Naturalis historia*, Berlin: Apud Weidmannos, 1867.

Posselt, Moritz C. *Peter der Grosse und Leibnitz*, Tartu: Friedrich Severin Verlag, 1843.

Ray, John. *The Wisdom of God Manifested in the Works of the Creation Being the Substance of Some Common Places Delivered in the Chappel of Trinity-College, in Cambridge*, London: Samuel Smith, 1691.

Raynal, Guillaume-Thomas. *Histoire des Deux-Indes*, Geneva: Pellet, 1780.

Ruini, Carlo. *Anatomia del cavallo, infermità, et suoi rimedii*, Venice: Appresso Fioravante Prati, 1618.

Sandoval, Alonso de. *Un tratado sobre la esclavitud*, ed. Enriqueta Vila Vilar, Madrid: Alianza, 1987 [1627].

Saxo Grammaticus. *Saxonis Grammatici Historia Danica, recensuit et commentariis illustravit Petrus Erasmus Müller; opus morte Mülleri interruptum absolvit Mag. Joannes Mattias Velschow*, Copenhagen: sumtibus Librariae Gyldendalianae, 1839–58.

Scheuchzer, Johann Jakob. *Physica Sacra*, Ulm, 1731.

Smith, John. *The Generall History of Virginia, New-England, and the Summer Isles*, London: Edward Blackmore, 1632 [1624].

Soranzo, Lazaro. *L'Ottomanno*, Ferrara, 1599.

Spinoza, Benedict de. *Opera*, ed. Carl Gebhardt, 4 vols., Heidelberg: C. Winter, 1925.

Sprat, Thomas. *History of the Royal Society*, London, 1667.

Stahl, Georg Ernst. *Georgii Ernestii Stahlii Negotium otiosum: Seu Skiamachia adversus positiones aliquas fundamentales* Theoriae varae medicae, Halle, 1720 [to appear in English translation as *The Leibniz-Stahl Controversy*, trans. and ed. François Duchesneau and Justin E. H. Smith, New Haven: Yale University Press, forthcoming].

Tacitus. *A Treatise on the Situation, Manners and Inhabitants of Germany*, ed. and trans. John Aikin, Oxford: W. Baxter, 1823.

Thucydides. *The Peloponnesian War*, trans. and ed. Richard Crawley, London: J. M. Dent, 1910.

Topsell, Edward. *The Historie of Foure-Footed Beastes, Describing the True and Lively Figure of Every Beast*, London, 1607.

Tumanskiĭ, F. V. *Sobranie raznykh zapisok o sochinenii, sluzhashchikh k dostavleniiu polnogo svedeniia o zhisni i deiatel'nosti Petra Velikogo*, Saint Petersburg, 1787.

Tyson, Edward. *Phocaena, or the Anatomy of a Porpess, Dissected at Greshame Colledge; with a Praeliminary Discourse Concerning Anatomy, and a Natural History of Animals*, London: Benjamin Tooke, 1680.

———. *Orang-outang, sive, Homo sylvestris, or, The Anatomy of a Pygmie Compared with That of a Monkey, an Ape, and a Man to Which Is Added, A Philological Essay Concerning the Pygmies, the Cynocephali, the Satyrs and Sphinges of the Ancients: Wherein It Will Appear That They Are All Either Apes or Monkeys, and Not Men, as Formerly Pretended*, London: Thomas Bennet, 1699.

Vanini, Lucilio. *Iulii Caesaris Vanini Neapolitani Theologi, Philosophi, & Iuris utriusque Doctoris, de admirandis naturae reginae deaeque mortalium arcanis*, Paris: A. Perier, 1616.

Vater, Christian. *Physiologia experimentalis*, Wittenberg, 1712.

Vega, Inca Garcilaso de la. *Commentarios reales de los lncas*, in P. Carmelo Saenz de Santa Maria (ed.), *Obras complétas*, Madrid, 1960 [1609, first French translation by Jean Baudouin, 1633; French translation: *Commentaires royaux sur le Pérou des Incas*, trans. René L. F. Durand, Paris: François Maspéro, 1982].

Verhandelingen uitgegeven door het Zeeuwsch Genootschap der Wetenschappen te Vlissingen, Negende Deel, Middelburg: Pieter Gillissen, 1782.

Vico, Giambattista. *The New Science of Giambattista Vico*, 5th ed., ed. and trans. Thomas Goddard Bergin and Max Harold Fisch, Ithaca, NY: Cornell University Press, 1994.

Voltaire. *Oeuvres complètes de Voltaire*, Paris: Renouard, 1819 [1756].

———. *The Works of Voltaire*, ed. and trans. Tobias Smollett, New York: E. R. DuMont, 1901.

Wallis, John. "A Letter to Edward Tyson," *Philosophical Transactions* 22 (1700): 772.

Wolff, Christian. *Oratio de Sinarum philosophia practica/Rede über die praktische Philosophie der Chinesen*, ed. Michael Albrecht, Hamburg: Felix Meiner Verlag, 1985.

Works Originally Published in 1900 or Later

Abraham, William E. "The Life and Times of Anton Wilhelm Amo, the First African (Black) Philosopher in Europe," in Molefe Kete Asante and Abu S. Abarry (eds.), *African Intellectual Heritage: A Book of Sources*, Philadelphia: Temple University Press, 1996, 424–40.

———. "Anton Wilhelm Amo," in Kwasi Wiredu (ed.), *A Companion to African Philosophy*, London: Blackwell, 2004, 190–99.

Abulafia, David. *The Discovery of Mankind: Atlantic Encounters in the Age of Columbus*, New Haven: Yale University Press, 2008.

Adams, Robert M. *Leibniz: Determinist, Theist, Idealist*, New York: Oxford University Press, 1994.

Appiah, K. Anthony. "The Uncompleted Argument: DuBois and the Illusion of Race," in L. Bell and D. Blumenfeld (eds.), *Overcoming Racism and Sexism*, Lanham, MD: Rowman & Littlefield, 1995, 21–37.

———. "Race, Culture, Identity: Misunderstood Connections," in Anthony Appiah and Amy Gutmann (eds.), *Color Conscious*, Princeton: Princeton University Press, 1996, 30–105.

Arnold, Günter. "'. . . der grösste Mann den Deutschland in den neuern Zeiten gehabt': Herders Verhältnis zu Leibniz," *Studia Leibnitiana* 37, 2 (2005): 161–85.

Atran, Scott. *The Cognitive Foundations of Natural History: Towards an Anthropology of Science*, Cambridge: Cambridge University Press, 1990.

———. *The Native Mind and the Cultural Construction of Nature*, Cambridge, MA: MIT Press, 2008.

Avramescu, Cătălin. *An Intellectual History of Cannibalism*, trans. Alistair Ian Blyth, Princeton: Princeton University Press, 2009.

Bailey, Eric J. *Medical Anthropology and African American Health*, Westport, CT: Greenwood, 2000.

Balibar, Étienne, and Immanuel Wallerstein. *Race, Nation, Class: Ambiguous Identities*, London: Verso, 1991.

Banton, Michael. "The Concept of Racism," in Sami Zubaida (ed.), *Race and Racialism*, London: Tavistock, 1970, 17–34.

Bartra, Roger. *El Salvaje en el espejo*, Mexico City: Ediciones Era, 1992.

Basso, Luca. *Individuo e comunità nella filosofia politica di G. W. Leibniz*, Soveria Mannelli: Rubbettino Editore, 2005.

Beltrán, Carlos López. "Hippocratic Bodies, Temperament and Castas in Spanish America (1570–1820)," *Journal of Spanish Cultural Studies* 8 (2007): 253–89.

Benz, Ernst. *Leibniz und Peter der Grosse*, Berlin: De Gruyter, 1947.

Berlin, Brent. *Ethnobiological Classification: Principles of Categorization of Plants and Animals in Traditional Societies*, Princeton: Princeton University Press, 1992.

Berlin, Ira. *Many Thousands Gone: The First Two Centuries of Slavery in North America*, Cambridge, MA: Harvard University Press, 1998.

Bernasconi, Robert. "Will the Real Kant Please Stand Up: The Challenge of Enlightenment Racism to the Study of the History of Philosophy," *Radical Philosophy* 117 (2003): 13–22.

———. "François Bernier and the Brahmans: Exposing an Obstacle to Cross-

Cultural Conversation," *Journal for the Study of Religions and Ideologies* 7, 19 (2008): 107–17.

———. "Kant's Third Thoughts on Race," in Stuart Elden and Eduardo Mendieta (eds.), *Reading Kant's Geography*, Albany: State University of New York Press, 2011, 291–318.

———. "Crossed Lines in the Racialization Process: Race as a Border Concept," *Research in Phenomenology* 42 (2012): 206–28.

Bernasconi, Robert, and Sybol Cook (eds.). *Race and Racism in Continental Philosophy*, Bloomington: Indiana University Press, 2003.

Bernasconi, Robert, and Tommy Lee Lott (eds.). *The Idea of Race*, Indianapolis: Hackett, 2000.

Bernasconi, Robert, and Anika Maaza Mann. "The Contradictions of Racism: Locke, Slavery, and the *Two Treatises*," in Valls, *Race and Racism in Modern Philosophy*, 89–107.

Bernheimer, Richard. *Wild Men in the Middle Ages*, Cambridge, MA: Harvard University Press, 1952.

Bethencourt, Francisco. *Racisms: From the Crusades to the Twentieth Century*, Princeton: Princeton University Press, 2013.

Bindman, David. *Ape to Apollo: Aesthetics and Race in the 18th Century*, Ithaca, NY: Cornell University Press, 2002.

Bindman, David, Henry Louis Gates Jr., et al. *The Image of the Black in Western Art, Volume I: From the Pharaohs to the Fall of the Roman Empire*, Cambridge, MA: Belknap, 2010.

———. *The Image of the Black in Western Art, Volume II: From the Early Christian Era to the "Age of Discovery," Part 1: From the Demonic Threat to the Incarnation of Sainthood*, Cambridge, MA: Belknap, 2010.

———. *The Image of the Black in Western Art, Volume II: From the Early Christian Era to the "Age of Discovery," Part 2: Africans in the Christian Ordinance of the New World*, Cambridge, MA: Belknap, 2010.

———. *The Image of the Black in Western Art, Volume III: From the "Age of Discovery" to the Age of Abolition, Part 1: Artists of the Renaissance and Baroque*, Cambridge, MA: Belknap, 2011.

———. *The Image of the Black in Western Art, Volume III: From the "Age of Discovery" to the Age of Abolition, Part 3: The Eighteenth Century*, Cambridge, MA: Belknap, 2011.

Blum, Lawrence. *I'm Not a Racist, But . . . : The Moral Quandary of Race*, Ithaca, NY: Cornell University Press, 2002.

———. "Stereotypes and Stereotyping: A Moral Analysis," *Philosophical Papers* 33, 3 (2004): 251–90.

Boulle, Pierre H. "François Bernier and the Origins of the Modern Concept of Race," in Sue Peabody and Tyler Stovall (eds.), *The Color of Liberty: Histories of Race in France*, Durham, NC: Duke University Press, 2003, 11–27.

Braudel, Fernand. "Histoire et sciences sociales: la longue durée," *Annales. Histoire, Sciences Sociales* 13, 4 (October–December 1958): 725–53.

Brecht, Martin. "August Hermann Francke und der Hallische Pietismus," in Martin Brecht (ed.), *Geschichte des Pietismus*, Band I: *Das 17. und frühe 18. Jahrhundert*, Göttingen: Vandenhoeck & Ruprecht, 1993–.

Brentjes, Burchard. "Ein afrikanischer Student der Philosophie und Medizin in Halle, Wittenberg und Jena (1727–1747)," in Wolfram Kaiser and Christine Beierlein (eds.), *In memoriam Hermann Boerhaave (1668–1738)*, Halle: Martin-Luther-Universität Halle-Wittenberg, 1969.

———. "Anton Wilhelm Amo, First African Philosopher in European Universities," *Current Anthropology* 16, 3 (1975): 443–44.

———. *Anton Wilhelm Amo: der schwarze Philosoph in Halle*, Leipzig: Koehler & Amelang, 1976.

Brodkin, Karen. *How the Jews Became White Folks*, New Brunswick, NJ: Rutgers University Press, 1998.

Brown, R. A., and George J. Armelago. "Apportionment of Racial Diversity: A Review," *Evolutionary Anthropology* 10 (2001): 34–40.

Cañizares-Esguerra, Jorge. *Nature, Empire, and Nation: Explorations of the History of Science in the Iberian World*, Palo Alto: Stanford University Press, 2006.

Carey, Daniel. *Locke, Shaftesbury, and Hutcheson: Contesting Diversity in the Enlightenment and Beyond*, Cambridge: Cambridge University Press, 2005.

Chuchmarev, V. I. "G. V. Leïbnits i russkaia kul'tura 18 stoletiia," *Vestnik istorii mirovoi kul'tury* 4 (1957): 120–32.

Clark, William. *Academic Charisma and the Origins of the Research University*, Chicago: University of Chicago Press, 2007.

Cook, Harold J. *Matters of Exchange: Commerce, Medicine, and Science in the Dutch Golden Age*, New Haven: Yale University Press, 2007.

Cooper, Alix. *Inventing the Indigenous: Local Knowledge and Natural History in Early Modern Europe*, Cambridge: Cambridge University Press, 2007.

Cooper, Rachel. "Why Hacking Is Wrong about Human Kinds," *British Journal for the Philosophy of Science* 55, 1 (2004): 73–85.

Cosmides, Leda, John Tooby, and Robert Kurzban. "Perceptions of Race," *Trends in Cognitive Science* 7 (2003): 173–79.

Crosby, Alfred W. *The Columbian Exchange: Biological and Cultural Consequences of 1492*, Westport, CT: Greenwood, 1972.

Curran, Andrew S. *The Anatomy of Blackness: Science and Slavery in an Age of Enlightenment*, Baltimore: Johns Hopkins University Press, 2011.

Dasgupta, Nilanjana. "Implicit Attitudes and Beliefs Adapt to Situations: A Decade of Research on the Malleability of Implicit Prejudice, Stereotypes, and the Self-Concept," *Advances in Experimental Social Psychology* 47 (2013): 233–79.

Da Silva, Denise. *Toward a Global History of Race*, Minneapolis: University of Minnesota Press, 2007.

Daston, Lorraine. "Type Specimens and Scientific Memory," *Critical Inquiry* 31, 1 (2004): 153–82.

Davis, David Brion. *Inhuman Bondage: The Rise and Fall of Slavery in the New World*, Oxford: Oxford University Press, 2006.

Descola, Philippe. *Par-delà nature et culture*, Paris: Gallimard, 2005.

DeSouza, Nigel. "Leibniz in the Eighteenth-Century: Herder's Critical Reflections on the *Principles of Nature and Grace*," *British Journal for the History of Philosophy* 20, 4 (July 2012): 773–95.

Devisse, Jean, and Michel Mollat. *L'image du noir dans l'art occidental*, Paris: Bibliothèque des Arts, 1979.

Dew, Nicholas. *Orientalism in Louis XIV's France*, Oxford: Oxford University Press, 2009.

Djuvara, Neagu. *Thocomerius-Negru Vodă. Un voivod cuman la începuturile Țării Românești*, Bucharest: Humanitas, 2007.

Doron, Claude-Olivier. "Race and Genealogy: Buffon and the Formation of the Concept of 'Race,'" *Humana Mente—Journal of Philosophical Studies* 22 (2012): 75–109.

Dreike, Beate Monika. *Herders Naturauffassung in ihrer Beeinflussung durch Leibniz' Philosophie*, Wiesbaden: Steiner, 1973.

Drescher, Seymour. *Abolition: A History of Slavery and Antislavery*, Cambridge: Cambridge University Press, 2009.

Dugatkin, Lee Alan. *Mr. Jefferson and the Giant Moose: Natural History in Early America*, Chicago: University of Chicago Press, 2009.

Durkheim, Émile. *Les formes élémentaires de la vie religieuse. Le système totémique en Australie*, Paris: Presses Universitaires de France, 1960 [1912].

Earle, Rebecca. *The Body of the Conquistador: Food, Race and the Colonial Experience in Spanish America, 1492–1700*, Cambridge: Cambridge University Press, 2012.

Edwards, A.W.F. "Human Genetic Diversity: Lewontin's Fallacy," *Bioessays* 25, 8 (2003): 798–801.

Eigen, Sarah, and Mark Larrimore (eds.). *The German Invention of Race*, Albany: State University of New York Press, 2006.

Ereshefsky, Marc, and Mohan Matthen. "Taxonomy, Polymorphism, and History: An Introduction to Population Structure Theory," *Philosophy of Science* 72 (2005): 1–21.

Eze, Emmanuel Chukwudi. *Race and the Enlightenment: A Reader*, London: Blackwell, 1997.

———. *African Philosophy: An Anthology*, London: Wiley-Blackwell, 1998.

Fanon, Frantz. *Peau noire, masques blancs*, Paris: Seuil, 1952; English translation: *Black Skin, White Masks*, trans. Richard Philcox, New York: Grove Press, 2008.

Faucher, Luc, and Edouard Machery. "Racism: Against Jorge Garcia's Moral and Psychological Monism," *Philosophy of the Social Sciences* 39 (2009): 41–62.

Faull, Katherine M. (ed.). *Anthropology and the German Enlightenment: Perspectives on Humanity*, *Bucknell Review* 38, 92 (1995).

Feingold, Mordechai (ed.). *Jesuit Science and the Republic of Letters*, Cambridge, MA: MIT Press, 2002.

Fenves, Peter. "Imagining an Inundation of Australians; or, Leibniz on the Principles of Grace and Race," in Valls, *Race and Racism in Modern Philosophy*, 73–88.

———. "What 'Progresses' Has Race-Theory Made since the Time of Leibniz and Wolff?," in Eigen and Larrimore, *German Invention of Race*, 11–22.

Foucault, Michel. *Histoire de la sexualité*, vols. 1–3, Paris: Gallimard, 1976–84.

Ganeri, Jonardon. *The Lost Age of Reason: Philosophy in Early Modern India*, Oxford: Oxford University Press, 2011.

Gannett, Lisa. "Questions Asked and Unasked: How by Worrying Less about the 'Really Real' Philosophers of Science Might Better Contribute to Debates about Genetics and Race," *Synthese* 177 (2010): 363–85.

Garber, Daniel. *Leibniz: Body, Substance, Monad*, Oxford: Oxford University Press, 2010.

Garcia, J.L.A. "Current Conceptions of Racism: A Critical Examination of Some Recent Social Philosophy," *Journal of Social Philosophy* 28 (1997): 5–42.

———. "Philosophical Analysis and the Moral Concept of Racism," *Philosophy and Social Criticism* 25 (1999): 1–32.

Geyer-Kordesch, Johanna. *Pietismus, Medizin und Aufklärung in Preussen im 18. Jahrhundert*, Tübingen: Niemeyer Verlag, 2000.

Ghiselin, Michael. "A Radical Solution to the Species Problem," *Systematic Zoology* 23 (1974): 536–44.

Gil-White, Francisco. "Are Ethnic Groups Biological 'Species' to the Human Brain? Essentialism in Our Cognition of Some Social Categories," *Current Anthropology* 42, 4 (2001): 515–54.

———. "The Cognition of Ethnicity: Native Category Systems under the Field Experimental Microscope," *Field Methods* 14, 2 (2002): 170–98.

Ginzburg, Carlo. *I benandanti. Stregoneria e culti agrari tra Cinquecento e Seicento*, Turin: Einaudi, 2002.

———. "Latitude, Slaves, and the Bible: An Experiment in Microhistory," in Angela N. H. Creager, Elizabeth Lunbeck, and M. Norton Wise (eds.), *Science without Laws*, Durham, NC: Duke University Press, 2007, 243–63.

Glasgow, Joshua. *A Theory of Race*. New York: Routledge, 2009.

Glasgow, Joshua, Julie L. Shulman, and Enrique Covarrubias. "The Ordinary Conception of Race in the United States and Its Relation to Racial Attitudes: A New Approach," *Journal of Cognition and Culture* 9 (2009): 15–39.

Gliozzi, Giuliano. *Adamo e il nuovo mondo. La nascità dell'antropologia come ideologia coloniale: dalle genealogie bibliche alle teorie razziali (1500–1700)*, Florence: Franco Angeli, 1977; French translation: *Adam et le nouveau monde. La naissance de l'anthropologie comme idéologie coloniale des généalogies bibliques aux théories raciales (1500–1700)*, trans. Arlette Estève and Pascal Gabellone, Paris: Théétète, 2000.

Glötzner, Johannes. *Anton Wilhelm Amo. Ein Philosoph aus Afrika im Deutschland des 18. Jahrhunderts*, Munich: Enhuber, 2002.

Godelier, Maurice. *Métamorphoses de la parenté*, Paris: Fayard, 2004.

Goldenbaum, Ursula. *Appell an das Publikum: Die öffentliche Debatte in der deutschen Aufklärung, 1687–1796*, Berlin: Akademie Verlag, 2004.

Goldenberg, David M. *The Curse of Ham: Race and Slavery in Early Judaism, Christianity, and Islam*, Princeton: Princeton University Press, 2003.

Gould, Stephen Jay. *Wonderful Life: The Burgess Shale and the Nature of History*, New York: Norton, 1989.

Gray, Chris. "Irish Travellers Gain Legal Status of Ethnic Minority," *Independent*, August 30, 2000, http://www.independent.co.uk/news/uk/this-britain/irish-travellers-gain-legal-status-of-ethnic-minority-710768.html.

Greenblatt, Stephen. *Marvelous Possessions: The Wonder of the New World*, Chicago: University of Chicago Press, 1991.

Gyekye, Kwame. *An Essay on African Philosophical Thought: The Akan Conceptual Scheme*, Cambridge: Cambridge University Press, 1995 [1987].

Hacking, Ian. "The Looping of Human Kinds," in Dan Sperber and A. J. Premack (eds.), *Causal Cognition*, Oxford: Oxford University Press, 1995, 351–83.

———. *Mad Travelers: Reflections on the Reality of Transient Mental Illness*, Charlottesville: University of Virginia Press, 1998.

———. *The Social Construction of What?*, Cambridge, MA: Harvard University Press, 1999.

———. *Historical Ontology*, Cambridge, MA: Harvard University Press, 2002.

———. "Why Race Still Matters," *Daedalus* 134, 1 (Winter 2005): 102–16.

———. "Genetics, Biosocial Groups and the Future of Identity," *Daedalus* 135, 4 (Fall 2006): 81–95.

Halbwachs, Maurice. *Leibniz*, Paris: Librairie Mellottée, 1907.

Hall, Stuart. "Race, Articulation, and Societies Structured in Dominance," in Philomena Essed and David Theo Goldberg (eds.), *Race Critical Theories: Text and Context*, Oxford: Blackwell, 2002, 38–68.

Hamblin, James. "'Rise of the Colored Empires.' White Babies Are No Longer the Majority in the US," *Atlantic*, June 13, 2013, http://www.theatlantic.com/health/archive/2013/06/rise-of-the-colored-empires/276844/.

Hannaford, Ivan. *Race: The History of an Idea in the West*, Baltimore: Johns Hopkins University Press, 1996.

Hanke, Lewis. *Aristotle and the American Indians: A Study in Race Prejudice in the Modern World*, Bloomington: Indiana University Press, 1970.

———. *All Mankind Is One: A Study of the Disputation between Bartolomé de las Casas and Juan Gines de Sepulveda in 1550 on the Intellectual and Religious Capacity of the American Indians*, DeKalb: Northern Illinois University Press, 1994.

Hardimon, Michael O. "The Ordinary Concept of Race," *Journal of Philosophy* 100 (2003): 437–55.

Haslanger, Sally. "Language, Politics and 'the Folk': Looking for 'the Meaning' of 'Race,'" *Monist* 93, 2 (April 2010): 169–87.

Hirschfeld, Lawrence. *Race in the Making: Cognition, Culture, and the Child's Construction of Human Kinds*, Cambridge, MA: MIT Press, 1998.

Hochman, Adam. "Against the New Racial Naturalism," *Journal of Philosophy* 110 (2013): 331–51.

Horton, Robin. *Patterns of Thought in Africa and the West: Essays on Magic, Religion, and Science*, Cambridge: Cambridge University Press, 1993.

Hountondji, Paulin J. *Un philosophe africain dans l'Allemagne du XVIIIe siècle*, Paris: Presses Universitaires de France, 1970.

———. *Sur la "philosophie africaine,"* Paris: Maspéro, 1976.

Hsia, Florence. *Sojourners in a Strange Land: Scientific Missions in Late Imperial China*, Chicago: University of Chicago Press, 2009.

Huddleston, Lee Eldridge. *Origins of the American Indians: European Concepts, 1492–1729*, Austin: University of Texas Press, 1967.

Hull, David. "A Matter of Individuality," *Philosophy of Science* 45 (1978): 335–60.

Hunter, Michael. *Boyle: Between God and Science*, New Haven: Yale University Press, 2009.

Ignatiev, Noel. *How the Irish Became White*, New York: Routledge, 1995.

Ingold, Tim. "Commentary on Gil-White," *Current Anthropology* 42 (2001): 541–42.

Isaac, Benjamin. *The Invention of Racism in Classical Antiquity*, Princeton: Princeton University Press, 2004.

James, Michael. "Race," in Edward N. Zalta (ed.), *The Stanford Encyclopedia of Philosophy*, Winter 2012., http://plato.stanford.edu/archives/win2012/entries/race/.

Jasmin, Claude. "Je suis fier de ma race," *Le Devoir*, May 30, 2013.

Jospe, Raphael. "Teaching Judah Ha-Levi: Defining and Shattering Myths in Jewish Philosophy," in Raphael Jospe (ed.), *Paradigms in Jewish Philosophy*, London: Associated University Presses, 1997, 87–111.

Kagame, Alexis. *La philosophie Bantu comparée*, Paris: Présence Africaine, 1976.

Kaplan, Jonathan Michael, and Rasmus Grønfeldt Winther. "Prisoners of Abstraction? The Theory and Measure of Genetic Variation, and the Very Concept of 'Race,'" *Biological Theory* 7 (2013): 401–12.

Karbowski, Joseph A. "Aristotle's Scientific Inquiry into Natural Slavery," *Journal of the History of Philosophy* 51, 3 (July 2013): 323–50.

Katasonov, Vladimir. "The Utopias and the Realities: Leibniz' Plans for Russia," in *Leibniz und Europa, Vorträge VI. Internationaler Leibniz-Kongress* 2 (1994): 178–82.

Kelly, Daniel, Luc Faucher, and Edouard Machery. "Getting Rid of Racism: Assessing Three Proposals in Light of Empirical Evidence," *Journal of Social Philosophy* 41 (2010): 293–322.

Kelly, Daniel, and Erica Roedder. "Racial Cognition and the Ethics of Implicit Bias," *Philosophy Compass* 3, 3 (2008): 522–40.

Kidd, Colin. *The Forging of Races: Race and Scripture in the Protestant Atlantic World, 1600–2000*, Cambridge: Cambridge University Press, 2006.

Kilgour, Maggie. *From Communion to Cannibalism: An Anatomy of Metaphors of Incorporation*, Princeton: Princeton University Press, 1990.

Kirsanov, Vladimir S. "Leibniz' Ideas in the Russia of the 18th Century," in *Leibniz und Europa, Vorträge VI. Internationaler Leibniz-Kongress* 2 (1994): 183–90.

Kitcher, Philip. "Species," *Philosophy of Science* 51 (1984): 308–33.

Kleingeld, Pauline. "Kant's Second Thoughts on Race," *Philosophical Quarterly* 57, 229 (2007): 573–92.

Koerner, Lisbet. *Linnaeus: Nature and Nation*, Cambridge, MA: Harvard University Press, 1999.

Kopec, Matthew. "Clines, Clusters, and Clades in the Race Debate," *Philosophy of Science* (forthcoming).

Lagier, Raphaël. *Les Races humaines selon Kant*, Paris: Presses Universitaires de France, 2004.

Landucci, Sergio. *I filosofi e i selvaggi, 1580–1780*, Bari: Einaudi, 1972.

Laqueur, Thomas. *Making Sex: Body and Gender from the Greeks to Freud*, Cambridge, MA: Harvard University Press, 1990.

Lebedev, Dmitriĭ. *Geografiia v Rossii petrovskogo vremeni*, Moscow: Izdatel'stvo Akademii Nauk, 1950.

Lestringant, Frank. *Le cannibale: grandeur et décadence*, Paris: Perrin, 1994.

———. *Le Brésil de Montaigne. Le nouveau monde des* Essais *(1580–1592)*, Paris: Persée, 2005.

Lestringant, Frank, Pierre-François Moreau, and Alexandre Tarrête (eds.). *L'unité du genre humain. Race et histoire à la Renaissance*, Paris: Presses Universitaires de Paris-Sorbonne, 2014.

Lévi-Strauss, Claude. *La pensée sauvage*, Paris: Plon, 1962.

Lewontin, Richard. "The Apportionment of Human Diversity," *Evolutionary Biology* 6 (1972): 391–98.

Livingstone, David. *The Pre-Adamite Theory*, Philadelphia: American Philosophical Society, 1992.

———. *Adam's Ancestors: Race, Religion, and the Politics of Human Origins*, Baltimore: Johns Hopkins University Press, 2008.

Lloyd, G.E.R. *Cognitive Variations: Reflections on the Unity and Diversity of the Human Mind*, Oxford: Oxford University Press, 2007.

Lovejoy, Arthur O. *The Great Chain of Being: A Study of the History of an Idea*, New Brunswick, NJ: Transaction, 2009 [1936].

Mabe, Jacob Emmanuel. *Anton Wilhelm Amo interkulturell gelesen*, Nordhausen: Traugott Bautz, 2007.

Machery, Edouard, and Luc Faucher. "Social Construction and the Concept of Race," *Philosophy of Science* 72 (December 2005): 1208–19.

———. "Why Do We Think Racially? Culture, Evolution and Cognition," in Henri Cohen and Claire Lefebvre (eds.), *Categorization in Cognitive Science*, Amsterdam: Elsevier, 2005, 1009–33.

Mallon, Ron. "Passing, Traveling and Reality: Social Constructionism and the Metaphysics of Race," *Nous* 38 (2004): 644–73.

———. "Race: Normative, Not Metaphysical or Semantic," *Ethics* 116, 3 (2006): 525–51.

Manning, Patrick (ed.). *Slave Trades, 1500–1800: Globalization of Forced Labour*, Brookfield, VT: Variorum, 1996.

Matthews, Gareth B. "Gender and Essence in Aristotle," *Australasian Journal of Philosophy* 64 (suppl., June 1986): 16–25.

Mayr, Ernst. *Systematics and the Origin of Species, from the Viewpoint of a Zoologist*, Cambridge, MA: Harvard University Press, 1942.

———. *The Growth of Biological Thought*, Cambridge, MA: Harvard University Press, 1982.

Mazzolini, Renato. "Skin Color and the Origin of Physical Anthropology (1640–1850)," in Susanne Lettow (ed.), *Reproduction, Race and Gender in Philosophy and Early Life Sciences*, Albany: State University of New York Press, 2014, 131–62.

Meijer, Miriam Claude. *Race and Aesthetics in the Anthropology of Petrus Camper (1722–1789)*, Amsterdam: Rodopi, 1999.

Memmi, Albert. *Le racisme: description, définition, traitement*, Paris: Gallimard, 1982.

Millikan, Ruth. "Historical Kinds and the 'Special Sciences,'" *Philosophical Studies* 95 (1999): 45–65.

Mills, Charles W. *The Racial Contract*, Ithaca, NY: Cornell University Press, 1997.

Mitton, Jeffry. "Genetic Differentiation of Races of Man as Judged by Single-Locus and Multilocus Analyses," *American Naturalist* 111, 978 (1977): 203–12.

Montagu, Ashley. *Statement on Race: An Extended Discussion in Plain Language of the Unesco Statement*, New York: Schuman, 1951.

Morrison, Toni. *Playing in the Dark: Whiteness and the Literary Imagination*, Cambridge, MA: Harvard University Press, 1992.

Mougnol, Simon. *Amo Afer. Un Noir, professeur d'université en Allemagne au XVIIIe siècle*, Paris: Harmattan, 2010.

Müller, Max. *Vorlesungen über die Wissenschaft der Sprache*, ed. Carl Böttger, Leipzig: Verlag Gustav Mayer, 1863.

Müller-Wille, Staffan, and Hans-Jörg Rheinberger. *A Cultural History of Heredity*, Chicago: University of Chicago Press, 2012.

Mungello, David E. (ed.). *The Chinese Rites Controversy: Its History and Meaning*, Monumenta Serica Monograph Series XXXIII, Nettetal: Steyler Verlag, 1994.

Nersessian, Nancy. "Opening the Black Box: Cognitive Science and History of Science," *Osiris* 10 (1995): 194–211.

———. *Creating Scientific Concepts*, Cambridge, MA: MIT Press, 2008.

Nkrumah, Kwame. *Consciencism: Philosophy and Ideology for De-Colonization and Development with Particular Reference to the African Revolution*, 2nd ed., New York: Monthly Review, 1970.

Nussbaum, Martha. "The Professor of Parody," *New Republic*, February 22, 1999.

Pagden, Anthony. *The Fall of Natural Man: The American Indian and the Origins of Comparative Ethnology*, Cambridge: Cambridge University Press, 1986.

Perkins, Franklin. *Leibniz and China: A Commerce of Light*, Cambridge: Cambridge University Press, 2005.

Pianigiani, Ottorino. *Vocabolario etimologico della lingua italiana*, Florence: Ariani, 1926.

Poole, William. "Seventeenth-Century Preadamism, and an Anonymous English Preadamist," *Seventeenth Century* 19 (2004): 1–35.

Popkin, Richard H. "The Philosophical Basis of Modern Racism," in Richard A. Watson and James E. Force (eds.), *The High Road to Pyrrhonism*, San Diego: Austin Hill Press, 1980, 79–102.

———. *Isaac La Peyrère (1596–1676): His Life, Work and Influence*, Leiden: Brill, 1987.

———. "Leibniz and Vico on the Pre-Adamite Theory," in Marcelo Dascal and Elhanan Yakira (eds.), *Leibniz and Adam*, Tel Aviv: University Publishing, 1993, 377–86.

Portuondo, María M. *Secret Science: Spanish Cosmography and the New World*, Chicago: University of Chicago Press, 2009.

Proust, Jacques. *Diderot et l'Encyclopédie*, 3rd ed., Paris: A. Michel, 1995.

Rasmussen, Knud. *Intellectual Culture of the Iglulik Eskimos*, Copenhagen: Nordisk Forlag, 1922.

Riley, Patrick. *Leibniz's Universal Jurisprudence*, Cambridge, MA: Harvard University Press, 1996.

Risch, Neil, Esteban Burchard, Elad Ziv, and Hua Tang. "Categorization of Humans in Biomedical Research: Genes, Race and Disease," *Genome Biology* 3, 7 (2002): 1–12.

Roberts, Sam. "Census Benchmark for White Americans: More Deaths than Births," *New York Times*, June 13, 2013, http://www.nytimes.com/2013/06/13/us/census-benchmark-for-white-americans-more-deaths-than-births.html?_r=0.

Robinet, André. *Le meilleur des mondes par la balance de l'Europe*, Paris: Presses Universitaires de France, 1994.

Rosenberg, Noah A., et al. "Genetic Structure of Human Populations," *Science* 298 (2002): 2381–85.

Rosenthal, Michael A. "'The Black, Scabby Brazilian': Some Thoughts on Race and Early Modern Philosophy," *Philosophy and Social Criticism* 31, 211 (March 2005): 211–21.

Rubiés, Joan-Pau. "Hugo Grotius's Dissertation on the Origin of the American Peoples and the Use of Comparative Methods," *Journal of the History of Ideas* 52, 2 (April–June 1991): 221–44.

———. "Christianity and Civilisation in Sixteenth-Century Ethnological Discourse," in Henriette Bugge and Joan Pau Rubiés (eds.), *Shifting Cultures: Interaction and Discourse in the Expansion of Europe*, Münster: Lit Verlag, 1995, 35–60.

Ruiu, Adina. *Les récits de voyage aux pays froids au XVII siècle: de l'expérience du voyageur à l'expérimentation scientific*, Montreal: Presses de l'université du Québec, 2007.

Salomon, Frank. *The Cord Keepers: Khipus and Cultural Life in a Peruvian Village*, Durham, NC: Duke University Press, 2004.

Samarin, Yu. F. *Stefan Iavorskiï i Feofan Prokopovich*, Moscow: Tipografiia A. I. Mamantova, 1880.

Saul, Jennifer. "Philosophical Analysis and Social Kinds: Gender and Race," *Proceedings of the Aristotelian Society* 80 (suppl., 2006): 119–44.

Schiebinger, Londa. *Plants and Empire: Colonial Bioprospecting in the Atlantic World*, Cambridge, MA: Harvard University Press, 2007.

Schimmel, Annemarie. *Im Reich der Grossmoguln: Geschichte, Kunst, Kultur*, Munich: C. H. Beck, 2000.

Schulenberg, Sigrid von. *Leibniz als Sprachforscher*, Frankfurt, 1973.

Schwartz, Stuart B. (ed.). *Implicit Understandings: Observing, Reporting, and Reflecting on the Encounters between Europeans and Other Peoples in the Early Modern Era*, Cambridge: Cambridge University Press, 1994.

Schwitzgebel, Eric. "Acting Contrary to Our Professed Beliefs; or, The Gulf between Occurrent Judgment and Dispositional Belief," *Pacific Philosophical Quarterly* 91, 4 (2010): 531–53.

Scupin, Raymond (ed.). *Race and Ethnicity: An Anthropological Focus on the United States and the World*, Upper Saddle River, NJ: Prentice Hall, 2003.

Sesardic, Neven. "Race: A Social Destruction of a Biological Concept," *Biological Philosophy* 25 (2010): 143–62.

Shapin, Steven. *A Social History of Truth: Civility and Science in Seventeenth-Century England*, Chicago: University of Chicago Press, 1994.

Sleigh, R. C. *Leibniz and Arnauld: A Commentary on Their Correspondence*, New Haven: Yale University Press, 1990.

Smith, Justin E. H. "'The Unity of the Generative Power': Modern Taxonomy and the Problem of Animal Generation," *Perspectives on Science* 17, 1 (Spring 2009): 78–104.

———. "'A Corporall Philosophy': Language and 'Body-Making' in the Work of John Bulwer," in Charles T. Wolfe and Ofer Gal (eds.), *The Body as Object and Instrument of Knowledge: Embodied Empiricism in Early Modern Science*, Dordrecht: Springer, 2010, 169–84.

———. *Divine Machines: Leibniz and the Sciences of Life*, Princeton: Princeton University Press, 2011.

———. "The Pre-Adamite Controversy and the Problem of Racial Difference in 17th-Century Natural Philosophy," in Marcelo Dascal and Victor Boantza (eds.), *Controversies within the Scientific Revolution*, Amsterdam: John Benjamins, 2011, 233–50.

———. "'Curious Kinks of the Human Mind': Cognition, Natural History, and the Concept of Race," *Perspectives on Science* 20, 4 (Winter 2012): 504–29.

———. "Leibniz on Natural History and National History," *History of Science* 50, 4 (December 2012): 377–401.

———. "'A Series of Generations': Leibniz on Race," *Annals of Science* 70, 3 (July 2013): 319–35.

———. "The Criminal Trial and Punishment of Animals: A Case Study in Shame and Necessity," in Andreas Blank (ed.), *Animals: New Essays*, Munich: Philosophia Verlag, forthcoming.

———. *A Global History of Philosophy*, Princeton: Princeton University Press, forthcoming.

Snowden, Frank M. *Blacks in Antiquity: Ethiopians in the Greco-Roman Experience*, Cambridge, MA: Harvard University Press, 1971.

———. *Before Color Prejudices: The Ancient View of Blacks*, Cambridge, MA: Harvard University Press, 1983.

Spencer, Quayshawn. "What 'Biological Racial Realism' Should Mean," *Philosophical Studies* 159 (2012): 181–204.

———. "The Unnatural Racial Naturalism," *Studies in History and Philosophy of Science Part C: Biological and Biomedical Sciences* 46 (June 2014): 38–43.

Sperber, Dan. "Why Are Perfect Animals, Hybrids, and Monsters Food for Symbolic Thought?," *Method & Theory in the Study of Religion* 8, 2 (1996 [1975]): 143–69.

Stanford, Dennis, and Bruce Bradley, "The North Atlantic Ice-Edge Corridor: A Possible Palaeolithic Route to the New World," *World Archaeology* 36, 4 (2004): 459–78.

Stock, Paul. "'Almost a Separate Race': Racial Thought and the Idea of Europe in British Encyclopaedias and Histories, 1771–1830," *Modern Intellectual History* 8, 1 (2011): 3–29.

Stoler, Ann Laura. *Race and the Education of Desire: Foucault's* History of Sexuality *and the Colonial Order of Things*, Durham, NC: Duke University Press, 1995.

Stuurman, Siep. "François Bernier and the Invention of Racial Classification," *History Workshop Journal* 50 (Autumn 2000): 1–21.

Suchier, Wolfram. "Ein Mohr als Student und Privatdozent der Philosophie in Halle, Wittenberg und Jena," *Akademische Rundschau* 4 (1916): 441–48.

———. "Weiteres über den Mohren Amo," *Altsachsen: Zeitschrift des Altsachsenbundes für Heimatschutz und Heimatkunde* 1, 2 (1918): 7–9.

Taylor, P. J. "Building on Construction: An Exploration of Heterogeneous Constructionism, Using an Analogy from Psychology and a Sketch from Socioeconomic Modeling," *Perspectives on Science* 3, 1 (1995): 66–98.

Tempels, Placide. *La philosophie bantoue*, Elisabethville: La Présence Africaine, 1945.

Thomson, Ann. "Diderot, le matérialisme et la division de l'espèce humaine," *Recherches sur Diderot et l'Encyclopédie* 26 (April 1999): 197–211.

Thornton, John K. *A Cultural History of the Atlantic World, 1250–1820*, Cambridge: Cambridge University Press, 2012.

Todorov, Tzvetan. *La conquête de l'Amérique: la question de l'autre*, Paris: Seuil, 1982.

Toulmin, Stephen. *Cosmopolis: The Hidden Agenda of Modernity*, Chicago: University of Chicago Press, 1990.

Tuplin, Christopher. "Greek Racism? Observations on the Character and Limits of Greek Ethnic Prejudice," in Gocha R. Tsetskhladze (ed.), *Ancient Greeks West and East*, Leiden: Brill, 1999, 47–75.

Utermöhlen, Gerda. "Leibniz im brieflichen Gespräch über Rußland mit A. H. Francke und H. W. Ludolf," *Leibniz und Europa, Vorträge VI. Internationaler Leibniz-Kongress* 2 (1994): 304–9.

Valls, Andrew (ed.). *Race and Racism in Modern Philosophy*, Ithaca, NY: Cornell University Press, 2005.

Van der Lugt, Maaike, and Charles de Miramon (eds.). *L'hérédité entre Moyen Age et époque moderne*, Florence: Sismel—Edizioni del Galluzzo, 2008.

Verlinden, Charles. *L'esclavage dans l'Europe médiévale, vol. 1: Péninsule ibérique—France*, Bruges: De Tempel, 1955.

———. *L'esclavage dans l'Europe médiévale, vol. 2: Italie—colonies italiennes du Levant—Levant latin—Empire byzantin*, Ghent: Philosophie et Lettres, 1977.

Vernant, Jean-Pierre, and Pierre Vidal-Naquet. *La Grèce ancienne 3: Rites de passage et transgressions*, Paris: Seuil, 1992.

Vidal-Naquet, Pierre. *Le chasseur noir*; English translation: *The Black Hunter: Forms of Thought and Forms of Society in the Greek World*, trans. Andrew Szegedy-Maszak, Baltimore: Johns Hopkins University Press, 1986.

Viveiros de Castro, Eduardo. *From the Enemy's Point of View: Humanity and Divinity in an Amazonian Society*, Chicago: University of Chicago Press, 1992.

———. "Os pronomes cosmólogicos e o perspectivismo ameríndio," *Mana* 2, 2 (1996): 115–44.

———. "Cosmological Deixis and Amerindian Perspectivism," *Journal of the Royal Anthropological Institute* NS 4 (1998): 469–88.

Voegelin, Eric. *Die Rassenidee in der Geistesgeschichte von Ray bis Carus*, Berlin: Junker und Dünnhaupt Verlag, 1933; English translation: *The History of the Race Idea, from Ray to Carus*, in Klaus Vondung (ed.), *The Collected Works of Eric Voegelin*, vol. 3, trans. Ruth Hein, Columbia: University of Missouri Press, 1989.

Wade, Nicholas. "East Asian Physical Traits Linked to 35,000-Year-Old Mutation," *New York Times*, February 14, 2013.

Wade, Peter. *Race, Nature and Culture: An Anthropological Perspective*, London: Pluto, 2002.

———. *Race and Ethnicity in Latin America*, 2nd ed., London: Pluto, 2010.

Wild, Markus. *Die anthropologische Differenz: der Geist der Tiere in der frühen Neuzeit bei Montaigne, Descartes und Hume*, Berlin: De Gruyter, 2006.

Wilson, Catherine. "Kant and the Speculative Sciences of Origins," in Justin E. H. Smith (ed.), *The Problem of Animal Generation in Early Modern Philosophy*, Cambridge: Cambridge University Press, 2006, 375–401.

Wilson, Robert. "Realism, Essence, and Kind: Resuscitating Species Essentialism?" in Robert Wilson (ed.), *Species: New Interdisciplinary Studies*, Cambridge, MA: MIT Press, 1999, 187–207.

Wiredu, Kwasi. "Amo's Critique of Descartes' Philosophy of Mind," in Kwasi Wiredu (ed.), *A Companion to African Philosophy*, London: Blackwell, 2004, 200–206.

———(ed.). *A Companion to African Philosophy*, London: Blackwell, 2004.

Wolfram, Herwig. *Geschichte der Goten: von den Anfängen bis zur Mitte des 6. Jahrhunderts*, Munich: Beck, 1979; English translation: *History of the Goths*, trans. Thomas J. Dunlap, Berkeley: University of California Press, 1988.

Wolfram, Herwig, and Walter Pohl (eds.). *Typen der Ethnogenese unter besonderer Berücksichtung der Bayern*, 2 vols., Vienna: Österreichische Akademie der Wissenschaften, 1990.

Zack, Naomi. "Mixed Black and White Race and Public Policy," *Hypatia* 10, 1 (1995): 120–32.

———. *The Philosophy of Science and Race*, London: Routledge, 2002.

Zammito, John. *Kant, Herder, and the Birth of Anthropology*, Chicago: University of Chicago Press, 2002.

Index